21st Century Biology and Agriculture

Ocean and Coastal Ecology

The Text book Series *21st Century Biology and Agriculture*

This series aims at fulfilling the knowledge needs of postgraduate learners in agriculture and biology, focussing on self-directed learning. Keeping this point in full view, the volumes published under this series would dilate on contemporary information in chosen themes in a pertinent, but brief backdrop of historical knowledge, with appropriate textual information laced with relevant illustrations. Most importantly, these books aim to cater to the self-learning needs in passionate and committed learners. By self-learning, the teacher's role turns into mentoring rather than tutoring. Vital details have been 'box'ed so that learners can internalize them easily, swiftly, and forever. Care has been exercised to integrate examples from the Indian subcontinent, so that the learners can relate to concepts and integrate principles quickly.

The most critical aspect is the inclusion of specific-case studies and interactive-mode of learning so that learners can learn about the day-today issues and application of theory that surround the nominated theme effectively. Every effort has been meticulously made to see that the books launched under this series are easily readable and user-friendly.

We are confident that post-graduate class teachers of biology and agriculture in the Indian subcontinent would find this material appropriate to prescribe this book as a learning resource.

Titles in this series:

- Growth and Development in Plants
- Medical and Veterinary Virology
- Ocean and Coastal Ecology
- Wildlife Ecology

21st Century Biology and Agriculture

Ocean and Coastal Ecology

– K. Kathiresan

Professor, CAS in Marine Biology,
Annamalai University,
Parangipettai - 608 502, Tamil Nadu, INDIA

SCIENTIFIC PUBLISHERS (INDIA)

Published by:

Scientific Publishers (India)
5 A, New Pali Road, P.O. Box 91
JODHPUR – 342 001 (India)

Scientific Publishers (India)
4806/24, Ansari Road, Daryaganj
NEW DELHI – 110 002 (India)

E-mail: info@scientificpub.com
Website: www.scientificpub.com

Cover illustration: This textbook series *21st Century Biology and Agriculture* launched by Scientific Publications, Jodhpur (India), celebrates the native, but critically threatened bird of the western semi-arid grasslands and scrubs of India, the great Indian bustard (GIB) *Ardeotis nigriceps* (Gruiformes: Otididae). GIB has been a part of Indian culture and tradition, known as *gonãdh* (*Sanksrit*), because the male GIB's call closely resembles the mooing of cows. Artist: Urmila Sharma.

ISBN: 978-81-7233-891-6

Typeset by Rajesh Parihar
Printed in India

Series Preface

With great pleasure, the Scientific Publishers (Jodhpur) and I launch the first volume of the postgraduate textbook series under the general title *21st Century Biology and Agriculture* for use in the Indian subcontinent.

This series aims at fulfilling the knowledge needs of postgraduate learners in agriculture and biology, focussing on self-directed learning. Keeping this point in full view, these volumes would dilate on contemporary information in chosen themes in a pertinent, but brief backdrop of historical knowledge, with appropriate textual information laced with relevant illustrations. Most importantly, these books aim to cater to the self-learning needs in passionate and committed learners. By self-learning, the teacher's role turns into mentoring rather than tutoring. Vital details have been 'box'ed so that learners can internalize them easily, swiftly, and forever. Care has been exercised to integrate examples from the Indian subcontinent, so that the learners can relate to concepts and integrate principles quickly. The most critical aspect is the inclusion of specific-case studies and interactive-mode of learning so that learners can learn about the day-today issues and application of theory that surround the nominated theme effectively. Every effort has been meticulously made to see that the books launched under this series are easily readable and user-friendly. Tanay Sharma of Scientific Publishers (Jodhpur) readily agreed to my demand to identify and recruit an efficient copy editor, who has done a neat job.

This book entitled *Ocean and Coastal Ecology,* executed by K. Kathiresan, a distinguished marine biologist of India, and a full professor at the Advanced Centre of Marine Biology, Annamalai University, Porto Novo, Tamil Nadu, commences this series. Many more similar titles have been enlisted and will be appearing in the near and far future. Scientific Publishers (Jodhpur) and I thank Kathiresan for readily and willingly accepting to our request by executing this task, and importantly doing it speedily.

India is a land of varied landscapes. One of the remarkable natural icons of India is the Great Indian Bustard, the populations of which are restricted to Western India. Because this series is being published from Jodhpur, I thought it will be appropriate to remind ourselves of the uniqueness of this splendid bird and make every effort of conserve its diminishing population.

All books to appear in the series *21st Century Biology and Agriculture* will celebrate this Indian icon.

I am confident that learners will benefit from the information and knowledge shared by Kathiresan; I am also confident that post-graduate class teachers of biology and agriculture in the Indian subcontinent would find this material appropriate to prescribe this book as a learning resource.

Anantanarayanan Raman
Editor-in-Chief, Textbook Series
21st Century Biology and Agriculture
School of Agricultural & Wine Sciences
Charles Sturt University
Orange, NSW 2800, Australia

Preface

Life on the Earth without ocean is impossible. Life originated on the Earth planet only in the oceans, some 500 million years ago. Oceans are unique in vastness, amazing in processes, astonishing in resources and remarkable in controlling global climate. Marine environment is the largest aquatic system on the earth and the biggest storehouse of mineral, food and energy resources. This supports 50% of the primary production of the earth, 85% of the global fish catch and more than 50% of the world's populations to live in the coastal areas. The oceans are of great ecological significance to have direct effect on global environment and human life.

Marine ecology is to understand on how organisms interact with each other and with their surrounding environment. There has been a growing interest in marine ecological studies in view of the global issues: losses of coastal habitats, serious declines in marine living resources, ever increasing human population, elevated levels of pollution, poor seawater qualities, overall deterioration of marine environment, natural calamities and climate change including sea temperature rise, sea level rise, and ocean acidification. It is a matter of necessity to understand scientific principles of the marine ecology in order to manage the marine environment and its resources.

Bearing this in mind, the present text book *entitled "Ocean and Coastal Ecology"* has been written in 10 chapters. Chapter-1 deals with basic principles and unique features of marine ecology. Chapter-2 is explaining various physical and chemical factors that influence the marine organisms. Chapter-3 is dealing with ecology of the ocean beyond 200 metres depth with a special note on hydrothermal vents and cold seeps. Chapter-4 is describing on various coastal ecosystems within 200 metres depth with additional information about seagrasses and kelp forests. Chapter-5 is dealing with ecological wonders of coral reef ecosystem. Chapter-6 is discussing about the ecobiology of estuaries with a special note on salt marsh community. Chapter-7 is explaining about ecological characteristics and importance of the mangrove ecosystem. Chapter-8 is on 'Functional marine ecology' discussing various oceanic processes: primary production, carbon sequestration, detritivory, herbivory, predation, parasitism, pathogenesis, fouling, boring, competition, succession, dispersal and settlement of larvae and marine food chains. Chapter-9 is dealing with different types of threats to various coastal and marine ecosystems.

Chapter-10 is discussing on the strategies for conservation and management of marine environment, in particular for promoting resilience of coral reefs and mangroves to climate change.

I am thankful to the authorities of Annamalai University, India for permitting me to write this book, to my team of research scholars particularly Dr. N. Rajendran, Mr. Sunil Kumar Sahu, Ms. Asmathunisha and Ms. V. Gomathi for helping me in preparatory work and to my wife Mrs. Sumathi Kathiresan for secretarial help.

I am extremely thankful to Anantanarayanan Raman, Editor-in-Chief, 21st Century Biology and Agriculture for having invited me to write this book and to Mr. Tanay Sharma, Managing Director, Scientific Publishers, Jodhpur, India for bringing out this book.

This book would be of immense help to students, scholars and all others who are interested in marine ecology.

K. Kathiresan, D.Sc.

Contents

Introduction to Marine Ecology 1

1.1. ECOLOGY AND MARINE ECOLOGY

Ecology is the science dealing with the relationships between organisms and their environments. No organism can exist by itself without an environment. Living organisms and their environment are inseparably interrelated, and they interact upon each other. All organisms interact with their own species, other species, and the surrounding environment. The physical and chemical conditions of the environment will influence the activities of living organisms; and, these organisms in turn, can alter the conditions of the environment. In short, ecology is the science of living environment or 'environmental biology'.

Marine ecology is the branch of ecology dealing with the relationships between marine organisms and their environments. In other words, the marine ecology is to understand on how marine organisms interact with each other and with their surrounding environment (Fig. 1.1). There are different interactions to be studied in the marine ecology as follow:

- Effects of environmental factors on the marine organisms;
- Effects of marine organisms on their environmental factors;
- Interactions among the marine organisms; and
- Interactions among the marine environmental factors

How does marine ecology differ from marine biology and oceanography?

The marine ecology is to study on how marine organisms interact with each other and with their surrounding environment, while the marine biology is the study of the marine organisms themselves. The oceanography is the study of oceans and their waves, currents and tides.

The organisms and their environments are grouped into different levels, such as species, population, ecological community and ecosystem. The 'species' is defined as a natural group of inter-breeding individuals. The 'population' is formed by all the individuals of a species in an area. The 'ecological community' is formed by several different species that occur together in a particular area. An 'ecosystem' is formed by communities and their surrounding environment. Thus the ecosystem is composed of living and non-living components that interact with each other, as a single functional unit. Populations, ecological communities and ecosystems are regulated by chemical and physical factors of the environment. Populations are also regulated by the biological factors such as competition, predation, parasitism and disease.

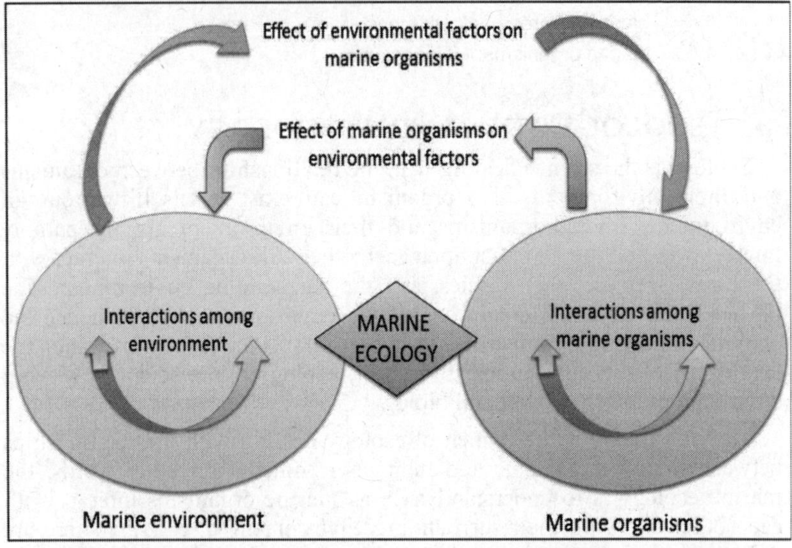

Fig. 1.1. Components of marine ecology

> The term 'ecology' is derived from the Greek words – '*oikos*' means 'home' and '*logos*' means 'study'.
>
> A **'Species'** is a natural group of inter-breeding individuals.
>
> A **'population'** is formed by all individuals of a species in an area.
>
> An **'ecological community'** is formed by several different species that occur together in a particular area.
>
> An **'ecosystem'** is formed by ecological communities and their surrounding environment that interact with each other, as a single functional unit.

In the marine environment, plants and animals of different species associate together to form the ecological communities. In the communities, their requirements to environmental conditions are almost the same. The communities are named based on the predominant associate (e.g. seagrass community) or the habitat (e.g. rocky shore community, estuarine community) or both (e.g. mangrove community). There are four important biological activities that produce changes in the community of the marine environment. These activities are: (1) Burrowing activities of worms, mollusks, crustaceans and others that live in the mud and sand, (2) Boring activities of several bivalves, crustaceans, sponges, worms and sea urchins into corals, rocks and woody structures, (3) Accumulation of calcareous or siliceous skeletal materials of organisms over the coral reefs and seafloor, bringing about changes in the environment, and (4) Deposition and accumulation of organic matter at the sea bottom, leading to the production of hydrogen sulphide and create anaerobic conditions that are favourable for anaerobic organisms.

How does 'habitat' differ from 'niche'?

A 'habitat' is the place where an organism is found, whereas a 'niche' is the role of an organism in a community. For an example, the habitat of the copepod crustacean, *Calanus* species is surface waters of the ocean, and its niche is herbivore feeding on the microscopic algae in the area.

Most organisms are divided into two types – opportunistic and equilibrium – based on their life history patterns in any habitat. The disturbed areas are colonized first by opportunistic species and later by equilibrium species. Equilibrium species are better competitors than opportunistic species. These two types show striking differences as shown in the Table 1.1.

Table 1.1. Differences between opportunistic and equilibrium species

Aspect	Opportunistic species	Equilibrium species
Life span	Short	Long
Time taken for reproductive maturity	Rapid	Long
No. of reproductive periods per year	Many	One or more
Death rate	High	Low
Size of organisms	Small	Large
Mobility	Mostly sessile and sedentary	Usually Mobile
Area of colonization	Frequently disturbed area	Not frequently disturbed area
Abundance	Changes over time	constant

- The marine ecology is to understand on how marine organisms interact with each other and with their surrounding environment.

1.2. IMPORTANCE OF MARINE ECOLOGY

1.2.1. Conservation of Marine Environment

Marine environment is of great ecological significance and economic importance. The world's oceans affect weather patterns through exchange of heat energy between atmosphere and the oceans. There has been a growing interest in marine ecology in view of the problems associated with conservation of marine environment. The problems are losses of coastal habitats, serious declines in marine living resources, ever increasing human population, elevated levels of pollution, poor water qualities, overall deterioration of marine environment, natural calamities and climate change including sea temperature rise, sea level rise and ocean acidification. The oceans have also become the ultimate sink for discharges of wastes from land-based sources. Other threats come from the transport of hazardous wastes, operational and accidental spillage of oil, discharge of radioactive materials and the transport of invasive species in the ballast water of ships. Harmful algal blooms are adversely affecting the recreational values of many coastal areas and in many cases reducing the fish populations and producing anoxia in the water column. Therefore, it is a matter of necessity to understand the ecological principles that govern marine life and environment.

1.2.2. Conservation of Marine Biodiversity

Biodiversity is the variability of life at its genetic, species and ecosystem levels of organization. It is the basis for human survival in meeting the needs of food, health and other needs of the growing world populations. Biodiversity is a measure of the total genetic and ecological diversity. In general, the biodiversity is used to summarize two community attributes, species richness (the number of species in a particular habitat) and relative species abundance. The high taxa (genera, families and phyla) are often used as an efficient measure in understanding biodiversity of different marine habitats. The value of biodiversity is scientifically proven. A biodiversity-based system produces one calorie of food from half a calorie of energy, whereas a homogeneous system produces one calorie of food from 10 calories of energy. Moreover, the biodiversity-rich system is resistant to all environmental stresses, whereas the less-diverse system is extremely vulnerable to environmental changes. Even today, we are not certain about how many marine species that exist in the ocean. Biodiversity is getting eroded globally at an alarming rate. Its losses are largely due to destructive activities of man. We are at present on the verge of mass extinction of countless species.

A vast majority of global human population has concentrated along the sea coasts. More than 50% of the world's populations live in the coastal areas and the total population is estimated to reach 6 billion by the year 2030. In East Asia, more than 70% of the population depends on coastal resources for food, employment and income generation. In south Asia, coastal populations are surviving on diminishing coastal resources. The western Pacific region suffers significant degradation of its coastal resources due to developmental activities. There is an inevitable need for biodiversity conservation due to erosion of biodiversity with increasing human pressure in the coastal areas. It is important to study marine ecology to find out the factors that are responsible for biodiversity loss and to undertake appropriate remedial measures to control the loss.

1.2.3. Conservation of Marine Bioresources

The marine bioresources provide a main source of food, medicines, fuels and livelihoods for millions of people worldwide. Coastal areas are highly vulnerable to man-made changes. This leads to degradation of ecosystems, destruction of species and reduction of overall productivity. In the middle of 1990's the level of seafood consumption in Asia and the Pacific regions exceeded that of the world's per capita seafood consumption. This was because fish prices were relatively lower than those of other sources of animal protein. An estimated 70% of the world's fish stocks are already being exploited at or beyond sustainable limits. The pressure on the oceans is not only due to over-harvesting but also due to the cumulative impact of land-based activities. This includes many of the effects of coastal development, especially destruction of wetlands, mangroves and coral reefs, sedimentation, dredging of sediments, damage to watersheds and the impounding of water supplies to support urban development in coastal areas. Nearly 50% of coastal mangroves that form vital nurseries for many fish species of commercial importance were already cleared in the last century. Because of siltation and other changes, nearly 10% of coral reefs have been degraded beyond recovery. Therefore, conservation of coastal habitats and their biodiversity deserves a top priority for preservation, sustainable utilization and ecological restoration. At least coastal habitats have adequate scientific studies, but the ocean ecology does not have much information. This is due to the reason that the coastal studies are easily accessible, whereas ocean studies are practically difficult due to lack of sophisticated oceanographic equipment.

1.2.4. Responsibility of Marine Ecologists

Over-exploitation and degradation of marine resources has two main consequences: (1) loss of marine economy through reduction in resource-based livelihood and income; and (2) destruction of marine environment through increased human pressures and natural disturbances (Fig. 1.2). It is essential to understand scientific principles of the marine ecology and to manage the marine environment and its resources. Marine ecologists attempt to promote conservation and protection through educating the general public about the importance of marine ecosystems and involving them in better protection of marine environment through the works related to coastal clean-ups, restoration of damaged shores, conservation of marine life and recovery of endangered marine species.

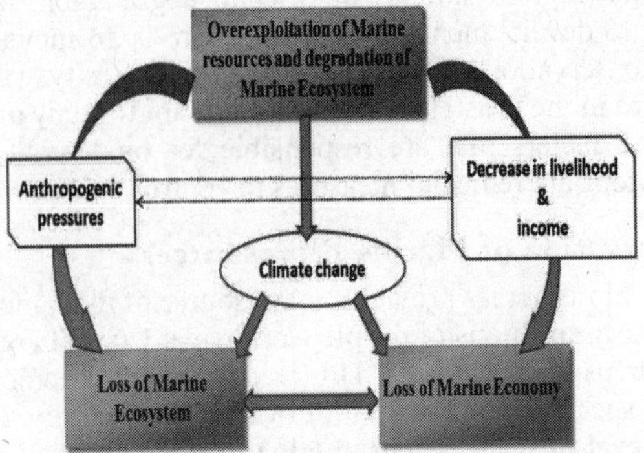

Fig. 1.2. A linkage between marine ecology and marine economy

- Marine ecological studies are important for sustainable management of marine environment and its biological resources.

1.3. UNIQUENESS OF MARINE ENVIRONMENT

Oceans are unique in occupying vast areas, amazing in processes, astonishing in resources and remarkable in controlling global climate. Marine environment is the largest aquatic system on the earth. The oceans occupy 71% of the globe, having an area of about 361 million square kilometres. The Earth planet is a 'water world'. About 95% of the volume of the global biosphere is in the oceans that have a total volume of about 1,347,000 million cubic kilometres. The average depth of the oceans is about 3,730 metres.

- The deepest part of the oceans is the Mariana Trench with a depth of 11,033 metres between Japan and Papua New Guinea in the Pacific Ocean.
- The highest sub-marine mountain is the Mount Kea in the Pacific Ocean with 10,203 metres above the seafloor. It is significantly higher than the Mount Everest which is the tallest mountain on land with a height of 8,849 metres.
- The mysterious place of the ocean is the 'Bermuda Triangle' which is a large area of the Atlantic Ocean that stretches between Bermuda, Puerto Rico and Florida where ships and aircraft have gone missing without a trace.

The sea offers 98% of the total space available to life. The oceans provide about 300 times more space for living organisms than that provided by the terrestrial and freshwater habitats. The marine living space is largely stable and uniform in physical and chemical factors as compared to that on the land. Most of the marine habitat is in deep water. Only about 3% of the world's waters lie over the continental shelf. Considering the entire volume of the biosphere, 79% of it is composed of marine water with depths greater than 1000 metres. Although the deep sea area is the largest habitat on earth, its biology is least known and explored. This is due to the difficulty of sampling and inaccessibility to human. About 2% of all known marine species are living in the water column (pelagic environment), while 98% inhabit the seafloor (benthic) environment.

The oceans are the earth's biggest storehouse of resources, such as minerals (chemicals, metals, oil and natural gas), food (fish, prawns, lobsters etc.) and energy (waves, water currents, tides). About 50% of the primary production of the earth takes place in the marine environment. Marine fisheries account for 85% of the global fish catch. Maritime shipping is involved in the transport of over 80% of the world's merchandise trade. The ice continent of Antarctica and the surrounding ocean, called the 'Southern Ocean', plays a vital role in global warming, sea level rise and ozone depletion, which have direct effect on global environment and human life.

Marine environment

- It occupies 71% of the globe, having an area of 361 million square kilometers.
- It is 95% of the volume of the global biosphere, having a volume of 1,347,000 million cubic kilometres.
- It has 79% of global water in the sea with depth greater than 1000meters.
- Average depth is about 3,730 metres.
- Deepest part of the oceans is the Mariana Trench with a depth of 11,033 metres.
- Highest Sub-marine Mountain is the Mount Kea in the Pacific Ocean with 10,203 metres above the seafloor.
- 2% of marine species live in the water column, while 98% inhabit the seafloor.
- 50% of the primary production of the earth takes place in the marine environment.
- Marine fisheries account for 85% of the global fish catch.
- Maritime shipping transports 80% of the world's merchandise trade.

- 'Southern Ocean', plays a vital role in environment and human life.

1.4. UNIQUENESS OF MARINE ORGANISMS

Oceans are the largest repository of life on the planet. The oceans are colonized by a variety of organisms ranging from microscopic algae to gigantic blue whales. In fact, life originated on the Earth planet only in the oceans, some 500 million years ago. The fish in the sea originated some 300 years ago. The whales originated in the oceans 65 million years ago. The biggest marine animal is the blue whale with the weight of 193 tons and the length of 31 metres. This big animal is feeding only on small animal zooplankton called 'krill' at a rate of 4 tons a day for its survival. The deepest diver is sperm whale, which can reach depths of 3,000 metres (Fig.1.4). Whales produce sound like singing, which can be heard several miles away. The largest fish is the whale shark measuring 18 metres long (Fig.1.5) and the smallest fish is the dwarf goby, measuring less than 1 cm long. The most feared animal in the sea is sharks especially the great white shark. The fastest fish is sailfish with the speed of 110 kilometres in an hour (Fig.1.6). Penguins are good swimmers and can stay underwater for about 18 minutes. The giant squid has the biggest eyes of any animal and they are 17 times bigger than human eyes, measuring 40 cm across. The biggest coral reef in the world is the Great Barrier Reef of Australia. The tallest seaweed is giant kelp with a length of 60 metres. Many marine communities are without autotrophic organisms because the greatest volume of the oceans does not have sunlight to perform the photosynthesis.

Marine microorganisms are unique. They are present even in extreme conditions of marine environment. The bacteria, archaea and protists account for the majority of the living biomass in the world's oceans. Over billion individual microorganisms are present in one litre of seawater or one gram of sediment. Two billion different bacteria may be present in the sea. There are 100 major phyla and 18 million DNA sequences are known for the marine microbes. This high diversity is due to mutations in the past 3.5 billion years of their existence.

Viruses are the most common biological units in the world's oceans. Interestingly, viruses that control marine environment were not known prior to 1989. They are 10 times greater than bacteria in terms of their population. They are extremely abundant (2.5×10^8 /ml of seawater) in nutrient-poor (oligotrophic) waters. However, viral infection is higher in nutrient-rich waters. The viruses are significant in ecological control of plankton and microorganisms. Viruses have a dramatic effect in regulating primary productivity. The viruses have a significant effect on natural communities of marine bacteria and phytoplankton. It is estimated that as much as 10-20% of marine bacteria are lysed daily with 2-3% of primary production loss through viral activity. The viruses are not only important regulators of bacterial and phytoplankton communities, but may also affect the grazers that consume them. The viruses can infect bacteriovorous flagellates which are ecologically important grazers in marine environment. The viruses infect all marine organisms. The viruses play significant roles in the sea: (1) enhance the oceanic respiration, (2) cause the diseases of marine organisms, (3) control the microbial species diversity, (4) exchange the genetic material among marine bacteria following viral attack, (5) help in the process of marine food chain and (6) disseminate the toxins by killing bacteria, such as cholera toxin. This leads to outbreak of cholera disease in human and it has a close correlation with sea surface temperature brought out by global warming.

Are marine microbes and viruses unique?

- Marine microorganisms account for majority of marine living biomass.
- High diversity of marine microorganisms is due to mutations in 3.5 billion years of existence.
- By number, marine microbes are 100 million times more than the stars. By total weight, the marine microbes are equated to 240 billion African elephants.
- Total number of bacterial and archaeal cells in the marine environment is 10^{29}.
- The total number of viral particles (Fig. 1.3) in the ocean is $4x10^{30}$, assuming average viral abundance of $3x10^{9}$/L and total volume of ocean is $1.3x10^{21}$ L.

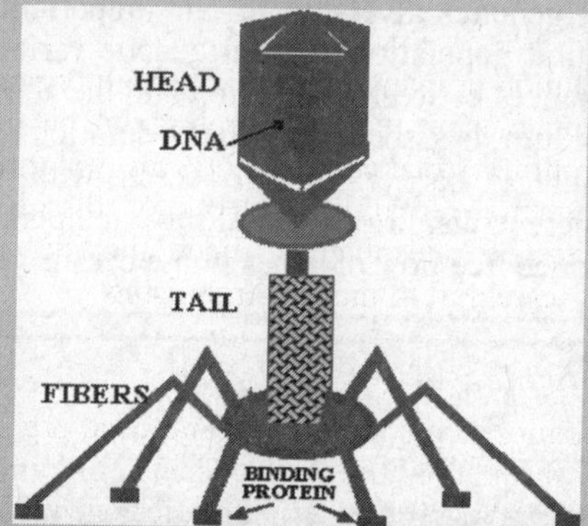

Fig. 1.3 : A typical viral structure

- Marine viruses are 10 times more in numbers than bacteria and archaeal cells in the oceans.
- Marine viruses store 200 million tons ($2x10^{11}$/g) of total carbon equivalent of 75 million blue whales.

- If all the marine virus particles are placed end to end they will span about 10 million light years (100 times the distance across) our own galaxy.

Marine environment is rich in biodiversity. The total number of marine species ranges between 10^5 and 10^7 species. There are 31 animal phyla in marine environment, compared with only 14 phyla on land. There are 15 endemic marine phyla that include 10 benthic, one pelagic and 2 both benthic and pelagic and two endoparasitic. Among the major groups of marine organisms, invertebrates are dominant (66%), followed by protozoa (13%), algae (13%), chordates (5%) and viruses and prokaryotes (3%) in the marine environment. The marine biodiversity is influenced by six major factors in the world's oceans, and they are: (1) latitude and longitude, (2) habitat size and complexity, (3) size of organisms, (4) depth of ocean, (5) grazing and predation, (6) genetics and dispersal of organisms. The biodiversity increases with the size and complexity of the habitats in terms of volume, biomass and area covered. Larger areas of a homogenous habitat are not supporting large number of species and therefore, habitat complexity is important for biodiversity. Regarding body size, there are a greater number of small organisms than larger organisms in a habitat. Species richness varies with depth and it exponentially declines from intertidal area to the deep sea. The interaction between grazers and predators in marine ecosystem may influence biodiversity. Feeding process by predators (starfish), and grazers (sea urchins) maintains the diversity of ecosystems. If the grazers and predators are not active, some prey organisms over-dominate, thereby reducing the biodiversity in the system. The genetic diversity is an important component of biodiversity in natural population. Spatial genetic variability is linked to the geographic distances between population and the ability of a species to disperse and this in turn, is associated with the duration of dispersal phase. There are geographical barriers in the ocean with varying physical structures and different conditions of salinity, temperature and current flow. These can reduce free flow of genes in the ocean.

Marine organisms

- Life originated only in the ocean, 500 million years ago.
- Total number of marine species ranges between 10^5 and 10^7 species.
- There are 31 animal phyla including 15 endemic phyla in marine environment, compared with only 14 phyla on land.
- Invertebrates are dominant group of marine organisms.
- Biggest marine animal is the blue whale with 193 tones weight and 31 metres length.
- Smallest fish is dwarf goby with less than 1cm length.

- Most feared marine animal is shark especially the great white shark.
- Giant squid has the biggest eyes with 40cm diameter - 17 times bigger than human eyes.
- Biggest coral reef in the world is the Great Barrier Reef of Australia.
- Tallest seaweed is giant kelp with 60 metres length.

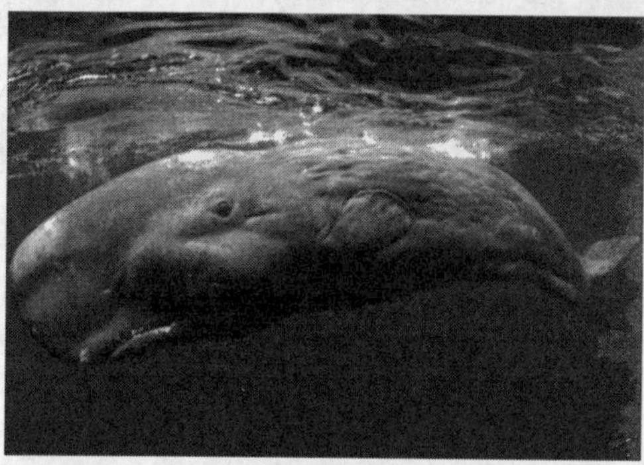

Fig. 1.4. Deepest diver is sperm whale, reaching depths of 3,000 metres (http://perfectdives.com/category/scuba-dive-essentials/conservation/page/4)

Marine wealth of India

India is rich in marine biodiversity with a record of over 15,000 species. This is due to diversified habitats. India has a long coastline of 7,516 km, a vast area of coastal ecosystem in 42,763 km^2 and an Exclusive Economic Zone (EEZ) of 2.02 million km^2. The long coastline, tropical climate and nutrients supplied by rivers have combined to produce a variety of productive marine habitats.

Most spectacular marine biological wealth in India are (i) Inter-tidal mudflats teeming with about 200 species of migratory birds, (ii) only dense mangrove forests at Sundarbans inhabited by the tiger in the world as well by globally threatened species such as Royal Bengal tiger, sea turtles, fishing cat, estuarine crocodile, the Gangetic dolphin and river terrapin. (iii) delicate seagrass meadows favoured by the seacow (dugong), (iv) most beautiful corals colonized with ornamental fishes, (v) sandy coast with the world's largest nesting site of sea turtles (Olive Ridley) at Orissa coast, (vi) rough sea migrated with the largest whale shark fish in Gujarat waters, (vii) occurrence of one of the two genetic paradise of mangroves in the world at Bhitarkanika, Orissa; (viii) Occurrence of all the three major types of coral reefs (atoll, fringing and barrier reefs) support about 11,000 species; (ix) highly productive seaweeds with 844 species, along the rocky shores; and (x) Mangroves with largest world record of about 4000 species; (xi) occurrence of the last population of one of five threatened sub-species of Asiatic wild ass (*Equus hemionus khur*) occurs in the salt marshes of the Little Rann of Katchchh in Gujarat.

Fig. 1.5. Largest fish is whale shark with 18 metres length
(http://wisecreatures.blogspot.in/2010/09/largest-animals.html)

Fig. 1.6. Fastest fish is sailfish with the speed of 110 km/hr
(http://www.ultimate-fishing-guide.com/sailfish.html)

HOW DO MARINE ORGANISMS DIFFER FROM TERRESTRIAL ORGANISMS?

Marine organisms are adapted to physical and chemical conditions of their surrounding environment. Seawater is over 800 times denser than air. This great density of water makes plankton, and even large organisms to float in the water. When water heats up, viscosity becomes less. Tropical marine animals encounter less resistance to movement than do temperate or polar animals. Moreover, the marine organisms are surrounded by seawater that is almost similar to their own body fluids in terms of salt content and hence the organisms do not spend much energy for maintaining the osmoregulation in the marine environment. In order to float in the seawater against gravity, the marine organisms do not have significant amounts of biomass. The marine organisms have less structural skeleton. They are also mostly rich in protein with lower energy storage. The seawater holds low level of oxygen and the marine organisms develop adaptations particularly in the form of integument respiration and small size body to provide adequate oxygen to the organisms. The speed of sound is four times-higher in seawater than air. The marine animals use shorter wave lengths of sound with higher frequency to detect their food prey and they also use the sound to monitor their environment and avoid predators. There are many striking differences between marine and terrestrial organisms as shown in table 1.2. However, the oceans and land have similar latitudinal gradients in biodiversity: (1) an increase in species diversity from Polar Regions to the tropics and (2) the greatest marine diversity within the tropics.

Table 1.2. Striking differences between marine organisms and terrestrial organisms

Aspect	Marine organisms	Terrestrial organisms
Size of plants	Mostly microscopic and single celled algae	Mostly large sized and multicellular plants
Photosynthetic area	Limited to the photic zone of ocean	Throughout the land
Herbivores	Mostly microscopic (copepods and protists); removing the autotrophs completely	Mostly macroscopic (cattle, sheeps and elephants); removing the autotrophs partially
Biodiversity of marine animals	Higher by number of animal phyla	Higher by number of animal species
No. of animal phyla (out of total of 34 phyla)	31	14

Aspect	Marine organisms	Terrestrial organisms
No. of animal classes	85	28
No. of endemic phyla	15	1
Feeding habit of large animals	Mostly carnivores	Mostly herbivores
Life history	Marine organisms shed both male and female gametes into the water	Terrestrial organisms do not shed female gametes
Life span of plants and animals	Short lived autotrophs and long lived animals	Long lived autotrophs and short lived animals
Food chain to reach top carnivores	Complex with 5 links (microscopic autotrophs, microscopic herbivores, sessile animals, small fishes and large fishes)	Simple with only 3 links (autotrophs, herbivores and carnivores)

1.5. MARINE ENVIRONMENT: DIVISIONS

Oceans are bigger than seas. There are 5 oceans and 54 seas. Some seas are parts of oceans, such as the Arabian Sea in the Indian Ocean. Other seas are enclosed by land, such as the Caspian Sea in Asia. All the oceans together are often just called the 'sea'. The five oceans are the Pacific Ocean, the Atlantic Ocean, the Indian Ocean, the Arctic Ocean and the Antarctic Ocean, in the order of decreasing size. The Pacific Ocean is the largest occupying 46% area of the five oceans. The Atlantic Ocean is the second largest occupying 23%, the Indian Ocean is the warmest covering 20% and the Arctic and Antarctic oceans are the smallest and coldest oceans occupying 11% of total oceans. The Arctic and Antarctic Oceans are frozen solid for most of the year. They have thousands of icebergs and only one ninth of an iceberg is visible above the sea's surface. The three major oceans (Pacific, Atlantic and Indian) are connected to each other, converging in the area around the Antarctic continent (Fig. 1.7). The oceans are not equally distributed over the earth. Oceans cover 81% of the Southern Hemisphere, but only 61% of the Northern Hemisphere. Most of the earth's land masses are present in Northern Hemisphere. About 170 million years ago, all the continents formed one piece of land called 'Pangaea' surrounded by only one ocean called 'Panthalassa'. As the Pangaea slowly split into today's seven continents, the Panthalassa became the five oceans. The Mount Everest has fossils of seashells and this reflects that the Himalayas were once part of the seafloor. Changes in the Earth's crust pushed them up into the air.

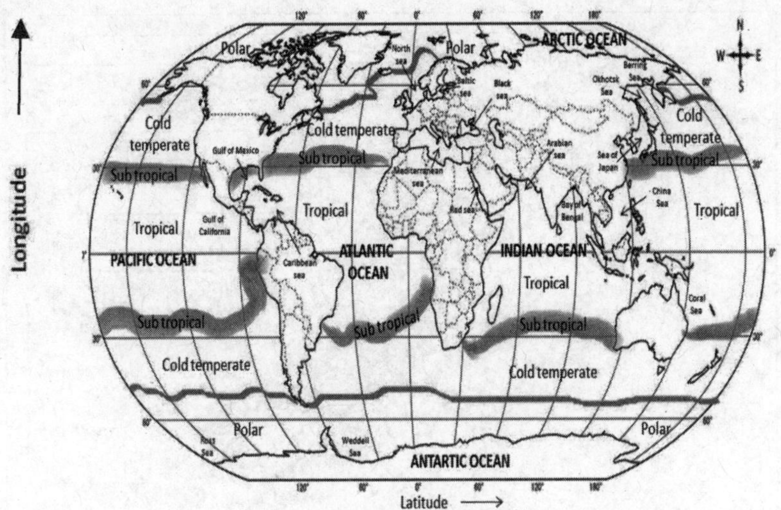

Fig. 1.7. Major oceans, seas and temperature-based bio-geographical regions of the world's oceans

The oceans are interconnected by the vertical and horizontal movement of water. There also exist differences in the intensity of temperature, pressure and light. These factors affect the distribution of organisms. Besides these factors, the topographical features of the ocean floor such as ridges, mount, canyons and trenches function as physical barriers in the distribution of organisms. On the basis of temperatures of sea surfaces and distribution of organisms, four major biogeographical zones can be categorized: Polar, Cold temperate, Warm temperate (sub-tropical) and Tropical (equatorial) (Fig. 1.7).

The marine environment is divisible into two major divisions, the 'benthic' and the 'pelagic'. The benthic division comprises the entire sea bottom, while the pelagic division comprises the whole water body that lies above sea bottom (Fig. 1.8).

The benthic division is made up of sea-shore and sea-bottom with three major zones: the littoral, the sub-littoral and the deep sea. The littoral zone is the territory between high water and low water marks of spring tides and this is the 'intertidal area'. Below the littoral zone is the sub-littoral zone which extends up to the end of the continental shelf or to a depth of 200 metres. The benthic division beyond the continental shelf is the deep sea zone which extends from 200 metres to deep ocean floor. The deep sea zone is further divided into upper (bathyal), middle (abyssal) and lower (hadal) zones. The bathyal zone extends down from the 200 to 2000 metres. The abyssal zone is below 2000 metres to 6000 metres. The hadal zone is beyond 6000 metres down to the deepest trenches (Fig. 1.8).

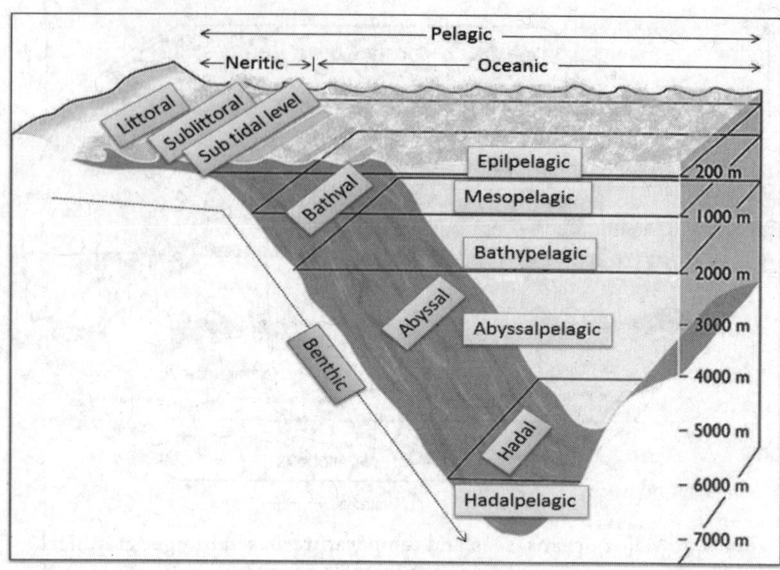

Fig. 1.8. Divisions of marine environment

The seafloor is also divided into three distinct topographic regions, the 'continental shelf', the 'continental slope' and the 'deep ocean basin'. The continental shelf is the land extension under seawater from all continents of the earth. It is usually very gentle slopes covered with sand, clay, mud and debris washed down from the land. The width of the continental shelf varies in different regions – 15 kilometres off Florida, 500 kilometres off the Gulf of Maine, 200 kilometres off Mumbai and 1000 kilometres off the Siberia. The Continental shelf bordering the Indian Ocean is narrower than those of the Atlantic. In the Bay of Bengal, the continental shelf is about 15 kilometres off Bengal, but narrows towards the south. Beyond the continental shelf, the seafloor slopes down sharply which is known as the 'continental slope'. In these steep slopes, there are V-shaped valleys called 'submarine canyons' with width ranging from 1.5 to 15 kilometres, and depth from few metres to several kilometers. The rivers extend hundreds of kilometers into the sea in the form of canyons.

Pelagic division is divided into a neritic zone (coastal) and an oceanic zone (offshore). Pelagic habitats are divided into five depth zones: epipelagic (0-200 metres), mesopelagic (200-1000 metres), bathypelagic (1000-2000 metres), abyssalpelagic (2000-6000 metres) and hadalpelagic (greater than 6000 metres). The shallow water over the continental shelf to a depth of 200 metres is the 'neritic zone'. The zone beyond a depth of 200 metres is the 'oceanic zone'. Both neritic zone and epipelagic zone of the ocean are abundant with larval stages of almost all the benthic animals. The mesopelagic zone is the area with temperature of 10°C. The bathypelagic zone is the area where temperature ranges between 10°C and 4°C. The abyssalpelagic zone is the deep aphotic zone, where the temperature is maintained constant at 4°C.

In the marine divisions, among the benthic component, the abyssal zone alone constitutes 75% by volume. This zone added with bathyal zone makes up 91% of the benthic domain. Among the pelagic component, the abyssal pelagic zone occupies 54%, followed by the mesopelagic (28%), the bathypelagic (15%), the epipelagic (3%) and the hadalpelagic (<1%) (Table 1.3).

Table 1.3. Marine divisions and their per cent volume

Zone	Depth (m)	Volume (%)
Benthic		
Hadal	>6000	1
Abyssal	2000-6000	75
Bathyal	200-2000	16
Sublittoral	0-200	8
Pelagic		
Hadalpelagic	>6000	<1
Abyssalpelagic	2000-6000	54
Bathypelagic	1000-2000	15
Mesopelagic	200-1000	28
Epipelagic	0-200	3

- There are 5 oceans and about 54 seas. The five oceans are the Pacific Ocean, the Indian Ocean, the Atlantic Ocean, the Arctic Ocean and the Antarctic Ocean, in the order of decreasing size.
- Oceans cover 81% of the Southern Hemisphere, but only 61% of the Northern Hemisphere. Most of the earth's land masses are present in Northern Hemisphere.
- On the basis of temperatures of sea surfaces and distribution of organisms, four major biogeographical zones can be categorized: Polar, Cold temperate, Warm temperate (sub-tropical) and Tropical (equatorial).
- The marine biodiversity increases from Polar Regions to the tropics.
- The marine environment has two major divisions: the 'benthic' (seafloor) and the 'pelagic' (water column).
- The benthic division is made up of sea-shore and sea-bottom with three major zones: the littoral, the sub-littoral and the deep sea. The littoral zone is the territory between high water and low water marks of spring tides and this is the 'intertidal area'. Below the

littoral zone is the sub-littoral zone which extends up to the end of the continental shelf or to a depth of 200 metres. The benthic division beyond the continental shelf is the deep sea zone which extends from 200 metres to deep ocean floor. The deep sea zone is further divided into upper (bathyal), middle (abyssal) and lower (hadal) zones. The bathyal zone extends down from the 200 to 2000 metres. The abyssal zone is below 2000 metres to 6000 metres. The hadal zone is beyond 6000 metres down to the deepest trenches. Among the benthic component, the abyssal zone is the largest.

- Pelagic division is divided into a neritic zone (coastal) and an oceanic zone (offshore). Pelagic habitats are divided into four depth zones: epipelagic (0-200 metres), mesopelagic (200-1000 metres), bathypelagic (1000-2000 metres), abyssalpelagic (2000-6000 metres) and hadalpelagic (greater than 6000 metres).
- The shallow water over the continental shelf to a depth of 200 metres is the 'neritic zone'. This zone is abundant with larval stages of almost all the benthic animals. Among the pelagic component, the abyssal pelagic zone is the largest.

Marine Living fossils

Living fossils in the sea are Coelacanth fish (405-345 million years old) and Horse shoe crab (200 million years old) (Fig. 1.9).

Coelacanth fish **Horse shoe crab**

Fig.1.9 : Marine living fossils: coelacanth fish and horse shoe crab

The most plentiful protozoa in the sea are the foraminifera. About 1000 live today and the rest being fossils. They are miniature seashells, made up of $CaCO_3$ (lime) extracted from seawater. Fossil foraminiferans are used to locate oil and natural gas deposits. A large number of foraminiferans are embedded in the lime stones, used for making the pyramids of Egypt.

1.6. MARINE ORGANISMS: MAJOR GROUPS

All marine organisms belong to one of the three domains: Archaea, Bacteria and Eukarya. Archaea have a group of simple microscopic bacteria-like organisms and their main functions are methane production and sulphur oxidation. The archaea prefers extreme temperature and pressure, and hence they are termed 'extremophiles'. They commonly occur in deep-sea vents and seeps. Bacteria are the microorganisms, lacking nucleus and they include purple bacteria, green non-sulphur bacteria and cyanobacteria. The archaea and bacteria are grouped under 'prokaryotes'. Eukarya are complex organisms having nucleus and they include protists, fungi and multicellular plants and animals. Although prokaryotes are small in size, they are a significant biomass component in aquatic ecosystems and play a major role in biogeochemical processes. It is generally assumed that the importance of prokaryotes to material and energy fluxes is high in low productive ecosystems and low in high productive ecosystems. Lagoon and coastal ecosystems are extremely dynamic and highly productive systems, dominated by macro-size eukaryotic organisms (both macrofauna and macrophytes).

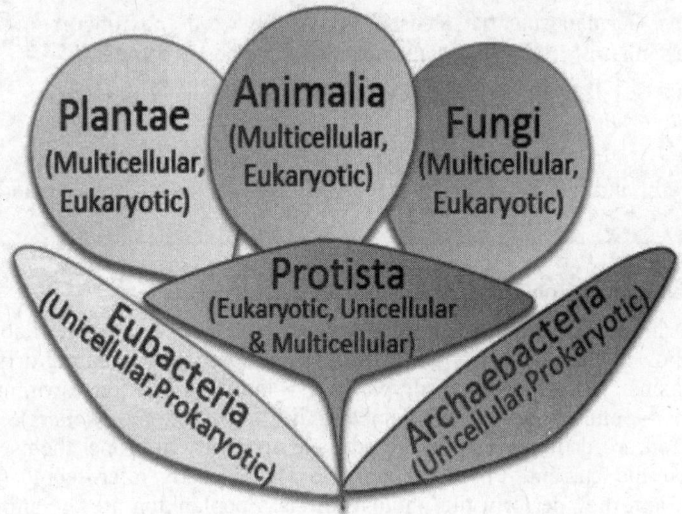

Fig. 1.10. Five kingdom classification of living organisms

The marine organisms belong to five major kingdoms: Monera, Protists, Fungi, Plantae and Animalia (Fig. 1.10). The Monera are single celled microorganisms without definite nucleus, and they include eubacteria and archaea. They play vital role in decomposition, synthesis of organic compounds and release of nutrients. Protists are single cellular organisms with definite nucleus and these include single celled algae, single-celled protozoa and multi-celled algae. They generate deep sea oozes of carbonates and silicates. Fungi include multicellular organisms, such as mold and lichen and they exist throughout the marine environment, in particular intertidal zone. The fungi are important decomposer in the marine environment and mineralize organic matter. Plantae (metaphyta) include multicellular plants and they are photosynthetic autotrophs. They are found attached in intertidal and shallow coastal environment. Examples are seaweeds, seagrass, saltmarsh and mangroves. Animalia (metazoa) include multicelled animals. They are mollusks (e.g. clams, oysters, snails and squids), arthropods (e.g. barnacles, crabs and shrimps), Annelids (e.g. polychaete worms), Ctenophores (e.g. comb jellies), Brachiopods (e.g. lamp shells), Echinoderms (e.g. starfish, sea urchins and sea cucumbers), Chordates (e.g. fishes, sea turtles, dolphins, seals and whales). Whales, seals and dolphins are mammals and warm-blooded animals. Sharks and fishes are cold-blooded animals.

Do you know?

In 1847, the Danish biologist (A.S. Orsted) was the first to see the presence of plankton in the seawater under a microscope.

The German scientist Hensen coined the word 'phytoplankton' (in Greek 'phytos' means plant and 'plankton' means to wander).

Red sea is so called because of blooms caused by blue-green alga, *Trichodesmium erythraeum.*

The Pacific Ocean means "peaceful" ocean. But it is not always so peaceful and is a home to many tsunami–generating earthquakes and El Nino.

Types of Marine Organisms based on Lifestyle

Marine organisms are classified according to where they live (habitat) and how they move (mobility). They are plankton (floating/drifting organisms; plankton – plural, plankter -singular), Nekton (swimmers; nekton – plural, nekter-singular) and benthos (bottom dwellers). The plankton are drifting organisms and have no ability to propel themselves against the current. Phytoplankton are algae mostly microscopic (Fig. 1.11), and they perform the photosynthesis. Zooplankton are the animals (Fig. 1.12) and they are heterotrophic, feeding on phytoplankton. The plankton are categorized into holoplankton and meroplankton, according to their life cycle. Holoplankton are permanent plankton, free-floating throughout entire life span, while meroplankton are temporary plankton, free-floating only during juvenile and larval stages of adults. The plankton are classified based on size as femtoplankton, picoplankton or ultraplankton, nanoplankton, microplankton or net plankton and macroplankton. All the plants such as phytoplankton, cyanobacteria and photosynthetic bacteria are restricted to photic zone of ocean for performing photosynthesis. Heterotrophic bacteria and animals survive at all depths of the oceans.

Types of plankton based on sizes

Femtoplankton (size 0.01-0.2 μm; viruses)

Picoplantkon (size 0.2-2 μm; Bacteria, archaea, cyanobacteria, nanoflagellates)

Nanoplankton (size 2-20 μm; Flagellates, diatoms, dinoflagellates, coccolithophores)

Microplankton of net plankton (size 20-200 μm; ciliates, diatom, dinoflagettates and other algae).

Macroplankton (>200 μm; jellyfish).

Abundance of plankton type increases with reduced size of organisms

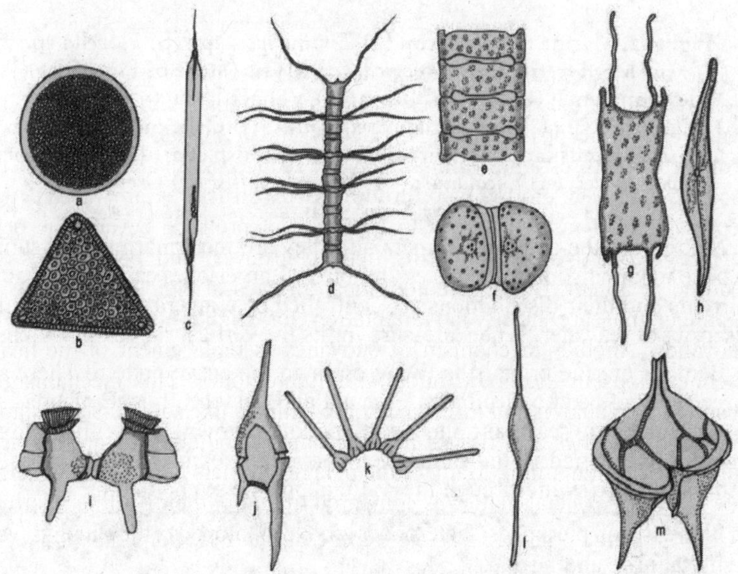

Fig. 1.11. Marine Phytoplankton : (a) *Coscinodiscus* sp.; (b) *Triceratium*; (c) *Rhizosolenia* sp.; (d) *Chaetoceros* sp.; (e) *Bellerochea malleus*; (f) *Hemidiscus hardmannianus*; (g) *Odontella sinensis* (h) *Pleurosigma* sp.; (i) *Dinophysis caudata*; (j) *Ceratium furca*; (k) *Aesterionellopsis glacialis*; (l) *Nitzschia* sp.; (m) *Protoperidinium* sp.

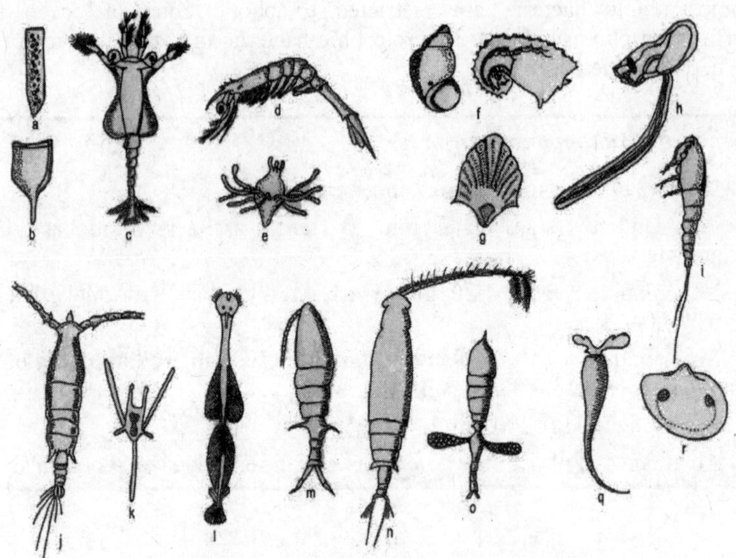

Fig. 1.12. Marine zooplankton (a) *Tintinnopsis* sp.; (b) *Flavella* sp.; (c) *Meganyctiphanes norvegica*; (d) Mysis (stage of Penaeid development); (e) Actinula larva; (f) Veliger larva; (g) *Cavolina* (h) *Oikopleura* sp.; (i) *Pseudocalanus elongatus*; (j) Calanoid (Acartia sp.); (k) Echinopluteus larva (Sea urchin); (l) *Sagitta* sp.; (m) *Pseudodiaptomus annandalei*; (n) *Eucalanus*; (o) *Oithona similis*; (q) *Creseis* sp.; (r) Veliger larva

Nekton are the active swimmers and they are most marine fish, squid, marine mammals marine reptiles and birds. Larger ones can swim against currents and their distributions are controlled by temperature, salinity and viscosity of seawater. Water pressure limits the vertical range of Nekton.

Benthos are the organisms living on or in the ocean bottom. There are three types of benthos: epifauna, infauna and nektobenthos. Epifauna live on surface of the sea floor, attached to rocks or moving along the bottom. Infauna live buried in the sand, shells or mud. Nektobenthos live on the bottom but also swim through the water above the sea bottom.

- All marine organisms belong to one of the three domains: Archaea, Bacteria and Eukarya. The marine organisms belong to five major kingdoms: Monera, Protists, Fungi, Plantae and Animalia.
- Marine organisms are classified into (1) plankton (floating/drifting organisms), (2) Nekton (swimmers) and (3) benthos (bottom dwellers).
- Phytoplankton are plant component of plankton, while zooplankton are animal component of plankton.

- Holoplankton are permanent plankton, free-floating throughout entire life span, while meroplankton are temporary plankton, free-floating only during juvenile and larval stages of adults.
- The plankton are classified based on size: macroplankton (>200 μm), microplankton (20-200 μm), nanoplankton (2-20 μm), picoplankton or ultraplankton (<2 μm) and Femtoplankton (<0.2 μm).
- Benthos are of three types: epifauna, infauna and nektobenthos.

Questions for Review/Exercise*

(1) Define 'Marine ecology' and explain its components.

(2) Tabulate the differences between opportunistic and equilibrium species.

(3) How are the marine ecological studies important?

(4) Explain the major divisions of marine environment.

(5) What are the striking differences between marine organisms and terrestrial organisms?

(6) Explain about the uniqueness of marine environment and marine organisms.

(7) Explain about the different groups and types of marine organisms.

* Frequent field visits to marine environment, marine museum and marine research laboratories.

Environmental Marine Ecology 2

2.1. PHYSICAL ENVIRONMENTAL FACTORS

The fundamental aspect of marine ecology is to study the relationship between the physical environment and marine life. Physical environmental factors such as temperature, light, salinity and pressure are important in the marine environment. These factors interact with each other and also with living organisms. The marine organisms adapt to withstand the physical factors for their survival.

2.1.1. Temperature

Temperature is one of the main factors in the marine environment. It is less than 0°C in the polar oceans and more than 30°C in the tropical oceans.Temperature varies from -1.9°C in the Arctic and the Antarctic oceans to 42°C in the Red sea and the Persian Gulf. The vertical profile of temperature is also not uniform in different latitudes. In the high latitudes, the difference between the surface water and the deep water of the sea is very little, when compared to the low and middle latitudes. In general, temperature is maintained more or less constant in deep sea, providing a stable environment for marine organisms. Thus, deep ocean waters are constant in temperature, ranging from -1.9°C in the Arctic to 4°C in equator. The temperature shows diurnal variations as the sunlight intensity varies during day time. The daily temperature variation is very small (0.25°C) in the seawater surface, since there is a continuous circulation of water. But the daily temperature fluctuates significantly in the shallow waters. Seawater surface temperature varies with latitude.

The temperature of seawater is mainly influenced by absorption of sunlight. The sunlight reduces when passing through the seawater and the light intensity reduces with increasing depth. Thus, the temperature decreases from sea surface to deep sea. Below the warm surface water, the temperature begins to fall and it undergoes a sudden decline at the depth range of 50-300 metres at an average of 200 metres. This depth zone of sudden decline in temperature is the 'thermocline' (Fig. 2.1). Below the thermocline, the water is cold and only a slight decrease in temperature towards bottom. The thermocline acts as a boundary layer between the warm surface water and cold deep water. The seawater zone with thermoclines is known as the 'discontinuity layer'. Thermocline is stable throughout the year in the tropical waters, while the thermocline occurs in the temperate waters during summer months. Thermoclines are absent in polar waters. There are different types of thermoclines – permanent thermoclines, seasonal thermoclines and diurnal thermoclines.

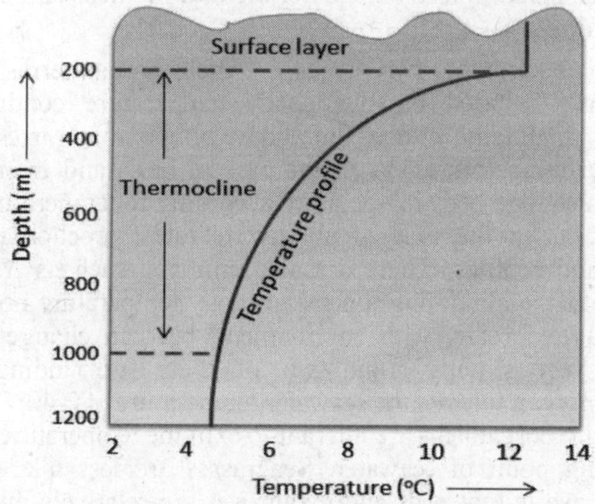

Fig. 2.1: Thermocline layer of sudden decline in temperature in the ocean

Temperature is important in changing the density and oxygen solubility in the seawater. Warm seawater is less dense than cold seawater of the same salinity. Oxygen solubility is more in dense seawater than that in the low dense water. The density and oxygen solubility in seawater get doubled, when the temperature of the seawater drops from 30°C to 0°C. The sudden decline in temperature that produces the thermocline also causes rapid increase in seawater density. This zone of rapid change in water density is termed as 'pycnocline'.

Temperature can control distribution, metabolism, growth and reproduction of an organism. For every 10 degree rise in temperature, the metabolic activity becomes doubled. Warm waters have more number of species, but lower total biomass production. In contrast, cold waters have less number of species, but higher total biomass production. Polar organisms grow slower, reproduce less frequently and live for a longer period than tropical organisms. Temperature can indirectly influence organisms by changing the activities of their predators or pathogens. Rise in temperature increases the predation and diseases in marine organisms. Therefore, the global warming is likely to increase the problems of predation and disease in marine organisms.

2.1.1.1. Temperature Tolerance

All living organisms are tolerant to different ranges of temperatures. Based on temperature tolerance, there are two types of organisms: 'Stenothermic' or 'eurythermic'. The stenothermic organisms can withstand only very small changes in temperature, whereas the eurythermic organisms can withstand large and rapid changes in temperature. The oceanic organisms are mostly stenothermic and the near-shore organisms are eurythermic.

The animals are of two types – the 'homiotherms' and the 'poikilotherms' – based on their body temperature conditions. The homiotherms maintain a constant body temperature irrespective of environmental temperatures, as in the case of birds and mammals. The poikilotherms change body temperature according to temperatures of their surroundings, as in the case of all invertebrates, prochordates, fishes, amphibians and reptiles. There are also animals, such as *Nerita, Uca,* known to neither maintaining a constant body temperature nor changing body temperature greatly with environment, but can change their body temperature very slightly from that of their surroundings. Marine microorganisms can tolerate the seawater temperature of 0-40°C. However, some of the microorganisms are tolerant to extreme temperatures from sub-zero to boiling point of seawater. Seagrasses are less tolerant to high temperatures, while seaweeds and mangroves are relatively more tolerant to high temperatures in the shallow waters and intertidal areas. Coral reefs and mangroves can not survive low temperatures and hence absent in the temperate waters, thereby restricted to tropical and subtropical waters.

2.1.1.2. Temperature and Reproduction

Temperature has a significant effect on reproduction of marine organisms. There are critical temperatures for the maturation and liberation of gametes. In general, marine organisms reproduce at narrow range of temperatures. For example, the eelgrass *Zostera marina* grows well between 10 and 20°C, but reproduces only between 15 and 20°C. In the Arctic and Antarctic regions, the polar species spawn only in spring and summer. Temperature requirement varies with stages of reproduction, such as gonad maturation and spawning. In marine invertebrates, these two processes are distinct. The optimal temperature required for gametogenesis is greater than spawning in the oysters. Temperature determines even the sex. For instance, sea turtle produces females at 31°C and males at 29°C.

2.1.1.3. Temperature and Activity

The respiratory activity increases with increased temperatures. In poikilotherms, the rate of metabolism increases about 2.5 times for each 10°C rise in temperature. In the polychaetes such as *Nereis* species, the swimming activity is increased from 13°C to 22°C, and there is no swimming below 13°C. The rate of beating of cirri in the barnacle *Balanus balanoids* increases regularly from 2 to 21°C and the beating activity is irregular below 2°C and above 21°C, but the activity is stopped completely at above 27°C. Boring rate by *Teredo* species increases with increased temperature from 5°C to 15°C.

2.1.1.4. Temperature and Growth

Temperature influences growth of marine organisms. The copepod *Calanus* species have two size groups: the larger ones are distributed in waters of low temperature, and the smaller ones occur in warm surface waters. However, diatoms are larger in warmer waters. Similarly the sea urchins and the gastropods are larger in warmer waters than their counterparts in the colder seas. Hydroids and algae grow luxuriantly in winter, while oyster, barnacle, bryozoans and other fouling organisms settle and grow fast in summer.

2.1.1.5. Temperature Periodicity

In many organisms, fluctuating temperatures favour the growth more than a constant temperature. Many animals except sessile forms, try to avoid unfavourable temperatures by moving away from them. The gastropod, *Littorina littorea* moves down from the intertidal zone to warmer conditions during winter. In most cases, migration takes place to avoid unfavourable temperature conditions or to get suitable temperature ranges for activities, such as mating and spawning. The periodicity is often related to other physical factors such as light and diurnal rhythms. Zooplankton descend deeper as the light increases but the descent is stopped when a low temperature is encountered. As the light decreases in the late afternoon, the zooplankters move up towards the sea surface.

Sea surface temperature (SST) is important to influence the whole structure of marine communities. The filter feeders, such as barnacles (*Balanus* spp.) and mussels (*Mytilus* spp.) increase in population linearly as mean SST increases, and this is linked to the increasing numbers of juveniles settling with increasing SST. However, the primary producers, such as seaweeds decrease with increasing SST and this is linked to increased numbers and activity of herbivores.

One of the most threatening physical factors in marine environment is the elevated SST on coral bleaching. The optimum temperature for most hard (Scleractinian) corals is between 25 and 29°C, while increases in water temperature to 31°C can cause serious loss of coral symbiotic algae namely zooxanthellae. Corals may die on a massive scale, if the symbionts are unable to recolonise or the warm conditions persist for a long time. Thus climate change and global warming are exerting pressures on the marine ecosystem. At the same time, some coral species and their algal components are getting adapted to increasing water temperature. However, it is predicted that the corals may not survive in the next 100 years.

- Temperature of seawater is mainly influenced by absorption of sunlight. The daily temperature variation is very small in the seawater surface, since there is a continuous circulation of water. But the daily temperature fluctuates significantly in the shallow waters. Deep ocean waters are constant in temperature.
- The temperature decreases from sea surface to deep sea. In the high latitudes, the difference between the surface water and the deep water of the sea is very little, when compared to the low and middle latitudes.
- There is a sudden decline of temperature at a depth range of 50-300 metres. This depth zone of sudden decline in temperature is the 'thermocline'. The thermocline is stable throughout the year in the tropical waters, while it occurs in the temperate waters during summer months. Thermoclines are absent in polar waters.
- Warm seawater is less dense with less oxygen solubility as compared to cold seawater.
- Warm waters have more number of species, but lower total biomass production. In contrast, cold waters have less number of species but higher total biomass production.
- The oceanic organisms are tolerant to a narrow range of temperature (stenothermic), while the near-shore organisms are tolerant to a wide range of temperatures (eurythermic).
- Temperature has a significant effect on respiration, growth and reproduction in each species.
- One of the most threatening physical factors in marine environment is the elevated sea surface temperature on coral bleaching.

2.1.2. Light

Light is an essential ecological factor to marine life. The natural sources of light are the sun, the moon, the stars and luminescence of organisms. Of these, sunlight is the largest source of energy for the photosynthetic process of marine plants, upon which the whole marine life depends. Light penetrates into the seawater and warms it. The intensity of light varies with time of the day, seasons, latitude of the area, cloudy condition and atmospheric absorption. Light penetration is high in the open ocean in the tropics, where turbidity and plankton density are low. The light penetration declines with depth of seawater. Of the light that enters the water, about 30% is lost in one meter depth, 50% in the upper five meters and only about 10% penetrates beyond a depth of 10 metres. Coastal waters are usually turbid than offshore ocean waters. Very little light penetrates more than a few metres into the coastal waters. In the waters with the suspended solids of 1 gram per litre, the light may penetrate less than 1 metre.

WHY IS THE SEA BLUE OR GREEN OR GREY?

The light penetration depends on its intensity on the sea surface and also transparency of water. The penetration of light also varies with different wave lengths of the light into the water. Infrared and ultraviolet rays are absorbed at the surface itself, while wave length in the blue green range may penetrate well beyond 100 meters. Blue light penetrates deep in clear water and greenish yellow light in turbid water. Thus colour of the sea varies with this differential penetration, which in turn depends on transparency of water. Blue waters are typical of the deep ocean, where there is low level of suspended matter. Green waters are characteristic of coastal shelf water, where there is high level of suspended matter. Sea can look dark grey in a storm.

The ocean can be sub-divided into three zones based on availability of light (Fig. 2.2). The euphotic zone extends only up to a depth of 100 metres. Below this zone, down to 200 metres, there is a dimly lighted area and it is referred to as 'disphotic zone'. In the water below 200 metres, there is little or no light, this zone is the 'aphotic zone'. The euphotic zone is the depth where light is sufficient for photosynthesis, and this zone is also known as photosynthetic zone. It varies with the depth of light penetration which in turn depends on solar altitude, turbidity of water, time of day, geographic position, and season.

Euphotic zone is 40-50 meters from the surface in middle latitudes, and more than 100 meters in low latitudes if the water is clear. Below this euphotic zone, there is only very dimly illuminated zone, insufficient for plant survival and it is known as 'disphotic zone' or 'twilight zone' which receives only less than 5% of light incidence. Below this disphotic zone, there is no light and it is known as the 'aphotic zone' or 'dark zone' (Fig. 2.2).

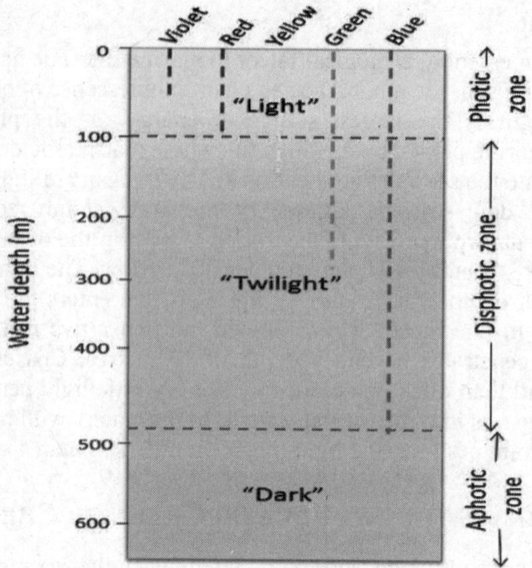

Fig. 2.2: Light penetration in relation to depth of ocean

During photosynthesis, oxygen is released. In contrast, oxygen is consumed during respiration. The light intensity at which oxygen production by photosynthesis and oxygen consumption by respiration equals is called the 'compensation point' and the depth at which the compensation point occurs is called the 'compensation depth'. Below the compensation depth, there will be no net primary production. This depth varies with changes in illumination, seasons of the year, time of the day, cloudiness and also with the clarity of the water. In the tropics, the compensation depth is well below 100 metres at noon in clear water throughout the year. But in high latitudes the compensation depth lies between 10 and 60 metres during summer. Sometimes, it decreases to zero during the winter months.

Coral bleaching is one of the most serious global threats to shallow marine tropical ecosystems. This is due to intense sun light with ultraviolet radiation. In such situation, the coral algae, zooxanthellae loss chlorophyll and may even die.

2.1.2.1. Light and Vertical Migration of Organisms

Each species has an optimum light requirement. Since the intensity of light at a particular depth varies throughout the day, the organisms are forced to perform vertical migrations. This behaviour of animals in relation to light intensity is performed by plankton and nekton, such as medusa,siphonophores,ctenophores,chaetognaths,pteropods,copepods,clad ocerans,amphipods, mysids, euphausids, pelagic decapods and others.

2.1.2.2. Light Period and Sexual Maturity

Light period delays maturity. Gonad development in female cuttle-fish *Sepia officinalis* is inhibited if the animals are illuminated for more than 12 hours per day. Similarly, light of more than 18 hours per day inhibits the formation of egg masses in the barnacle *Balanus balanoides*.

2.1.2.3. Light and Colour of Organisms

Light influences the colour of organisms. Colour of the bottom dwelling organisms is related to the colour of the background. Many littoral crustaceans associated with seaweeds acquire the colour of the substratum on which they live. Most of the pelagic organisms have transparent, bluish, greenish or brownish backs with silvery sides and bellies. The colour for organisms living in the surface waters of the sea is blue, as seen in the flying fish, the mackerel and the tuna. Colourless animals are characteristic of the open sea from near the surface to depths of about 150 metres, as in the case of Sagitta, ctenophores, copepod *Eucalanus* and fish larvae. This colourlessness is advantageous to escape from predators. The zone between 150 and 500 metres is colonized by silvery and grayish fishes. In the region below 500 metres of the ocean where there is no light the animals are reddish as seen in red-violet prawns, reddish copepods, bright red medusa and red cephalopods. Deep-sea fishes are mostly black, violet or brownish.

2.1.2.4. Light and Size of Eyes

Light has impact on the eyes of marine organisms. Many organisms depend on their eyesight to detect their prey. There is a close relationship between the nature of the eyes and the depth in which animals live. The size of the eyes is larger in the surface waters and then it decreases in the deeper lightless zone. Many of the deep-sea organisms have small eyes. In the dimly lighted areas, the eyes are enlarged due to the overgrowth of the lens, to provide telescopic vision to the fishes. The small eyes of deep sea fishes contain only the rod cells which are more sensitive. In some deep water species, the eyes are tubular structures to provide a binocular vision to assist perception of distance.

HOW DO MARINE ORGANISMS ADAPT THEMSELVES TO DARK CONDITIONS OF THE OCEAN?

2.1.2.5. Adaptations for Life in Darkness

Marine organisms have unique adaptation to live in the dark conditions of the ocean. There are three important adaptations: (i) development of tactile structures, (ii) special food procuring devices, and (iii) light production or bioluminescence. In general, deep-sea animals produce elongated thread-like tails which function as tactile organs. These organs are produced either by fin rays or from head portion. The main food procuring device of the deep sea organisms is the presence of a larger mouth and elastic stomach for swallowing large amount of food stuff, as well curved teeth to prevent the prey from escaping once it is caught.

2.1.2.6. Bioluminescence

Bioluminescence or 'living light' occurs mostly in deep sea environment. Bioluminescence is the production and emission of light by a living organism. It is a hybrid word, derived from the Greek word "*bios*" for "living" and the Latin word "*lumen*" for "light". It is more common in tropical waters than in higher latitudes. The light is produced by a biochemical reaction in which the enzyme luciferase breaks down the luciferin and emits light. This light is predominantly blue as well as yellow and green, but free from ultraviolet and infrared rays. Certain bacteria like bioluminescent bacteria are able to produce light and the bacteria are present in marine organisms, such as dinoflagellates, radiolarians, hydroids, ctenophores, bryozoans, polychaetes, brittle stars, many crustaceans, gastropods, bivalves, cephalopods, chordates and fishes. The bioluminescent bacteria are present on the body surface or within the cells of the marine organisms. Many mollusks do have luminescent organs. In the dinoflagellate, *Noctiluca*, light is produced by the granules present in the periphery of the cell. In hydromedusae, the luminous cells are grouped beneath the endoderm. Ctenophores have glandular structures within the radial canals. Certain penaeid shrimps have photophores over the appendages. Bioluminescence is the only source of light for many deep sea organisms in locating predators and prey, attacking prey and finding mates for reproduction. Some organisms use the bioluminescence in courtship behaviour. For example, the male and female partner of polychaete *Odontosyllis* emits light during spawning.

- The light penetration depends on its intensity on the sea surface and also transparency of water. The light penetration declines with depth of seawater. Blue waters are typical of the deep ocean, where there is low level of suspended matter. Green waters are characteristic of coastal shelf water, where there is high level of suspended matter.
- The ocean can be sub-divided into three zones based on availability of light: euphotic zone, disphotic zone (twilight) and aphotic zone (no light).
- The light intensity at which oxygen production by photosynthesis and oxygen consumption by respiration equals is called the 'compensation point' and the depth at which the compensation point occurs is called the 'compensation depth'. Below the compensation depth, there will be no net primary production. This depth varies with changes in illumination, seasons of the year, time of the day, cloudiness and also with the clarity of the water.

- Some organisms perform vertical migrations in order to get an optimum light requirement. Light period delays maturity.
- Colour of the bottom dwelling organisms is related to the colour of the background. The dominant colour for pelagic organisms living in the surface waters of the sea is blue. Colourless animals are characteristic of the open sea. Where there is no light the animals are reddish.
- The size of the eyes is larger in the surface waters and then it decreases in the deeper lightless zone.
- To live in the dark conditions of the ocean, marine organisms have unique adaptation: (i) development of tactile structures, (ii) special food procuring devices and (iii) light production or bioluminescence.
- Bioluminescence occurs mostly in deep sea environment. It is more common in tropical waters than in higher latitudes. The light is produced by a biochemical reaction in which the enzyme luciferase breaks down luciferin and emits light. This light is predominantly blue, but free from ultraviolet and infrared rays. Certain bacteria are able to produce light and they are present in marine organisms. Bioluminescence is important for animals in locating predators and prey, attacking prey and finding mates for reproduction.

2.1.3. Salinity

Salinity is the most important physical factor of seawater. Total amount of dissolved material is termed as 'salinity'. Ocean water contains 35 grams of dissolved compounds in one litre of seawater in other words seawater contains 3.5% dissolved salts, and the remaining 96.5% water. The term used for the salinity is practical salinity units (psu) or parts per thousand (ppt) and the salinity of seawater is 35psu or ppt. Mixtures of seawater and freshwater are termed 'brackish' and the salinity ranges between 8 and 33ppt.

The salinity of sea surface depends mainly upon two important factors: (1) evaporation and (2) precipitation. These are closely related to the atmosphere. Salinity is high in the seas where surface evaporation is high, and freshwater input is low. The Red Sea in the Indian Ocean has the saltiest seawater in the world; Part of it is known as the Dead Sea. The Mediterranean Sea has a surface salinity of about 38-39ppt. Surface water salinity varies across the globe. It is lowest towards the Polar Oceans and highest in the tropical latitudes where surface evaporation is greatest. Low salinity usually occurs in the polar seas through the melting of ice during summer and also in the coastal belts and in estuaries where there is an influx of fresh water. Low salinity is also recorded during high rainfall or where there is dilution through river discharge into the sea. Salinity also varies with depth. Lower saline water is less dense, and will lie over higher saline water.

In India, the salinity range of the surface waters exhibits a great seasonal variation. The Bay of Bengal has a salinity of about 33-34ppt, which is lower than that of the Arabian Sea. This is because of the large freshwater discharge from six large rivers namely the Ganges, the Brahmaputra, the Mahanadi, the Godavari, the Krishna and the Cauvery from Indian side and four rivers, namely the Irrawadi, the Sittang, the Salween and the Tennasserim from the eastern side. The Arabian Sea has a salinity of about 36ppt throughout the year. This is because only three large rivers namely the Indus, the Narmada and the Tapti flow into the Arabian Sea. Another reason is the high evaporation and the influx of high saline water with 50ppt from the Red Sea and Persian Gulf.

2.1.3.1. Salinity Tolerance and Distribution of Organisms

Living organisms are tolerant to different range of salinity. Based on this tolerance, two types of organisms are classified as 'Stenohaline' and 'Euryhaline'. The stenohaline organisms are tolerant to narrow range of salinity, whereas the euryhaline organisms are tolerant to a wide range of salinity. The oceanic organisms are mostly stenohaline and the nearshore organisms are euryhaline. Most stenohaline organisms are isotonic with seawater and they have not developed any mechanism of osmoregulation. Salinity in body fluid of stenohaline forms is equivalent to that in ocean and there is no net transfer of water. Most of the euryhaline forms have the ability of osmoregulation. In hypertonic osmoregulation, the body fluid is more saline than surrounding seawater and the water passes in to the cells; in order to overcome excessive entry of water, the organisms do not drink water, absorb salt through cells and excrete large volume of much diluted urine (Fig. 2.3a). In hypotonic osmoregulation, the body fluid is less salty rather than the surrounding seawater and the water passes from cells into the seawater. In order to avoid this, organisms drink seawater, excrete salt through special chloride releasing cells, and produce low urine by kidney (Fig. 2.3b).

Most of the fishes are stenohaline, but many of them acclimatize to slow salinity changes. The milk fish *Chanos chanos* maintains an internal concentration of chloride within a range of 0.5%, although the external medium ranges from nearly freshwater to about 50ppt. Euryhaline fishes such as the eel *Anguilla* (Fig. 2.4), when in seawater they drink seawater as they are unable to absorb external water against a concentration gradient and the fishes also excrete salt through the gills. There is an efficient osmoregulation in more specialized crustaceans. *Metapenaeus monoceros* maintains the same level of sodium chloride in the blood while the salinity fluctuates from 6 to 30ppt.

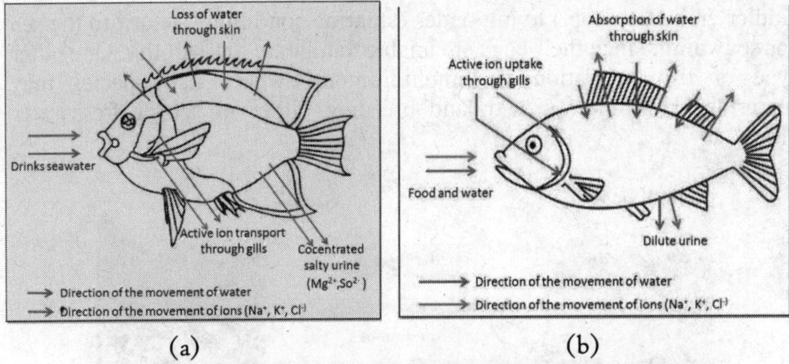

Fig. 2.3 : In hypertonic osmoregulation (a) and hypotonic osmoregulation (b) processes in fishes with their surroundings

Fig.2.4 : *Anguilla* sp., an euryhaline fish

Organisms are classified by their ability to withstand salinity variation. The 'obligate freshwater organisms' do not live in waters that exceed 8 ppt, whereas the 'obligate marine organisms' which will not tolerate salinities below 30ppt. Echinoderms, such as starfish and sea-urchins are stenohaline and their water vascular systems function only when their internal body fluids are isotonic having an equal osmotic pressure with the surrounding seawater. Many of the crustaceans (shore crabs) and fish (salmon, flounder and eel) are euryhaline. There are two types of euryhaline species: (1) anadromous and (2) catadromous. Many fish (e.g. *Tenulosa ilisa*) species undertake their growth in the sea and return to freshwater for breeding. These species are termed 'anadromous' species. In contrast to this, species such as eels (e.g *Anguilla* spp.) undertake their growth in the freshwater and return to the seawater for breeding. These species are termed 'catadromous'. The blue crab (*Portunus* spp.) and the fiddler crab (*Uca* spp.) living under estuarine conditions, return to the sea for spawning, since their eggs are unable to tolerate low salinity. One other type of fish migration is 'amphidromous' where fish species may undertake both their growth and breeding either in sea or freshwater (Fig. 2.5).

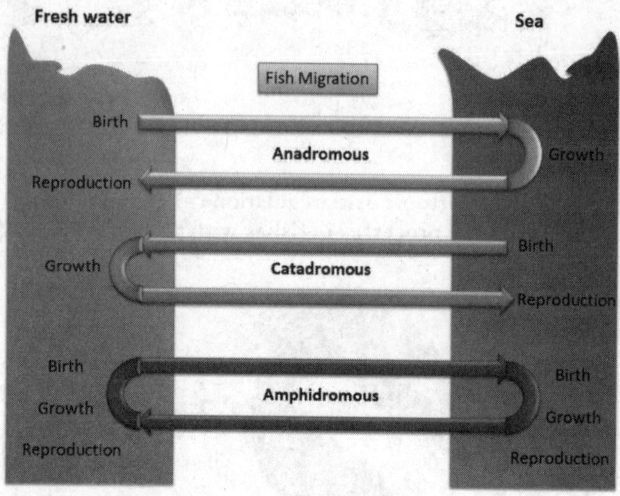

Fig. 2.5 : Types of fish migration in relation to freshwater and seawater

2.1.3.2. Salinity as Related to Size

Euryhaline animals are smaller in size when they live in reduced salinities, while the animals are larger in size when they live in high saline waters. The smaller size of marine species in lesser salinities can be attributed to that the organisms spend a considerable part of their energy for the maintenance of the water balance and also to the non-availability of food.

- Total amount of dissolved material is termed as 'salinity'. Ocean water contains 35ppt that is 35 grams of dissolved compounds in one litre of seawater. Mixtures of seawater and freshwater are termed 'brackish' and the salinity ranges between 8 and 33ppt.
- The salinity of sea surface depends mainly upon two important factors: (1) evaporation and (2) precipitation. Salinity is high in the seas where surface evaporation is high, and freshwater input is low.
- Stenohaline organisms are tolerant to narrow range of salinity.
- Euryhaline organisms are tolerant to a wide range of salinity.
- Stenohaline organisms have no osmoregulatory mechanism.

- Euryhaline organisms have osmoregulatory mechanism.
- In stenohaline organisms, salinity of body fluid is equivalent to that in seawater and there is no net transfer of water.
- In hypertonic osmoregulation, the body fluid is more saline than seawater and the water passes into cells. To avoid this, organisms absorb salts and excrete urine.
- In hypotonic osmoregulation, the body fluid is less saline than seawater and the water passes from cells. To avoid this, organisms excrete salts and produce less urine.
- The oceanic organisms are tolerant to a narrow range of salinity (stenohaline) and the near-shore organisms are tolerant to a wide range of salinity (euryhaline).
- Euryhaline animals are smaller in size when they live in low salinities, while the animals are larger in size when they live in high salinity.
- Based on migration pattern, there are different types of euryhaline species: (1) anadromous, (2) catadromous and (3) amphidromous.
- Anadromous fish grow in the sea and migrate to freshwater for breeding.
- Catadromous fish grow in the freshwater and migrate to seawater for breeding.
- Amphidromous fish grow and breed in either freshwater or in seawater.

2.1.4. Pressure

Pressure is one of the significant physical factors influencing the marine organisms. Pressure increases with the depth of seawater. The pressure at the surface of the sea is 1 atmosphere, and increases by 1 atmosphere for every 10 metres increase in depth. The average depth of the oceans is about 3800 metres and so the average pressure of the ocean bottom is about 380 atmospheres. Thus in the deeper oceans exceeding 10,000 metres, the pressure may even reach one ton atmospheres on each square centimeter of the ocean bottom. Organisms in particular bacteria are adapted to live in very deepwater at very high pressures are termed 'piezophilic'.

Pressure influences the distribution of animals. It acts as a limiting factor for vertical distribution of the animals. When the fish is moving downward the swim bladder is compressed and when it moves upward, the bladder expands. The fish has to remove or add the gas to the swim bladder in order to maintain buoyancy equilibrium. The levels of pressure and carbon dioxide are higher in the deep sea than in surface waters. This condition makes calcium carbonate more soluble in deep waters and so incorporation of calcium carbonate into shells and bones is very difficult. This is reflected in most of the deep sea organisms in which bony structures are very poorly developed.

Eurybathic animals are tolerant to a wide range of varied pressure and performing vertical migration, while stenobathic animals are restricted to a narrow range of pressure and they do not perform vertical migrations. Many plankton and nekton organisms move vertically down the seawater with increasing pressure. These eurybathic organisms are undergoing varied pressures daily. Some organisms such as air breathing animals and fishes with air bladders are stenobathic organisms, living in the surface waters and restricted to a particular vertical depth range.

- Pressure increases with the depth of seawater. The pressure at the surface of the sea is 1 atmosphere and it increases by 1 atmosphere for every 10 metres increase in depth. Organisms in particular bacteria are adapted to life in very deepwater at very high pressures are termed 'piezophilic'.
- Eurybathic animals are tolerant to a wide range of varied pressure and performing vertical migration, while stenobathic animals are restricted to a narrow range of pressure and they do not perform vertical migrations.

2.1.5. Oceanic Currents

Currents are the water movements that are caused by the combined effects of the wind action on the sea surface and the density difference of water masses. The currents are also caused by tides. There are three types of currents: (1) wind currents or horizontal currents, (2) density currents or vertical currents, and (3) tidal currents. Currents are like rivers in the oceans. They can carry heavy amounts of water with a greater speed. The Antarctic Circumpolar Current can transport 100 million cubic metres of seawater per second. Some currents travel at the speed of 4 metres per second. One drop of seawater may take 5,000 years to travel through all the world's oceans.

Ocean currents are induced mainly by wind acting on the sea surface. The rotation of the Earth rotates current flow to the right of the wind direction in the northern hemisphere and to the left of the wind direction in the south, through a force. This force is known as 'Coriolis force', and it is zero at the Equator (Fig. 2.6).

Horizontal currents circulate in paths called 'gyres', which rotate in a clockwise direction in the northern hemisphere and an anti-clockwise direction in the southern hemisphere. Horizontal currents also occur in the deep-sea. Upwelling occurs in the areas where adjacent surface currents diverge towards different directions. In low latitudes, divergence between the equatorial currents and counter-current causes upwelling close to the

Equator and this helps in maintaining a high productivity for the equatorial waters.

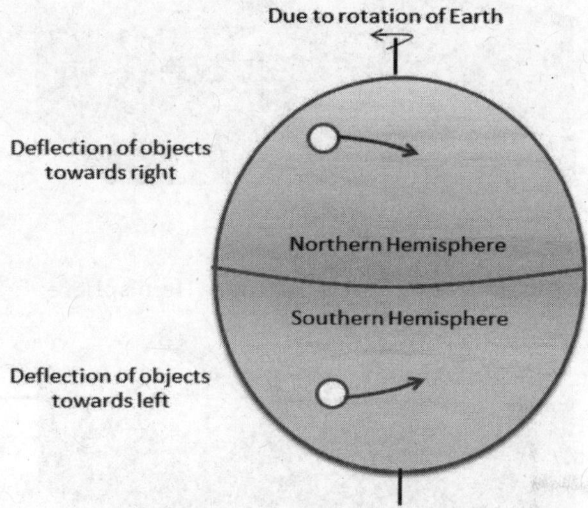

Fig. 2.6: Coriolis force through which current flows to right of the wind direction in the northern hemisphere and to left of the wind direction in southern hemisphere due to rotation of Earth.

Vertical currents in the ocean are driven by small differences in water density due to differences in salinity and temperature. These water movements are termed 'thermohaline circulation'. Increased salt content increases the density of water and the density of water decreases with increasing temperature. However, below 4°C the density of pure water starts to decline with decreasing temperature and this property is very important in Arctic and Antarctic waters.

Tidal currents cause only very little water movement and do not bring about large-scale transport of water to long distances. The maximum speed of tidal currents at spring tide in the Andhra coast in India is estimated to be only 3-4 cm per second.

What are upwelling and down-welling?

Upwelling is the movement of water offshore resulting in replacement by cooler, nutrient-rich water originating from the bottom. This results in the transport of large amounts of nutrients from the bottom and great biological productivity in the surface waters. Reverse to upwelling is down-welling. In this process, surface water with low nutrients falls down to the sub-surface level of the sea (Fig. 2.7).

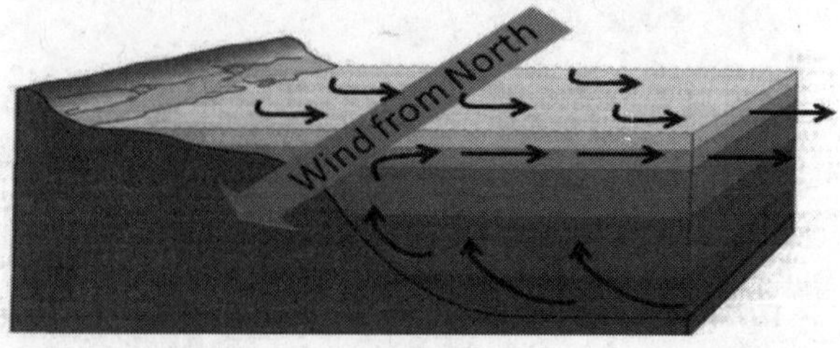

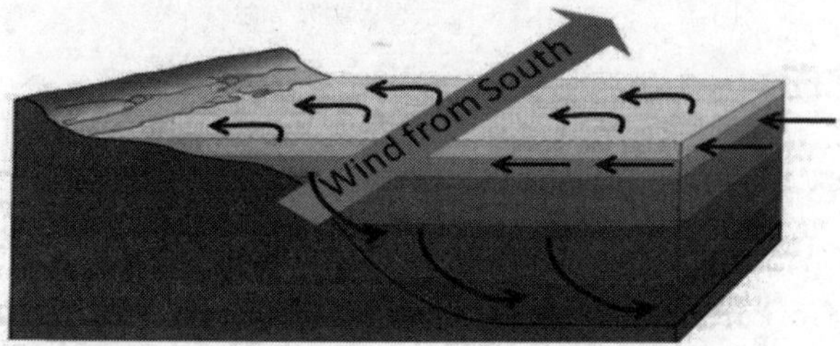

Fig. 2.7: Processes of upwelling and down-welling in Northern hemisphere

ARE WATER CURRENTS CRUCIAL FOR MARINE LIFE?

Water currents are crucial for marine life. Currents keep the water well-mixed, influence the distribution of salinity and heat, bring the nutrients to the surface, necessary for primary productivity and carry oxygen to the depths of the oceans. Currents help in the dispersal of organisms by transporting them to different areas, favourable for settlement, growth and reproduction. Besides dispersal, the benthic organisms depend on currents for respiration, waste disposal and food requirement. The dependence of sessile, benthic organisms to water current is great. The currents help them to get food particles. The sperms are carried to the ova mainly by water movement or the gonad elements of both male and female are discharged into the surround water where currents bring them together to help fertilization. The European eel fish *Anguilla vulgaris* travels 2000 to 3000 miles to reach the sea near Bermuda for spawning.

Many marine organisms are adapted to tidal currents. The lugworm, *Arenicola marina* lives in a U-shaped tube in sand or mud. The waste opening of the tunnel is raised above the seabed by only a centimeter or so, but the other opening is flushed with the sand surface. When the tide is flowing, this height difference creates a pressure difference, which, in turn, causes water to flow into the lower opening and so helps to supply the worm with fresh seawater and hence food and respiratory gases.

The current velocity influences the distribution of filter feeders that live close to the seabed. Active filter feeders are animals that create their own feeding currents and they are most abundant closer to the seabed with slow current speed. Examples of active filter feeders are bivalves, sponges and sea squirts. Passive filter feeders like cnidarians are animals that do not create their own feeding currents and they are abundant in the water column with faster current speed.

- Currents are water movements that are caused by the combined effects of the wind action on the sea surface and the density difference of water masses. The currents are also caused by tides.
- There are three types of currents: (1) wind currents or horizontal currents, (2) density currents or vertical currents and (3) tidal currents.
- Ocean circulation is induced by wind acting on the sea surface. The rotation of the earth rotates current flow to the right of the wind direction in the northern hemisphere and to the left of the wind direction in the south, through a force. This force is known as 'Coriolis force', and it is zero at the Equator.
- Horizontal currents circulate in paths called 'gyres', which rotate in a clockwise direction in the northern hemisphere and an anti-clockwise direction in the southern hemisphere.
- Upwelling is the movement of water offshore resulting in replacement by cooler, nutrient-rich water originating from the bottom. This results in the transport of large amounts of nutrients from the bottom and great biological productivity in the surface waters. Reverse to upwelling is down welling.
- Vertical currents in the ocean are driven by small differences in water density due to differences in salinity and temperature. These water movements are termed 'thermohaline circulation'.
- Tidal currents cause only very little water movement and do not bring about large-scale transport of water to long distances.
- Currents keep the water well-mixed, influence the distribution of salinity and heat, bring the nutrients to the surface necessary for primary productivity and carry oxygen to the depths of the oceans. Currents help in the dispersal of organisms by transporting them to different areas, favourable for settlement, growth and reproduction. Besides dispersal, the benthic organisms depend on currents for respiration, waste disposal and food requirement.

2.1.6. Tides

Tides are the periodic rise and fall of the sea. When sea level is at its highest height, the tide is called 'high tide'. When sea level is at its lowest height, the tide is called 'low tide'. The most important tidal waves are caused by the gravitation interaction between the Earth and the Moon (lunar waves) and also between the Earth and the Sun (Solar waves). The height of the average solar tide is about 50% of the average lunar tide. Thus moon has strong influence on tides. The gravitation attraction of the moon causes the oceans to bulge out in the direction of the Moon. Another bulge occurs on the opposite side (Fig. 2.8). As the earth is rotating, there are about two high tides per day, but the moon actually takes about 24 hours and 50 minutes to return to the same position in the sky from one day to the next day. Thus in general, each high tide is 12 hours and 24 minutes. Spring tides are strong tides that occur when the Earth, the Sun and the Moon are aligned in a straight line and the gravitational pull of the moon and the sun are working together. This arrangement occurs at the full and new moons, so that there are two spring tides every month. The smallest tidal ranges over the lunar cycle are called 'neap tides' and these occur when the sun and moon are placed in opposite directions. These events occur during the first and last quarters of the moon.

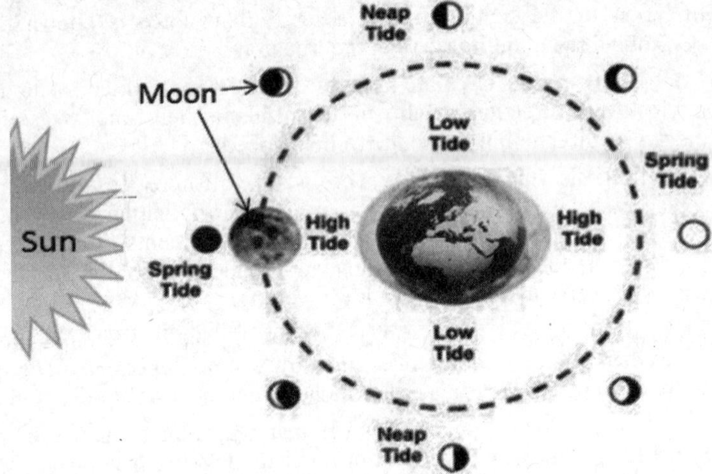

Fig. 2.8 : Formation of different tides

2.1.6.1. Types of Tides

There are three types of tides: (1) diurnal tides, (2) semi-diurnal tides and (3) mixed tides – based on the geometric relationship of moon and sun to locations on the Earth's surface. The 'Diurnal tides' have one high tide and one low tide per day. These are common in parts of the northern Gulf of Mexico and Southeast Asia. The 'Semi-diurnal tides' have two high tides and two low tides per day. They occur in many parts of the world. The 'mixed tides' have highest high tide and lowest high tide as well as highest low tide and lowest low tide. The mixed tides occur around the west coast of Canada and the United States. The biggest tide is in the Bay of Fundy, Canada with the height difference between high and low water is 16.3 metres and this is even taller than a three-story building.

The tides provide two conditions – exposure and immersion – to the intertidal zone and also to marine organisms living there. The duration of exposure or immersion controls the duration of adverse factors, such as desiccation and insulation, as well as beneficial factors, such as feeding and other physiological activities.

WHAT ARE THE BIOLOGICAL EFFECTS OF TIDES?

Currents formed by the tides are of advantage especially to sedentary forms in the intertidal zone as they function as carriers of oxygen and nutrients, also as removers of waste products. They may also adversely affect organisms by damaging them mechanically by disturbing the sediments from soft bottom and making the area unsuitable for colonization.

Many marine organisms synchronize their behaviour to the tides. The tides govern the mass spawning of corals, the swarming of bioluminescent worms and the release of spores by seaweeds. The marine alga *Dictyota* produces its gametes at the time of the full moon spring tide. Similarly spawning occurs during spring tides in the case of gastropod, *Littorina* species living at the intertidal zone. In the coral *Pocillopora bulbosa*, spawning occurs at spring tides at new moon in summer and full moon in winter. In the case of mussel, *Mytilus edulis,* maturation of gonads takes place during the period of new moon spring tides and spawning during the subsequent first quarter neaps, full moon springs and third quarter neaps.

- Tides are the periodic rise and fall of the sea. The most important tidal waves are caused by the gravitation interaction between the Earth and the Moon (lunar waves) and also between the Earth and the Sun (Solar waves).
- The height of the average solar tide is about 50% of the average lunar tide.
- Spring tides are strong tides that occur when the Earth, the Sun and the Moon are aligned in a straight line. This arrangement occurs at the full and new moons, so that there are two spring tides every month.

- The smallest tidal ranges during the lunar cycle are called 'neap tides' and these occur when the sun and moon are placed in opposite directions. These events occur during the first and last quarters of the moon.
- The tides provide two conditions - exposure and immersion - to the intertidal zone and also to marine organisms living there.
- Many marine organisms synchronize their behaviours like spawning to the tides.

2.1.7. Waves

Oceanic waves are caused by winds blowing over the sea surface. The waves do not move horizontally, they only move up and down. The wind pushes the water upwards and then gravity pulls it down again. When the wind is strong, the waves are large. The waves range from ripples of less than a centimeter in height to giant waves of more than 30 metres in height. In addition to the ordinary sea waves, internal waves occur in the deep sea. These are formed at the boundary between two distinct types of water of differing density and also when strong winds blow over the areas.

What are 'storm surges' and 'tsunamis' or 'seismic waves'?

In addition to wind, change in atmospheric pressure over the sea causes the waves associated with storm termed the 'storm surges'. The storm surges are significant in the shallow areas, occurring usually in the North Sea between Iceland and Scotland and in the Bay of Bengal. Seismic waves or tsunamis are destructive giant waves caused by underwater earthquakes and sub-marine volcanic eruptions. The seismic waves occur in the open sea and also near the coasts.

Wave strength depends on the speed and duration of the wind and the distance of sea surface over which it moves. The wave height is the distance between the wave crest (the highest point of the wave) and trough (the lowest point of the wave). The distance between wave crests is termed the 'wave length' (L). The time taken by two successive crests to pass a fixed point is the 'wave period' (T). Wave speed (V) is calculated as

$$V = L/T$$

The maximum depth at which the wave speed is experienced is half the wave length (Fig. 2.9). Therefore organisms and habitats will be less affected by the wave in the deeper water. As water approaches the shore, some of it is moved parallel to shore in the form of long shore currents and some moves rapidly offshore in concentrated rip currents. Long shore currents are responsible for the transport of sediment parallel to the shore.

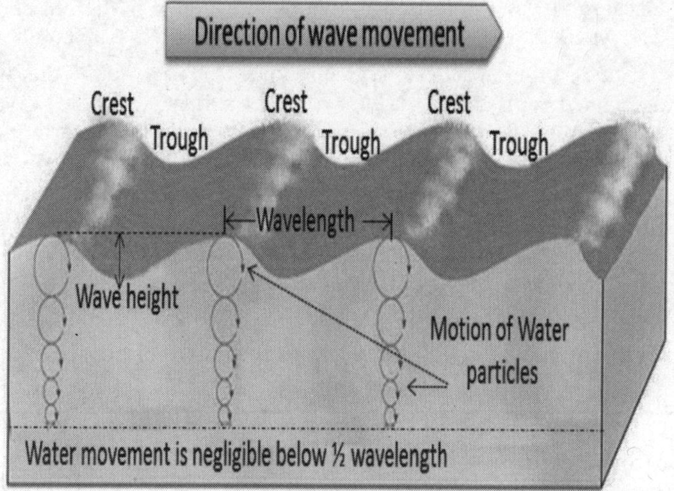

Fig. 2.9 : Crest and trough of waves and reduction of waves in shallow water

WHAT ARE THE BIOLOGICAL EFFECTS OF WAVES?

The wave action of coast determines the nature of the substrate and the community of plants and animals. The wave action causes effective mixing of atmospheric gases with the surface waters and promotes gaseous exchange in marine organisms. The organisms of intertidal zone are greatly influenced by waves. All living organisms are adversely affected by the abrasive action of waves. Along rocky shores with wave action, the flora and fauna of the upper side of a rock differ greatly from those of the underside. Algae, barnacles and limpets dominate the upper side of rock, while sponges and ascidians dominate the underside of rock. Most of the organisms of rocky shores that are subjected to wave action have adaptations for adhesion. Actinians with sucking discs, echinoderms with tube feet and lamellibranchs with byssal threads attach themselves firmly on the substratum and thrive in the intertidal zone. Waves influence the vertical zonation of intertidal organisms. The gastropod *Littorina* species and the brown alga *Fucus* species, increase their zone of occurrence with increasing wave exposure. Barnacles have better growth rates with high wave exposure through the increased food supply brought within their reach. The incidence of larvae is much greater in wave exposed areas when compared with sheltered regions. Wave action modifies the shape of reef corals. Flat shape is the better adaptation for withstanding wave action. Soft corals are better able to withstand the wave action than hard corals.

- Oceanic waves are caused by winds blowing over the sea surface. The waves do not move horizontally, they only move up and down.
- Changes in atmospheric pressure over the sea cause the waves associated with storm result in 'storm surges'. Seismic waves or tsunamis are destructive waves caused by underwater earthquakes, submarine volcanic explosions or land slides.
- Wave strength depends on the speed and duration of the wind and the distance of sea surface over which it moves. The maximum depth at which the wave speed is experienced is half the wave length. Therefore organisms and habitats will be less affected by the wave in the deeper water.
- The wave action of coast determines the nature of the substrate and the community of plants and animals.

2.1.8. Substratum

Substratum is required for organisms for attachment, shelter and nourishment. There are two types of substrates available to the marine organisms: solid substrates (rock, wood, metal, shell) and particulate substrates (sand and mud). Living organisms also provide substrates for animals (epizoic forms), plants (epiphytes), symbionts and parasites. The underwater structures provide substrates for the attachment of fouling organisms, such as algae, barnacles, mussels, ascidians and tube worms. Underwater wooden structures serve as a substratum to bacteria, fungi and wood boring organisms such as *Limnoria, Sphaeroma, Martesia* and shipworms.

The organisms depend on the nature of the substratum, its nutritive content and stability. Holothurians and other bottom animals of muddy habitats pass the mud through their alimentary canal and extract the nutrients from it. Many algae require substratum for only attachment and not nutrients. In the sandy substratum, only few animals which are capable of quick burrowing are present. When this sand is mixed with mud especially in intertidal areas, there are high populations of mollusks, worms and crustaceans. The mud bottom favours growth of the rooted plants, such as the eelgrass and occurrences of marine organisms such as sea cucumbers, brittle stars, sea-urchins and different kinds of worms. Deposit feeders prefer a substratum with finer particle size with high organic content. There is also a possible presence of toxic hydrogen sulphide in muddy substrate. To overcome this toxicity, many organisms living in such substrates are adapted for anaerobic life and other organisms maintain sufficient water circulation in their surrounding environment.

A suitable substratum is selected by animals mostly in their larval stages. Barnacles settle in greater numbers on rough surface than on smooth surface. Animals, such as Pholads and the sea-urchins can bore into even very hard rocks and concretes. Timber-borers such as *Martesia, Teredo, Limnoria* and *Sphaeroma* destroy the substratum on which they settle. These organisms deteriorate substrates. On the other hand, corals and calcareous algae deposit calcium carbonate and built up the substrates. Also, shells of oysters and mussels provide substratum for many organisms like *Balanus*.

Why is the sandy beach with pebble stones colonized with very few organisms? How do the organisms survive in such a beach?

The pebble stones in a sand beach suffer abrasion through grinding action between the stones, and therefore, such a beach is colonized with only very few organisms. In order to withstand the abrasive action of waves in the littoral zones, marine organisms have hold-fasts, cementing organs, suckers, as well as a remarkable ability for regeneration. Most of the invertebrates and fishes on the shoreline possess abrasion-resistant outer surfaces, such as hard shells of mollusks and crustaceans, as well as tough skin and thick scales of fishes. Seaweeds are flexible with leathery texture to adjust with the abrasive action on them.

Current speed determines the quantity of suspended particulate matter in the water column, when the seabed is composed of soft sediments. This is because faster flows of water lead to greater turbulence and higher rates of vertical mixing and resuspension. As currents slow down, particles suspended within the water column will drop out of the suspension. This process therefore, produces and maintains physical habitats in the sea, which in turn, determine the types of organisms to live in the habitats. Slow currents will create soft sediments and support burrowers and detritus feeders, whereas fast currents support encrusting species that filter feed.

Can electricity be generated in marine sediments?

Marine sediments contain large amount of organic material and this can provide a source of energy. If a graphite rod electrode (anode) is placed in anoxic marine sediment and connected to a cathode in the O_2 containing substrate, electricity is generated. It is sufficient to power oceanographic monitoring equipment. This has been experimentally proved using the sediment with anaerobic sulphur reducing bacteria (*Desulfuromonas acetoxidans*) in coupling with the oxidization of acetate to reduction of sulphur.

The amount of suspended particles in the water column has a significant influence on the community of plants and animals living in an area. The ecological effects of suspended matter depend mainly on two factors: the size range of the sediment particles and the food content of the suspended sediment. The organic content of suspended particulate matter (SPM) decreases with increasing load of the suspended solids. The suspended solid material may reduce feeding efficiency and affect the function of gills and other respiratory surfaces. Deposited sediments may simply cause death by suffocation of sessile and sedentary animals and plants. Some organisms such as copepods survive well in conditions of high SPM and elevated turbidity. However, high turbidity reduces photosynthetic efficiency of sub-littoral plants through reducing the light penetration.

- Substratum is required for organisms for attachment, shelter and nourishment.
- Solid substrates (rock, wood, metal, shell), particulate substrates (sand and mud) and living organisms themselves provide substrates for marine organisms.
- A suitable substratum is selected by animals mostly in their larval stages.
- In order to withstand the abrasive action of waves in the littoral zones, the marine organisms are well-adapted through hold-fasts, cementing organs, suckers, fast regeneration, hard shells (mollusks and crustaceans), tough skin and thick scales (fish), flexible with leathery texture of body (seaweed).
- Slow currents will create soft sediments and support burrowers and detritus feeders, whereas fast currents support encrusting species that filter feed.

2.1.9. Climate Change

Climate change is the cause of physical environmental changes especially increased temperature. Two of the most important aspects of climate change are: (1) sea level rise; and (2) El Nino changes. The annual mean global temperature has risen steadily and is now about 0.5°C warmer than it was 100 years ago. This is reflected in three fold increase in the rate of sea level change. The rising sea levels will have serious problems for low-lying coastal areas on land, but the marine species will be little affected. Temperature changes can have greater impact on marine organisms. The best example is 'El Nino'. In addition to the El Nino, there are deep-jets of water travelling from the sea floor up above 3,000 metres. These jets of water cause anomalies in wind, rainfall and sea temperature across the tropical Atlantic.

What are 'El Nino', 'La Nino' and 'ENSO'?

'El Nino' is an unusual condition of warm ocean temperatures, whereas 'La Nino' is an unusual condition of cold ocean temperatures.

El Nino is an oscillation of the ocean-atmosphere system in the equatorial Pacific. This results in short-term weather changes around the world. There is an increased rainfall associated with destructive flooding across the southern part of the US and in Peru, and there is also an increased drought in the West Pacific and Southeast Asia.

One of the El Nino's biggest influence is on primary productivity of the sea. The nutrient-rich water is prevented from upwelling in various parts of the world with serious negative impact on marine productivity. The most significant effect of El Nino is on commercial fishing stocks. Anchovy suffered huge declines in stocks in Peruvian coast, whereas sardine increased in the Peruvian coast. These fluctuations can have serious impacts on the socio-economics of the region.

The most rapid climate change of this El Nino is taking place in the southern ocean which comprises much of the Antarctic Sea. This El Niño effects are called "El Nino-Southern Oscillations" (ENSO) for their variations. ENSO will have serious impacts, such as the reduction in krill populations in the Antarctic with all the associated food chain effects on marine fauna.

El Nino is a Spanish name for 'little boy' referring to the birth of Christ, since it begins around Christmas.

Ocean acidification is one of the most serious physical changes for the future of our oceans. The pH of seawater falls, when the level of carbon dioxide rises. This problem is aggravated when acid rains to surface seawater. For example, the surface seawater pH in Bermuda of the Atlantic Ocean is decreasing by 0.0017 pH unit per year. It has been experimentally proved that rate of calcification is reduced by 20% in the hard coral *Montipora capitata* treated with diluted hydrochloric acid at the acidity expected in the ocean during the 21st century.

- The most important aspects of climate change are: (1) sea level rise, and (2) El Nino changes.
- The rising sea levels will have serious problems for low-lying coastal areas on land, but the marine species will be little affected.
- Temperature changes can have greater impact on marine organisms and the best example is 'El Nino'.
- El Nino prevents the nutrient-rich water from upwelling in various parts of the world with serious negative impact on marine productivity.
- The most rapid climate change of this El Nino is taking place in the southern ocean which comprises much of the Antarctic Sea. This El Niño effects are called "El Nino-Southern Oscillations" (ENSO) for their variations.

- One of the most serious physical changes for the future of our oceans is ocean acidification. This will reduce the rate of calcification in shellfishes, coral reefs and calcareous algae.

2.2. CHEMICAL ENVIRONMENTAL FACTORS

2.2.1. Oxygen

Oxygen is essential for living organisms including marine life. It is available in seawater as dissolved oxygen. The amount of oxygen dissolved in seawater is about 3.5 millilitres in one litre of seawater, as against 200ml per litre in the air. The level of oxygen in the oceans is influenced by several processes, such as exchange across the sea surface and atmosphere at the water-air interface, circulatory and mixing processes, photosynthetic product-ion of oxygen and respiratory consumption of oxygen.

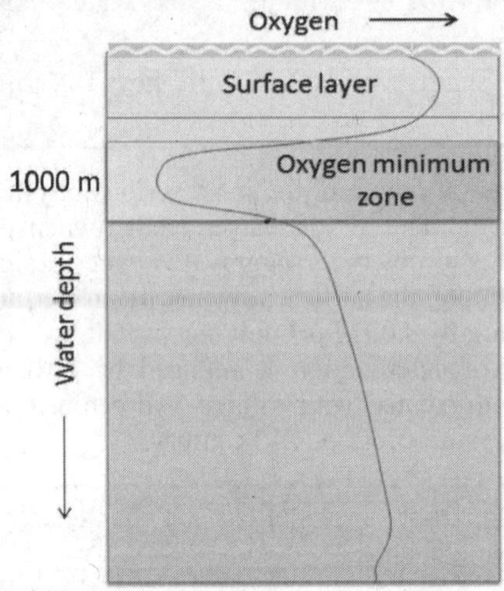

Fig. 2.10 : Dissolved oxygen changing with depth of ocean and the oxygen minimum zone

The surface layer of the seawater receives oxygen sufficiently from the air. In the open ocean, the oxygen content at the surface is high (6 ml per litre). The oxygen content decreases to a minimum value with increasing depth until at about 1000 metres. The reason for the decrease in oxygen level is the consumption by bacteria for decomposition of the organic matter, the very little water circulation and the high concentrations of nitrates and phosphates. This 'oxygen minimum layer' is known to reduce the abundance and diversity of life in the mid-water region (Fig. 2.10). Minimal values of 0.15 ml/litre has been recorded in the Arabian Sea.

HOW DOES TEMPERATURE INFLUENCE THE OXYGEN SOLUBILITY IN SEAWATER?

Solubility of oxygen in seawater is a function of temperature: lower the temperature, greater is the solubility. Warm water holds less oxygen, whereas cold water holds more oxygen. Therefore, deep ocean waters are not anoxic, because the cold water has maximum oxygen to support the populations of animals in the deep waters. The deeper layers in the oceans are also well oxygenated through circulatory and mixing processes. Cold water in the Antarctic surface water is higher (7.5ml O_2/litre) than warm surface water in the tropics (4ml O_2/litre). It is due to the circulation of water, when the cold and highly oxygenated surface water in the high latitudes sinks to greater depths and then flows towards equator. The tropical surface water containing only about 4ml oxygen per litre in the tropics, gradually increases in a concentration of up to 8ml/litre at the polar regions and when this water reaches the deep sea by sinking, it may hold as much as 5ml/litre after losing some amount of oxygen on the way as a result of the respiration of organisms and decomposition of organic matter. Thus the bottom of the sea receives oxygen through the direct flow of water from the enriched polar sea. This process of circulation is naturally very slow. The movements of Arctic surface water to deep layers at mid latitudes may require more than 1000 years.

WHAT ARE THE BIOLOGICAL EFFECTS OF OXYGEN IN THE SEA?

Oxygen is one of the key factors that determines the abundance and diversity of marine life especially in estuaries and shallow coastal waters. Oxygen availability determines hatching of eggs in marine organisms. When the level of oxygen decreases in the eggs toward end of their development, carbon dioxide increases and pH gets decreased in egg fluids. This causes significant loss of egg hatching in marine animals especially in cephalopods. When the shoals of fish enter the oxygen minimum zone, mass mortality may occur owing to asphyxiation. Organic particles sink to the deeper layers and anaerobic decomposition of organic matter by anaerobic bacteria produces a great amount of hydrogen sulphide. Reduction of oxygen level leads to hypoxic ($<$2 ml O_2/litre) and then anoxic (0 ml O_2/litre) conditions. This occurs in several places that may likely lead to loss of fisheries, loss of biodiversity and change of food webs.

Oxygen is not a major limiting factor in the distribution and movements of most marine organisms. The lower limit of dissolved oxygen, required for normal respiration of marine animals is 1-2ml per litre. However certain bacteria and some animals can survive anaerobically. In fact, marine organisms are adapted to low oxygen. The respiratory pigments provide an oxygen reserve for some animals. Glycogen is stored to nearly half the dry weight of the tissues and it is used as a source of oxygen supply in animals such as *Teredo*. The mussel *Mytilus edulis* (Fig. 2.11) lives in polluted waters of low oxygen content, and it can survive without oxygen for several weeks; during that period it remains inactive and does not feed. Many active animals such as mackerel fish move rapidly through water, keeping open their mouths, in order to get a fast flow of water over their gills and oxygen from the water is extracted by the gills. Many sessile organisms produce a respiratory current of water which removes the metabolic wastes. Plants can tolerate greater oxygen deficiency than animals as their tissues have lower oxygen requirement. The oxygen consumption varies with animals. It is 2.2ml/gram of wet weight for protozoans, 0.005 for coelenterates (jelly fish and ctenophores), 0.026 for echinoderms, 0.017 for annelids and 0.181 ml/gram for crustaceans (shrimps) at a temperature range of 17 to 25°C. In general the differences in the amount of oxygen used per gram of wet weight are related to the water content of the living organisms.

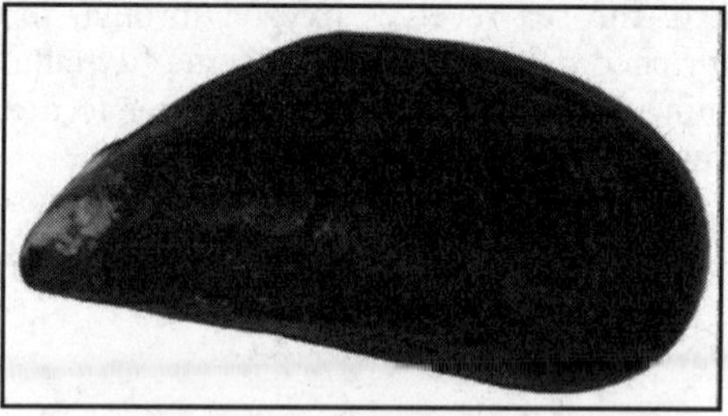

Fig.2.11 : *Mytilus edulis*, a mussel tolerant to low oxygen condition

Level of oxygen is related to photosynthetic and respiratory activities of marine organisms which in turn, are influenced by light and temperature. Light influences the rate of photosynthetic release of oxygen, but it does not influence the rate of oxygen consumption of animals. Light increases photosynthetic release of oxygen and increases oxygen content in the water. Temperature increases the respiratory consumption of oxygen by animal and reduces oxygen content in the water. For example, the oxygen content of seawater is 8.03ml per litre at 0°C and only 4.5ml/litre at 30°C.

- The amount of oxygen dissolved in seawater is about 3.5 millilitres present in one litre of seawater.
- The level of oxygen in the oceans is influenced by exchange across the sea surface and atmosphere at the water-air interface, circulatory and mixing processes, photosynthetic production of oxygen and respiratory consumption of oxygen.

- In the open ocean the oxygen content at the surface is high (6ml per litre). The oxygen content decreases to a minimum value with increasing depth until at about 1000 metres. This 'oxygen minimum layer' is known to reduce the abundance and diversity of life in the mid-water region.
- The deeper layers in the oceans are well-oxygenated through circulatory and mixing processes. The bottom of the sea receives oxygen through the direct flow of water from the oxygen-enriched polar sea. This process of circulation is naturally very slow.
- Oxygen determines the abundance and diversity of life especially in estuaries and shallow coastal waters. Oxygen availability determines hatching of eggs in marine organisms. Many ecosystems that are now severely stressed by hypoxia may lead to loss of fisheries, loss of biodiversity and change of food webs.
- Oxygen is not a major limiting factor in the distribution and movements of most marine organisms. The lower limit of dissolved oxygen, required for normal respiration of marine animals is 1-2ml per litre. However certain bacteria and some animals can survive anaerobically. In fact, marine organisms are adapted to low oxygen. The respiratory pigments provide an oxygen reserve for some animals. Glycogen is stored to nearly half the dry weight of the tissues and it is used as a source of oxygen supply in animals.
- Level of oxygen is related to photosynthetic and respiratory activities of marine organisms which in turn, are influenced by light and temperature. In the seawaters, increasing light increases oxygen content due to photosynthetic release of oxygen, while increasing temperature reduces oxygen content due to respiratory consumption.

2.2.2. Carbon Dioxide and Carbonates

The ocean is a large reservoir for carbon dioxide. Carbon dioxide is more soluble in seawater than oxygen. Seawater contains about 47ml of carbon dioxide per litre which is about 15 times more than in air. Carbon upon entering seawater reacts with water to produce carbonic acid which dissociates into a bicarbonate ion and hydrogen ion. Bicarbonate may further dissociate into another carbonate ion and hydrogen ion, as given below :

$$CO_2 + HOH \longrightarrow H_2CO_3 \text{ (carbonic acid)}$$

$$H_2CO_3 \longrightarrow H^+ + HCO_3^-$$

$$HCO_3^- \longrightarrow H^+ + CO_3^{2-}$$

At the pH of the ocean water (7.5-8.4), carbon dioxide – carbonic acid - bicarbonate system functions as a buffer to keep the pH of the seawater more or less constant.

In seawater, carbon dioxide is present mainly in the form of bicarbonate ions, which decrease with increasing temperature and salinity. The dissolved carbon dioxide is removed in large amounts from the surface water by photosynthetic plants and also by plants and animals for synthesis of calcium carbonate. This loss is compensated by atmospheric carbon dioxide through atmosphere-ocean interface. At the deep water, carbon dioxide is produced by the decomposition of organic matter by bacterial activity using up the available dissolved oxygen. This results the increase in carbon dioxide content and decrease in dissolved oxygen with seawater depth.

The carbon dioxide content of the polar region is low, when compared to the tropical region. The atmospheric circulation carries carbon dioxide from low latitudes to high latitudes, where it again dissolves in the seawater, which in turn, brings it back towards the equator.

The fundamental role of carbon dioxide in the marine environment is in the photosynthesis of plants. Besides this role, carbon dioxide together with calcium plays an essential role in the metabolism of many marine organisms. Bacteria, algae and several groups of invertebrates deposit calcium carbonate in their body. It occurs under the conditions of high temperature, high salinity and low carbon dioxide concentrations especially in shallow waters. In the tropics, conditions are favourable for rapid calcium precipitation. The calcareous formations in the marine environment are the result of the activity of calcareous algae and other marine organisms, such as foraminiferans, pteropods, protozoans, coelenterates, annelids, crustaceans and mollusks. The photosynthetic activity removes carbon dioxide from seawater and increases the concentration of carbonate ions. In tropical waters, there is always super-saturation of calcium carbonate, owing to high temperature and pH. These conditions are favourable for the precipitation of calcium carbonate. In the deep sea, conditions are different just the opposite of those that prevail at the surface especially in the tropics. The high pressure prevailing in the abyssal zone favours the accumulation of more carbon dioxide which makes calcium carbonate more soluble in the deep waters. The skeleton of star fishes of the deep sea is considerably reduced and the shell of the barnacles is poorly calcified. Many deep sea fishes have uncalcified or weakly calcified skeletons.

- Seawater contains about 47ml of carbon dioxide per litre. At the pH of the ocean water (7.5-8.4), carbon dioxide – carbonic acid - bicarbonate system functions as a buffer to keep the pH of the seawater constant.

- The fundamental role of carbon dioxide to play in the marine environment is in the photosynthesis of plants. Besides this role, carbon dioxide together with calcium plays an essential role in the metabolism of many marine organisms.
- In tropical waters, there is always super-saturation of calcium carbonate, owing to high temperature and pH. These conditions are favourable for the precipitation of calcium carbonate. In the deep sea, conditions are different just the opposite of those that prevail at the surface especially in the tropics.
- The high pressure prevailing in the abyssal zone favours the accumulation of more carbon dioxide which makes calcium carbonate more soluble in the deep waters. Many deep sea fishes have uncalcified or weakly calcified skeletons.

2.2.3. Hydrogen Sulphide

The presence of hydrogen sulphide in seawater is associated with deficiency or absence of oxygen. In areas with little or no circulation of water near the bottom and with an accumulation of organic detritus, the heterotrophic bacteria and sulphur-bacteria produce large amounts of hydrogen sulphide. The muddy shores and sediments, where aeration is very poor, bacterial activity will lead to the formation of hydrogen sulphide by decomposition of organic matter. The presence of hydrogen sulphide is indicated by the black-colour of the sediment. This black colouration is the result of formation of ferrous sulphide which gets converted into the stable mineral known as 'pyrite'. The bacteria involved in the process are of the genus *Desulfovibrio*. Most of the hydrogen sulphide present in the Black Sea is derived from *Desulfovibrio*. Hydrogen sulphide may be formed through a process of reduction of the proteins by heterotrophic bacteria or reduction of inorganic compounds of sulphur, such as sulphates and sulphides by heterotrophic bacteria under anaerobic conditions. The process of hydrogen sulphide formation involves two steps: sulphate is reduced to sulphide and then hydrogen sulphide is released. The hydrogen-rich substrates are colonized mainly with anaerobic bacteria and presence of a few fauna with special adaptations. Polychaetes and lamellibranches inhabiting such areas maintain a circulation of normal water through their burrows or siphons. In the case of the tube-dwelling polychaete, *Arenicola marina* the mucous lining of the burrow prevents the diffusion of hydrogen sulphide from the mud into the water in the burrow.

- The muddy shores and sediments, where aeration is very poor, bacterial activity will lead to the formation of hydrogen sulphide by decomposition of organic matter.

- The presence of hydrogen sulphide is indicated by the black-colour of the sediment as the result of formation of ferrous sulphide which gets converted into the stable mineral known as 'pyrite'.

2.2.4. Hydrogen Ion Concentration (pH)

Seawater is slightly alkaline and its pH varies between 7.5 and 8.4. Seawater is a highly buffered solution. If the pH is around 7, most of the carbon dioxide will be present as bicarbonate ion. At high pH values, carbon dioxide occurs as carbonate ions while at low pH values, carbon dioxide exists in the free condition. Thus, the bicarbonate and carbonate ions in the seawater resist changes in pH. This stability of pH favours marine organisms for getting carbon dioxide, essential for photosynthetic activity of primary producers and also by providing alkaline condition for formation of shells with calcium carbonate. During peak periods of diatom production, there may be a marked rise in pH, as a result of the rapid utilization of carbonic acid for photosynthesis by the diatoms. This rise in pH may act as a natural check for high proliferation of the diatoms.

- Seawater is slightly alkaline and its pH generally varies between 7.5 and 8.4.
- Seawater is a highly buffered solution. This stability of pH favours marine organisms for getting carbon dioxide, essential for photosynthetic activity of primary producers and also by providing alkaline condition for formation of shells with calcium carbonate.

2.2.5. Inorganic Salts

Seawater is a solution of many inorganic salts. Inorganic salts (chloride, sodium, sulphur, magnesium calcium, potassium bicarbonate, bromide, boric acid and strontium) make up 99.99% by weight of the dissolved substances. Among the remaining 0.01% of dissolved substance in seawater are – nitrates, phosphates, silicon dioxide, iron, manganese, cobalt and copper – essential to marine organisms.

What is Redfield ratio?

The elemental ratios of phytoplankton and dead organic matter in the oceans are similar throughout the world. The C:N:P ratio is known as the "redfield ratio", averaging 106:16:1 in atomic composition. This ratio is stable in the deep ocean and the N:P ratio can vary with species composition of phytoplankton growing with limiting concentration of nutrients.

Calcium is an essential building block for exoskeletons and potassium is an important fertilizer for marine primary production, while boron is used for cellular process by plants such as seagrasses. Nitrates and phosphates are required by plants to synthesize the organic material during photosynthesis. Silicon dioxide required by diatoms and radiolarians to construct skeletons. In the ocean, the nitrate and phosphate are in low concentrations and this short supply limits the plant growth. In some parts of the world oceans, nitrate and phosphate are present adequately; even then plant production is low due to lack of iron in the water column.

How do organisms adapt to iron-deficient Ocean?

Iron is essential element required for photosynthesis and nitrogen fixation. But it is very little in the open ocean. Organisms have an efficient mechanism of scavenging the low levels of iron – in the form of iron containing proteins.

The importance of iron in the marine environment was experimentally proved. Moss Landing Marine Laboratory tested the effect of iron in the equatorial Pacific waters. Half a ton of iron when added to 100 km^2 of waters, the plant biomass increased by 30 times, and the carbon dioxide depleted by 60% and these resulted in greening of the ocean.

Not only increasing oceanic productivity, iron helps organisms to overcome the problem of wave action. For example, the sand dollar (*Dendraster excentricus*; Fig. 2.12) accumulates iron compounds in a special area of their digestive tracts. This serves as a 'weight belt' to keep them down in the presence of wave action

Fig. 2.12 : *Dendraster excentricus* (http://bss.sfsu.edu/holzman/courses/fall02%20projects/sandollar/sanddollar.html)

The inorganic salts are regenerated through different biogeochemical cycles and these salts never get depleted to cause serious ecological problems. However, silicate and calcium carbonate are permanently removed in large amounts from the seawater by foraminiferans, coral polyps, calcareous algae, diatoms, mollusks, siliceous sponges and radiolarians. Similarly, strontium and lithium are also removed and added on to the bottom deposits.

- Inorganic salts (chloride, sodium, sulphur, magnesium calcium, potassium bicarbonate, bromide, boric acid and strontium) make up 99.99% by weight of the dissolved substances. Among the remaining 0.01% of dissolved substance in seawater are – nitrates, phosphates, silicon dioxide, iron, manganese, cobalt and copper - essential to marine organisms. Calcium is an essential building block for exoskeletons and potassium is an important fertilizer for marine primary production, while boron is used for cellular process by plants such as seagrasses.
- The inorganic salts are regenerated through different biogeochemical cycles and these salts never get depleted to cause serious ecological problems. However, silicate and calcium carbonate are permanently removed in large amounts from the seawater by foraminiferans, coral polyps, calcareous algae, diatoms, mollusks, siliceous sponges and radiolarians. Similarly, strontium and lithium are also removed and added on to the bottom deposits.

2.2.6. Dissolved Organic Matter

Organic matter is in dissolved form in seawater. The sources of the soluble organic matter are death and decay of marine organisms, excretion by marine organisms, run-off from the land, diffusion from sediments and secretions of marine organisms. The dissolved organic matter in seawater is a source of food to many organisms, such as saprophytic bacteria and microflagellates. Several organic compounds, such as vitamins, auxins and hormones have growth promoting effect. Man-made organic compounds, such as detergents and pesticides which enter the sea also affect the organisms. Traces of DDT are present even in the seals and penguins of the Antarctic.

- The sources of the soluble organic matter are mainly death and decay of marine organisms.
- The dissolved organic matter (vitamins, auxins and hormones) in seawater is a source of food and growth promoters to saprophytic bacteria and microflagellates.
- Thus chemical substances dissolved in the seawater are inorganic salts, organic compounds, and dissolved gases (oxygen and carbon dioxide).

Changing chemical environment

Chilka Lake exhibits contrasting levels of dissolved inorganic carbon (DIC) and organic carbon (DOC) in wet and dry seasons. DIC is lower than DOC in dry season. But, the reverse is recorded in wet season. The DIC/DOC ratio is higher in wet season than dry season, indicating intense respiration of organic carbon in wet season. This suggests that Chilka Lake is undergoing the biological transformation of organic carbon to inorganic carbon and its export to the atmosphere especially during wet season.

In Cochin estuary, along west coast of India also exhibits a shift from autotrophy to heterotrophy, during wet season. The terrestrial organic matter input through rivers reduce primary production but enhance bacterial production and respiration leading to oxygen undersaturation and exceptionally high CO_2. The CO_2 supersaturation caused by increased bacterial respiration (in excess of PP) is a result of bacterial degradation of organic matter. During the periods, zooplankton population reduces and hence phytoplankton population increases.

Questions for Review/Exercise*

(1) Define the terms – Salinity, Thermocline, bioluminescence, Pycnocline, El Nino, Oxygen minimum layer, Piezophils, neap tide, waves, Coriolis force, anadromous, stenohaline

(2) How do marine organisms respond to temperature?

(3) Explain the zonation of ocean based on availability of light. Add a note on the influence of light on marine organisms.

(4) Give an account of currents, tides and waves and their influence on marine organisms.

(5) How do the marine organisms adapt to salinity?

(6) "Oxygen is not a major limiting factor in the distribution and movements of marine organisms" – Discuss.

(7) How is the pH of the seawater maintained constant?

(8) What is "El Nino-Southern Oscillations"? Explain their impacts on climate change.

(9) Write about the importance of inorganic salts and dissolved organic matter for marine organisms.

(10) Discuss the substrate as a physical factor of marine environment.

* Frequent field visits to various marine environment during wet and dry seasons and analyze various physical and chemical parameters at different day times.

Ocean Ecology 3

Ocean Ecology deals with the interrelationships between marine organisms and their ocean environment that occurs beyond the continental shelf of 200 metre depth (Fig. 3.1). The ocean has two major components: water column (pelagic zone) and sea bottom (benthic zone). The pelagic environment is quite uniform, while the benthic environment varies with different habitats such as rocky, sandy, muddy, flat, sloped, irregular and mixed bottom. The pelagic organisms do not experience extreme environmental changes, and there is no need to adapt significantly for their survival. In contrast, the benthic organisms are strongly adapted to the environmental changes. In general, marine species are diverse in the benthic environment due to structural complexity and diversity of habitats in the sea.

The deep sea is the largest habitat with cold and dark environment on earth. The deep sea is a part of the marine environment that lies below the level of light penetration and beyond 200 metres depth. The deep sea constitutes 85% of the water area and 90% of the water volume of the entire earth. About 79% of deep sea is composed of marine water with depths greater than 1000 metres. This vast deep sea is the least known and unexplored area. This is due to the difficulty of sampling and the inaccessibility to humans.

- Ocean Ecology deals with the interrelationships between marine organisms and their ocean environment that occurs beyond the continental shelf of 200 metre depth.
- The pelagic environment is quite uniform, while the benthic environment varies with different habitats.
- The pelagic organisms do not adapt significantly for their survival. In contrast, the benthic organisms are strongly adapted to the environmental changes.
- Marine species are diverse in the benthic environment due to structural complexity and diversity of habitats in the sea.
- The deep sea is the largest habitat with cold and dark environment on earth. The deep sea lies below the level of light penetration and beyond the continental shelf that occurs beyond 200 metre depth.

3.1. PELAGIC ZONE OF OCEAN

The pelagic zone comprises the water column that is overlying seafloor. The pelagic waters are blue in colour and transparent with little suspended matter. Environmental changes influence only the upper layers of the water column, and the rest of this pelagic zone is stable and uniform. The pelagic zone is divided into five vertical biozones : (1) epipelagic, (2) mesopelagic, (3) bathypelagic, (4) abyssalpelagic and (5) hadalpelagic. All the biozones, except epipelagic zone, are together known as 'deep sea' environment.

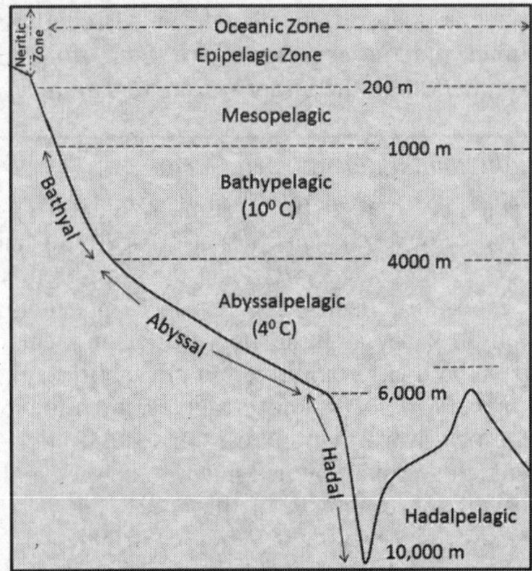

Fig. 3.1: Divisions of oceanic zone, beyond 200 m depth

3.1.1. Epipelagic Zone

The 'epipelagic' zone is the uppermost layer of the ocean, and it falls from the seawater surface to 200 metres depth. Upper part of epipelagic zone has sufficient light to support photosynthesis. The epipelagic zone is the photic zone dominant with phytoplankton. Many of the larger planktonic protozoa, particularly the Foraminifera and Radiolaria, contain intracellular and extracellular photosynthetic symbionts. Many planktonic ciliates and small mollusks retain the active chloroplasts of their algal food for a short time. The contribution of this type of symbiosis to pelagic plankton productivity is also important. Other organisms include in this zone are zooplankton, mostly holoplanktonic organisms and nekton, in particular fishes, such as tunas (Scomberidae), flying fish (Exocoetidae), sharks and sail fish.

Phytoplankters of the epipelagic zones are dominated by smaller photosynthetic organisms such as nanoplankton (2-20μm in diameter) and picoplankton (<2μm in diameter). However, they are less dominant in coastal waters and upwelling areas, where net phytoplankters with diameter of 20-200μm are abundant. Dominant groups of nanoplankton and picoplankton include coccolithophores, prochlorococcus, small flagellates and several kinds of prokaryotes and eukaryotic cells. The nanoplankton and picoplankton account for more than 80% of all photosynthetic activity and 75% of the phytoplankton biomass in open ocean waters especially in tropics and subtropics. The reason for dominance of the very smaller photosynthetic organisms is that the low nutrients of ocean environment. The smaller photosynthetic organisms that have greater surface area absorb nutrients from surrounding environment. The smaller photosynthetic organisms are more tolerant to seasonal changes of environment than the larger net phytoplankton.

A predominant phytoplankter in Open Ocean

Prochlorococcus is smaller (0.6-1 μm in diameter) than coccoid cyanobacteria, and they reach densities of 10^5 cells in one millilitre of oceanic water. *Prochlorococcus* is the differently pigmented cyanobacteria. *Prochlorococcus* species lack phycobiliproteins, but possess chlorophyll b as a light harvesting pigment, whereas the cyanobacteria have both chlorophyll-a and phycobiliproteins. Because of its abundance, *Prochlorococcus* has received attention for its role in climate change, global warming, and carbon sequestration in the surface ocean. At present, this prokaryote is known to be the smallest and most abundant photosynthetic organism in the ocean, but not reported from freshwater or terrestrial ecosystems. They contribute 10-20% of global primary productivity. Globally *Prochlorococcus* is abundant in tropical waters and much lower abundances at temperate waters. *Synechococcus* species are unicellular cyanobacteria (Fig. 3.2) and are typically responsible for 5-25% of primary productivity in temperate and tropical oceanic environments.

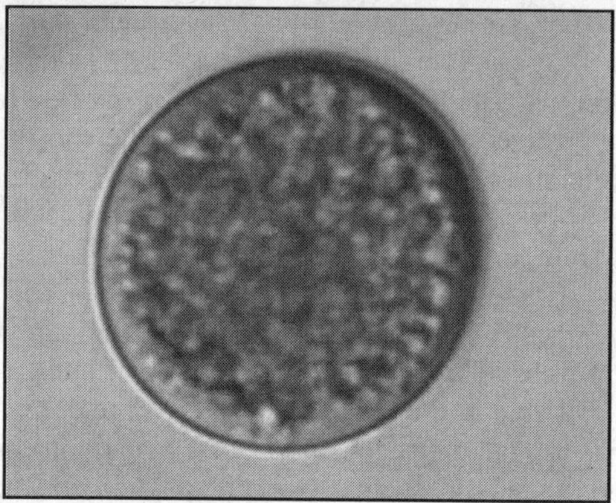

Fig. 3.2: *Synechococcus* species

The picoplankton including *Prochlorococcus* are consumed by small heterotrophs, such as non-photosynthetic flagellates and various ciliates belonging to the kingdom, Protista. The flagellates act like amoeba and engulf the entire cells of picoplankton and nanoplankton, while the ciliates filter the cells from the water column. Respiration rate of nanozooplankton exceeds that of net plankton by a factor of 10. This indicates that nanozooplankters consume most of the energy fixed by autotrophs. This initial link is thus from nanoplankton and picoplankton to ciliates and small flagellates, and this link is important in the open ocean. Therefore, the classical model of diatom or dinoflagellates being grazed by copepods is not the first step in the epipelagic food chains of the ocean (Fig. 3.3).

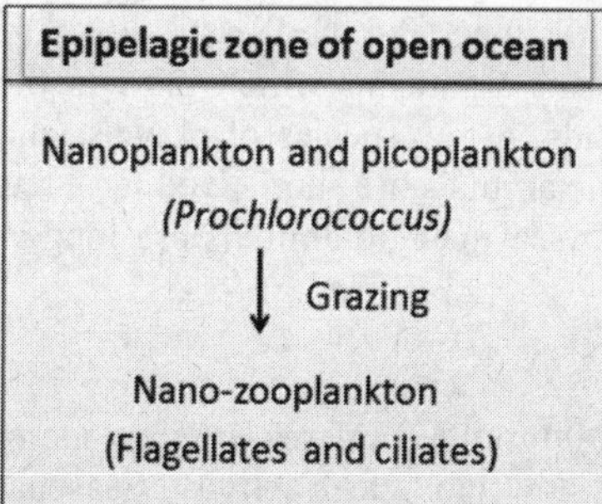

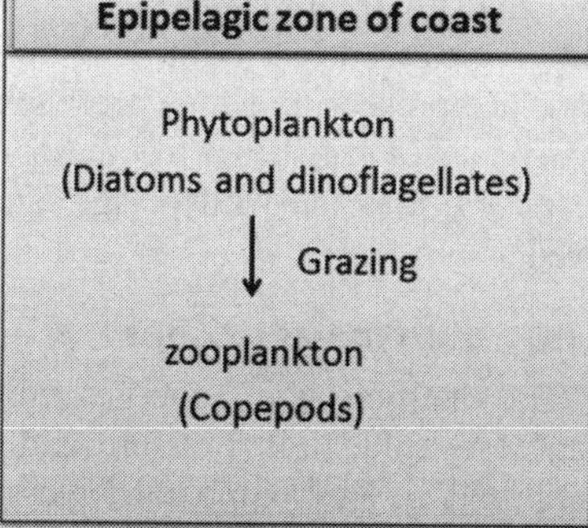

Fig. 3.3. The first step involved in the epipelagic food chain of open ocean and in that of coastal water

Dissolved organic carbon in the waters of the world's ocean is one of the largest reservoirs of organic carbon on the earth. Bacterial counts are more in nutrient-rich coastal waters (10^6 cells/ml) than that in nutrient-poor open seawaters (10^5 cells/ml). However, bacterial biomass exceeds phytoplankton biomass in nutrient-poor seawaters. Dissolved organic matter is utilized by bacteria directly. The bacteria also attack both living and dead organic particles, thus releasing more dissolved organic matter. Bacteria have a very high assimilatory efficiency, converting a considerable amount of dissolved organic carbon in to particulate organic carbon in the form of bacterial cells. The bacterial cells are lysed by viruses thereby releasing dissolved organic matter which in turn supports the bacterial growth and reproduction. The bacteria are consumed by nanozooplankters, and can channel a significant amount of energy into planktonic food webs. This is termed as 'microbial loop'. This microbial loop plays a major role in the total carbon flux of the oceans. The bacteria are also responsible for the regeneration of nutrients in the photic zone, helping the continued productivity of the phytoplankton, even in the nutrient-poor waters.

3.1.2. Mesopelagic Zone

The mesopelagic zone falls from 200 to 1000 metres. It is the disphotic zone of decreasing light intensity. Only 1% of the light rays reach the upper level of the mesopelagic zone, to a depth of about 800 to 950 metres. Characteristics of the mesopelagic zone are decreased temperature, low level of dissolved oxygen, increased hydrostatic pressure and high levels of nutrients. The environmental conditions are almost static. Many of the invertebrates of this disphotic zone are transparent. The layer of water column with minimum level of dissolved oxygen is termed "Dissolved oxygen minimum layer" and is found at depth about 700-1000 metres. This is the most preferred zone for zooplankters. Most of the fishes belonging to this zone do not normally move out of this zone, and the fish usually have larger eyes to detect light. Bioluminescent organisms which glow in the dark are present and they are euphausids, a few species of prawns, and many species of fishes. Most mesopelagic animals are short-lived for a few years, whereas the bathypelagic and abyssal pelagic animals are longer-lived.

3.1.3. Bathypelagic Zone

The 'bathypelagic' zone lies from 1000 to 4000 metres. It is a zone of darkness, cold, high pressure and very low biological activity. Seasonal variations in environmental factors are almost non-existent. This zone is inhabited only by a few organisms such as amphipods and Foraminifera. The important nekton animals that occur in bathypelagic zone are the fishes such as angler fishes. The fishes are characterized by complete blindness and black body colouration often with luminescence.

3.1.4. Abyssalpelagic Zone

The 'abyssalpelagic' zone falls from 4000 to 6000 metres. This zone is characterized by darkness, high pressure, and intense cold (less than 4°C). Planktonic forms are rare in this zone. Some amphipods and fishes such as deep water eels and deep-sea anglers are present in this zone.

3.1.5. Hadalpelagic Zone

The Hadalpelagic zone lies from 6000 metres down to the deepest trenches. Similar to abyssalpelagic zone, planktonic forms are rare in this zone and, some amphipods and fishes such as deep water eels and deep-sea anglers occur in this zone.

3.1.6. Marine Organisms of Pelagic Zone of Ocean

The organisms living in the pelagic zone are grouped into two broad categories – (1) 'plankton' and (2) 'nekton' based on their ability of movement through the water. The plankton consist of floating plants and animals which drift with water movement. Nekton include all the organisms capable of powerful swimming and remaining localized in spite of the water movement. The nekton are composed of bony fishes, sharks and rays as well as lesser number of marine mammals, reptiles and birds. The only invertebrates that occur in ocean as nekton are the cephalopod mollusks. The significant environmental conditions to nekton are the transparency of the ocean, lack of shelter to escape from predators, and the absence of physical boundaries or structure.

In trophic structure and food webs, the pelagic community varies in different geographical areas of the oceans. In the tropics, the top carnivores are fishes. In cold temperate waters the top carnivores are sharks, marine birds and mammals. In Antarctic waters, the top carnivores are marine mammals.

The fish are the significant component of nekton in the pelagic zone of the ocean (Fig. 3.4). There are two types of epipelagic fishes - holoepipelagic and meroepipelagic – based on their life history in the zone. Holoepipelagic fish spend their entire lives in the epipelagic zone. Examples for holoepipelagic fish are sharks, most flying fish, tunas and sword fish. These fish often lay floating eggs and have epipelagic larvae, and they are most abundant in the surface waters of the tropics and subtropics. Meroepipelagic fish spend only part of their life cycle in the epipelagic zone, as adults but spawn in inshore waters. Examples for meroepipelagic fish are herring, whale, shark and dolphin. Nekton reptiles are turtle and sea snakes; only true nekton birds are the flightless penguins of the southern Hemisphere; but, cormorants and other seabirds do dive up to 100 m depth for food and spend much time swimming in the ocean.

Fig. 3.4: Common fish of the pelagic zone of the ocean

Marine mammals are the second component of oceanic nekton. The marine mammals are whales, dolphins, porpoises (order – Cetacean), seals, elephant seals, sea lions and walruses (order – Carnivory, suborder – Pinnipedia), manatees and dugongs (order - Sirenia) and sea otters (order – Carnivory) (Fig. 3.5). Marine mammals are ecologically significant in the open-oceans. Antarctic whales have been reduced by 90% by hunting. This reduction of the whales has led to large increases in populations of Antarctic birds and pinnipeds due to the availability of krill which is not consumed by the whales. In the Antarctic sea, one baleen whale population consumes 190 million metric tonnes of krill (*Euphausia superba*) per year. Dead whales provide food for the deep sea benthic organisms.

Fig. 3.5: Marine mammals of pelagic zone of the ocean

3.1.7. ADAPTATIONS OF PELAGIC ORGANISMS OF OCEAN

3.1.7.1. General Adaptive Features

The pelagic zones of deep sea include mesopelagic (disphotic), bathypelagic, abyssal pelagic and hadalpelagic. In other words, the deep sea is that part of the ocean environment that lies below the level of light penetration in the ocean. In the deep sea, the major environmental factors that affect animals are temperature, pressure, absence of sunlight, food and oxygen content. The pressure is very high in the deep sea, and has a strong effect on the metabolism of the animals, including enzyme systems, membranes and secretion of calcium carbonate. Food is scarce in the deep sea, and it decreases with depth and distance from land. This low food availability results in small size of deep-sea organisms. Oxygen is adequate for life in most of the deep sea environment. Mid-water organisms are dominated by small fishes and crustaceans. These organisms have their own niche to avoid competition for food, and this allows diverse numbers of species.

Adaptations observed in mesopelagic animals include red colour in crustaceans, black colour in many fishes, tubular eyes in fishes, large mouths with curved teeth to enable the animal to feed on prey. Most deep-sea animals are small, but some crustacean groups exhibit gigantism. The anatomical organization of the animals becomes simplified and they show increased water content with increasing depth.

Bioluminescence is widespread, particularly in the mesopelagic zone. Many animals emit light often through light producing stalk called 'photophores'. Flash-light fish have two light organs under each eye, made up of billions of microscopic glowing bacteria. The angler fish have a glowing light in front of their mouths to attract prey. Bioluminescence is important for animals in locating predators and prey, attacking prey and finding mates for reproduction.

In deep sea, there are two reproductive patterns. In one type, the animals live as both adults and young stages in the deep sea. In another type, the animals live in the surface waters at their young stages and after metamorphosis or maturity the animals migrate to the deep sea.

3.1.7.2. Adaptation for Floatation

The planktonic organisms are adapted morphologically for floatation. This is brought about by increasing the surface area through the development of long spines or feather-like projections or by extreme flattening of the body. For example, diatoms *Chaetoceros* species have elaborate branching (Fig. 3.6) and *Coscinodiscus* exhibits flattening. The copepod has long spines. Some larvae of squilla and lobster have extremely flattened body. Gelatinous tissues are developed in the case of the medusae, the siphonophores, the ctenophores and copepods such as *Eucalanus*. The oil globules are present in fish eggs and diatoms to keep them suspended in water. Gas is present in the cyanobacteria, *Trichodesmium,* and the pneumatophores of siphonophores to assist in floatation. The barnacles like *Lepas* secrete mucus and foamy substance for supporting floatation.

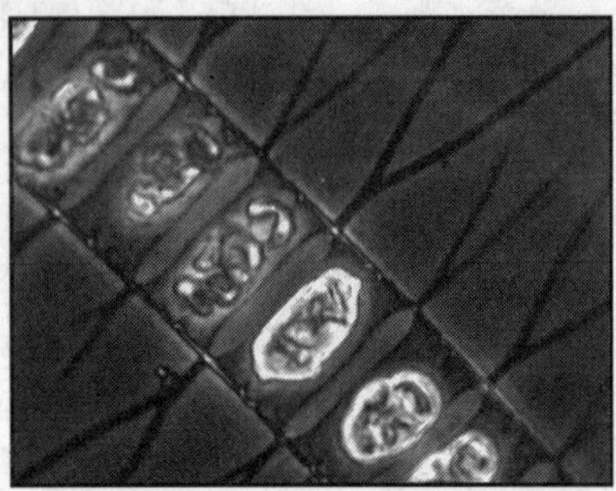

Fig. 3.6: *Chaetoceros* species, a diatom with branching structures helping in floatation
(http://cimt.ucsc.edu/habid/phytolist_diatoms/13chaetocerosspeices.html)

Buoyancy adaptations to nekton are the presence of a gas bladder in most fishes, air sacs and blubber in marine mammals, air channels in the bones of marine birds, and lipid-filled livers in sharks. In addition, the projected structures of the animal body are reduced as an adaptation. In the case of tuna fish, normally protruding body structures are located into depressions. In fast moving fish, eyes do not protrude beyond the body surface. Pectoral and pelvic fins fit into grooves, and body scales are reduced or absent. Birds have additional air sacs in their bones in the air channels that are mostly an adaptation for flight, but can also offer buoyancy. Another mechanism of buoyancy is replacement of the heavy chemical ions in the body fluids with lighter ones. This mechanism is common in plankton and in few nektonic animals like squids. Some squids have body cavities and tissues in which heavy sodium ions are replaced with lighter ammonium. Many of the pelagic animals have reduced weight due to less of skeletal matter as compared to that of their benthic relatives. In the case of pteropods, heteropods and cephalopods, the shell is very light, delicate and fragile.

3.1.7.3. Adaptation for Osmoregulation

Marine mammals and birds have lower concentration of salts in their body fluids, than surrounding seawater. This situation may cause dehydration of the body fluids, but the animals avoid dehydration either by drinking seawater or by obtaining water from their food. Marine birds and turtles eliminate excessive salt through salt secreting glands in the orbit of the eyes. Marine mammals do not have such glands or sweat gland, and the only organ for salt excretion in marine mammals is kidney.

3.1.7.4. Adaptation to Cold Temperature

Marine birds and mammals have special adaptation to survive in the cold water. They have fat layers to reduce heat loss and circulatory systems to conserve heat. Heat loss is prevented by the arteries that bring the warm blood out to these extremities are surrounded by a number of smaller veins that bring blood back to the central core of the mammals. Because of this arrangement, the heat of the blood in the arteries can be absorbed by the cooler blood returning in the veins before it is lost to the external water through the thin flesh of the outer extremities. This conserves heat. If animals are too warm they dissipate heat by waving their flippers in the air. Seals and sea lions open their mouths and pant like a dog.

Many birds and marine mammals are diving animals. They have large blood volume than their terrestrial relatives to transport more oxygen. Slowing the heart beat ('bradycardia') during submersion in the ocean is yet another adaptation. This cuts off blood supply to various organs including muscles. The muscular system is rich in oxygen-containing compound called 'myoglobin'. If the oxygen of the myoglobin is depleted in the muscle, the animal continues to survive in anaerobic conditions. When the animal reaches the surface and begins breathing again, the blood is recharged with oxygen.

3.1.7.5. Adaptation for Fast Movement

Marine animals are adapted for fast movement in the ocean by two ways: (i) by creating the propulsive force, and (ii) by reducing the resistance of the body to passage through the water. Most fishes, whales and sea snakes propel themselves using undulating movements of the body, whereas pinnipeds, turtles and birds paddle using the limbs. Most nekton have tear drop-shaped bodies and reduced forms of protrusions from the body that provide less resistance for body movement in the water. Marine mammals have little or lack of hairs; otherwise, hairs will give much resistance to movement of the mammals. Mammary glands are flattened and reproductive organ of the males and nipples of females do not protrude beyond the skin, when they are not used.

3.1.7.6. Adaptation of Vertical Migration

Vertical migrations are performed by many zooplankters such as copepods, euphausids and chaetognaths and some pelagic fishes like lantern fishes. The vertical migration is mainly stimulated by light and it varies diurnally and seasonally. The vertical migration helps in getting food and favourable environmental conditions for the animals.

3.1.7.7. Adaptation for Preying

Nektonic animals have well-developed sensory systems reacting to light (vision), sound (hearing) or chemicals (olfaction). Sound reception is well developed in marine mammals to navigate, locate and identify food. The nekton animals defend themselves from predators by means of alteration of body shape and also camouflage through colour change.

Plankton feeders consume larger zooplankton. Examples of plankton feeders are flying fish, sardine fish and the baleen whales (e.g. blue whale). The baleen whales have large mouths devoid of teeth, but sheets of baleen or whale bone which hang down from the roof of the mouth cavity. The baleen consists of a comb-like structure, filtering plankton. However, the grey whale is not plankton feeder but a bottom feeder. They filter the sediment through their baleen and retain small invertebrates like amphipods. The largest carnivore in the oceans is the sperm whale and is the only animal which preys upon the largest squid. The second largest carnivore of the nekton is the killer whale which preys upon fishes, penguins, porpoises, seals and sea lions. Smaller toothed whales, seals and sea lions feed on squids. The fish eating whales and dolphins are feeding on lantern fish. These dolphins have a shorter, slender rostrum with many teeth. Porpoises feed on squids, and they have reduced number of teeth, but increased ridges in the palate to hold squid. Other squid eaters are pilot whales.

Sea birds such as Albatross possess the tubular nostrils with an olfactory epithelium located within the nose and also large olfactory bulbs in the brain. They travel long distances for food. The birds respond quickly to the presence of the gas (Dimethyl sulphide) in the air. When krill feed upon phytoplankton, their prey releases the gas. The gas attracts the seabird for feeding the krill.

3.1.7.8. Adaptation for Reproductive Behaviour

Reproductive behaviour varies with animal groups. Pelagic bony fishes produce a large numbers of eggs that are subjected to heavy loss by predation. In contrast, pelagic sharks produce only few eggs that hatch into large juveniles. Marine birds and reptiles retain the reproductive characteristics of their terrestrial relatives producing and laying eggs on land. There are two types of marine mammals: those that give birth on land and in the water. Whales give birth in the water, while pinnipeds give birth on land.

The pinnipeds, such as seals, sea lions and walruses give birth to young ones on land or on floating ice. Males are much larger in size than the females. Many are polygamous and territorial in behaviour on the breeding grounds. The largest male contracts few numbers of females, and protects the females from other males. In general, they migrate long distances for breeding. In contrast to pinnipeds, the cetaceans, such as whales give birth to young ones in the water. The juvenile whales remain close to their mothers for protection from predators, while pups of sea lions are left on the breeding grounds for several days, while the mother forages for food in the open ocean. Like pinnipeds, certain cetaceans may migrate for breeding purposes especially thousands of miles from the feeding areas in cold water to warmer waters. The young ones of whales and pinnipeds grow very fast. Blue whales can grow 3 tonnes at birth to 23 tonnes in seven months. The reason for the fast growth is due to the milk that is rich in fat about 10 times as much fat as in cow's milk. Nursing lasts 15 months in sperm whales, but several years in pilot whales. Because of the high amount of energy is spent on milk production, one young at a time is produced by whales. The female manatee floats upright while nursing her young, using her front feet to cradle it.

Marine mammals are long lived: 46 years in grey seals, 36 years in harbour seals, 32 years in bottle nose porpoise, 77 years in sperm whale and 80 years for fin whale. The sexual maturity is attained only after 9-10 years, but sperm whales after 20 years. Many nekton undertake extensive migrations for breeding or feeding. The means of navigation in marine mammals and turtles are linked to an ability to detect the earth's magnetic field.

- The pelagic zone of ocean is divided into (1) epipelagic, (2) mesopelagic, (3) bathypelagic, (4) abyssalpelagic and (5) hadalpelagic. All the zones, except epipelagic zone, are together known as 'deep sea' environment.
- Environmental changes influence only the upper layers of the water column, and the rest of this pelagic zone is stable and uniform.
- The oceanic nekton are bony fishes, sharks, rays, marine mammals, reptiles and birds. The only invertebrates that live as oceanic nekton are the cephalopod mollusks.
- The fish are the major component of nekton in the pelagic zone of the ocean. The marine mammals are the second important component of oceanic nekton.
- The marine mammals are whales, dolphins, porpoises (order - Cetacean), seals, elephant seals, sea lions and walruses (order - Carnivory, suborder - Pinnipedia), manatees and dugongs (order - Sirenia) and sea otters (order - Carnivory).
- In the tropics, the top carnivores are fishes. In cold temperate waters the top carnivores are sharks, marine birds and mammals. In Antarctic waters, the top carnivores are marine mammals.

- The planktonic organisms are adapted morphologically for floatation by increasing the surface area through the development of long spines or feather-like projections or by extreme flattening of the body.
- Buoyancy adaptations to nekton include the presence of a gas bladder in most fishes, air sacs and blubber in marine mammals, air channels in the bones of marine birds and lipid-filled livers in sharks. Many of the pelagic animals have reduced weight due to less of skeletal matter.
- Most deep-sea animals are small, but some crustaceans exhibit gigantism. The anatomical organization of the animals becomes simplified and they show increased water content with increasing depth.
- The animals are adapted for fast movement in the ocean, either through creating the propulsive force, or reducing the resistance of the body to passage through the water.
- Vertical migrations are performed by many zooplankters and some pelagic fishes.
- Nektonic animals have well-developed sensory systems reacting to light (vision), sound (hearing) or chemicals (olfaction).
- Marine mammals and birds avoid dehydration either by drinking seawater or by obtaining water from their food. Many diving marine mammals and birds have large blood volume to transport more oxygen. Slowing the heart beat ('bradycardia') during submersion in the ocean is another adaptation. The muscule is rich in oxygen-containing compound called 'myoglobin'.
- There are two reproductive patterns that exist in the deep-sea: (i) the animals live as both adults and young stages in the deep sea, and (ii) the animals live in the surface waters at their young stages and after metamorphosis or maturity the animals migrate to the deep sea.

3.2. BENTHIC ZONE OF OCEAN

The benthic zone of the sea extends from about 200 metres depth to the deep ocean bottom. This zone is further divided into an upper 'bathyal' zone, a middle 'abyssal' zone and a lower 'hadal' zone (Fig. 3.1). This benthic zone has dark and cold conditions with slow growing organisms. The animals of the zone are dominated by detritus feeders, represented with low density of carnivores and herbivores. There is a rare presence of suspension feeders due to the low density of food in the water. Deep-sea diversity particularly in the benthos is high due to stability and extensive area of the habitat.

Is ocean floor flat plain?

The ocean basins are mostly flat plains. In some places, the plains are cut by deep trenches. Deep sea trenches are longitudinal depressions having a depth of more than 6000 metres, mostly found in the Pacific Ocean. In most cases the trenches are troughs as continuations of the continental slope. In other places, the ocean basin elevates as oceanic rises, oceanic islands, seamounts, guyots and mid-oceanic ridges (Fig. 3.7). The oceanic rises are present in non-seismic areas, as in Bermuda of the North Atlantic Ocean. Oceanic islands are formed by volcanic rocks derived out during the eruption of submarine volcanoes. The islands are found above the sea level as long chains, up to 2700 kilometres long in the Hawaiian region. The seamounts and guyots are also formed by submarine volcanic eruptions but below sea level. Seamounts have sharp cone peaks, while guyots are flat-topped mountains. The mid-oceanic ridges are sub-marine mountains with irregular slopes and extend continuously through the Atlantic, the Indian and South Pacific oceans. The mid Atlantic ridge has the height of 1-3 kilometres and width of over 1000 kilometres above the ocean basin. The Mid-ocean ridges are of high volcanic activity, and are estimated to produce 75% of the total annual output of molten volcanic rock, the 'magma' on earth. There are more than one million submarine volcanoes, and as many as 75,000 of these volcanoes rise over 1 kilometre above the ocean floor. The Galapagos archipelago in the Pacific Ocean off the coast of South America is well known example of a volcanic island group.

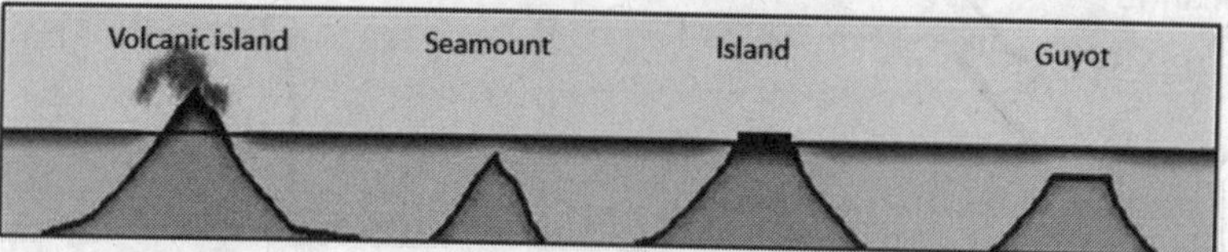

Fig. 3.7: Ocean basin, showing volcanic island, seamount, island, guyot

Is Ocean floor the site of volcanic and seismic or earthquake activity?

The outermost layer of the earth is the crust. Its thickness is about 10 kilometres beneath the oceans and 30-35 kilometres beneath the continents. The crust of ocean and continent is composed of about a dozen tectonic plates that are floating on the underlying hottest mantle. The plates are bounded by ridge and trench (troughs) systems. Mid-oceanic ridges are centres of sea-floor spreading where new crust is created by volcanic activity. This new crust continuously spreads away from the mid-oceanic ridges and the crust descends or gets pushed down into deep trenches. Thus, a new crust is formed at the mid-oceanic ridges in one side; and in the opposite side, the margin of one plate dives beneath the margin of the adjoining plate at the trench system, a process called 'subduction'. Both the oceanic ridge system and the subduction zones of the trenches are the sites of volcanic and seismic or earthquake activity, in other words, most earth quakes and volcanoes occur where plates collide (Fig. 3.8). All along the trenches, lava comes out at temperatures of about 800°C and builds up a chain of volcanic islands, followed by earthquakes. This zone is also known as 'island arc' or 'Ring of Fire'. This occurs around the Pacific Ocean.

The oceans were believed to be formed by massive clouds of water vapour rising from volcanoes and hot rocks on the earth's surface. As the earth surface cooled, the water vapour turned to rain which fell and filled up dips in the earth's surface and taking with it any mineral salts on the land and so formed the oceans.

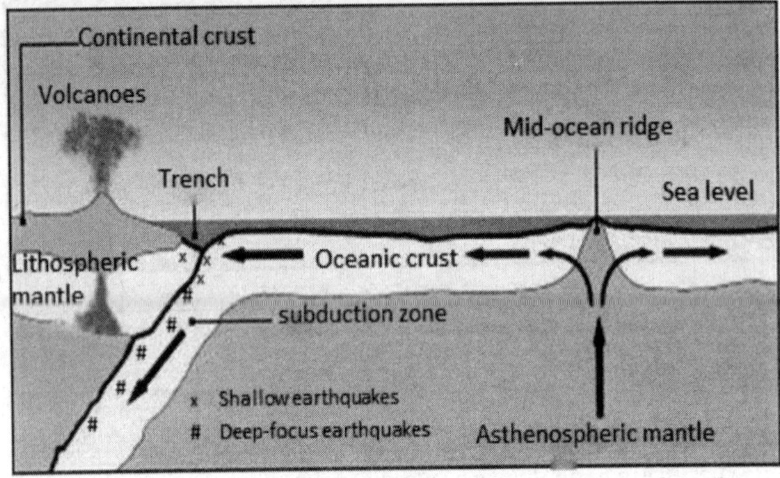

Fig. 3.8: Sites of volcanic and seismic or earthquake activity

Definitions

A trench is a type of excavation or depression in the ground. They are generally more deeper than their width.

Lithosphere is the outer solid part of the earth, including the crust and uppermost mantle. It is about 100 km thick.

Asthenosphere is a portion of the upper layer just below the lithosphere that is involved in plate tectonic movement and the mantle coming out of asthenosphere is known as asthenospheric mantle.

What are compositions of the deep sea floor?

The deep sea is red clay made up of fine sediment, rich in iron and manganese. The seafloor is covered with sediments derived either from terrestrial or marine sources. The organic materials of terrestrial origin are mainly derived from plant materials, whereas those of marine origin consist of seaweeds, diatoms, sponge spicules, coral skeletons and the remains of crustaceans, echinoderms and mollusks. The rate of sedimentation is relatively faster in the shallow coastal areas where about 30 centimetres is deposited within a period of 1000 years, whereas it is slow in the abyssal depths beneath the open ocean where only about 1 millimeter is deposited in a period of 500 years. Large areas of the seafloor are covered by black, potato like objects called 'manganese nodules'. These contain metals such as copper, nickel, cobalt, iron and manganese that are precipitated as oxides or hydroxides from seawater. They are growing at slow rates- a few millimeters in 1 million years.

HOW ARE THE ANIMALS DISTRIBUTED ON THE SEAFLOOR?

There are significant observations in distribution of animals on the seafloor (Fig. 3.9): (1) Sea-anemones, starfishes, and polychaetes decline while holothurians are increasing between depths of 400 and 6000 metres, (2) A sharp decline in the presence of sponges from 1000 to 5000 metres, (3) Decapod crustaceans are rare or absent beyond a depth of 5000 metres, (4) Starfishes are absent beyond 8000 metres, (5) Amphipods, gorgonians, and isopods are rare between depths of 8000 and 9000 metres, (6) Sea anemones and mollusks decline sharply at depths greater than 9000 metres.

The two Polar Regions exhibit striking differences in their seafloors. The Antarctic region has higher in species diversity of benthic organisms, and it has an extremely high degree of endemism. Benthic biomass in the Antarctic is about 2-times higher than those of Arctic. This is attributed to nutrient-rich water and high offshore productivity. Antarctic seas have 120 species of fish. Most of them are endemic because they are isolated by deep water from all the continents. These fishes have special mechanisms to keep their body fluids from freezing, and some lack respiratory pigments. Such specialized endemic fish are not present in the Arctic. In the Antarctic seas, all birds are adapted to forage in the marginal ice area or near the edge of the ice. Only few such birds forage in the Arctic.

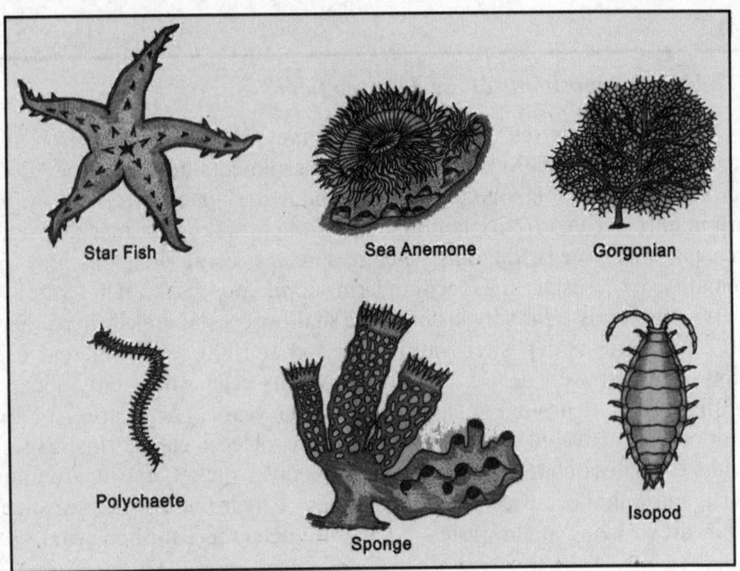

Fig. 3.9: Important marine invertebrates on the seafloor

3.2.1. Bathyal Zone

The bathyal zone is the sea floor environment of the continental slope and it extends down from 200 to 4000 metres depth of ocean. It is a broad zone lying between the sub-littoral and the true deep sea zone. It is also referred to as 'archibenthic zone'. The temperature of the bathyal zone ranges from 15 to 5°C in the low and middle latitudes, while it varies from 3 to -1°C in the higher latitudes. The bathyal zone is aphotic, without sunlight penetration. In low and mid-latitudes, the bottom waters of the bathyal zone are almost stagnant at about 1000 metres depth. In contrast to this, the bottom waters are often flowing with current movement in high latitudes. The bottom sediments of the bathyal zone consist of clay and silt with a large amount of decomposing organic debris. The sediments are mostly derived from terrestrial sources and then from the continental shelf, down the continental slope by currents and gravitational movement. The sediments are also derived from pelagic sources of the ocean and the pelagic sediments consist of shells of pelagic organisms, such as Foraminifera, Coccolithophoridae and Pteropoda.

Benthic organisms of the bathyal zone include mostly meiofauna and microflora. In India, 128 species of meiofaunal organisms have been identified in the Bay of Bengal continental slope of 200-1000 m depths with population density of 4 - 434 individuals per 10 square metre area of bottom. Density decreases with the increasing depth. Average meiobenthic biomass has been weighed at 1094 kilograms in an area of one square kilometre. Nematodes are most dominant followed by foraminiferans. Regarding microbial diversity, in the same depth of 200-1000 m depth, there are 11 genera of bacteria dominating with Gram negative bacteria such as *Pseudomonas* rather than Gram positive bacteria and 16 fungal genera dominating with the genus *Aspergillus* and class deuteromycotina. The microbial counts also reduce with increasing depth.

The characteristic deep sea fauna such as sponges, coelenterates, crustaceans, echinoderms and fishes, are present at depths below 200-300 metres. Crustaceans are represented by decapods, amphipods, isopods and cirripeds in the bathyal zone. Among decapods, Anomuran crabs and deep sea shrimps are abundant. Mollusks are rarely present as the bathyal benthos. Echinoderms, especially holothurians, brittle stars and starfishes are abundant in the bathyal zone. The holothurians are important substrate ingesters of the deep sea, and they help in the reprocessing and redistribution of marine bottom sediments.

Most of the deep sea fishes are pelagic, and only few of them are present as benthic forms in the bathyal zone. The fish have large head, very large eyes, almost wing-like pectoral fins and whip-like tail. There is a decrease in the number of species with increasing depth, and this is attributed to reduced availability of food materials in the deep sea.

3.2.2. Abyssal Zone

This zone is extending from 4000 to 6000 metres depth. It occupies about 83% of the ocean bottom. The environmental conditions of the abyssal zone are uniform. The temperature ranges from 4 to -1°C with no significant diurnal or seasonal changes. There is total absence of sunlight, and the only light available is emitted by bioluminescent organisms. The salinity of water below 2000 metres is around 34 ppt. The pressure in the abyssal depth is strong, increasing with depth at the rate of 1 atmosphere per 10 metres. In the abyssal zone, the sea bottom is covered by soft clay soil along with dead plankton and calcareous shells of coccolithophores. In addition, manganese nodules of various sizes and volcanic ash are also present. The manganese nodules are formed by the precipitation of manganese oxides and other mineral salts from seawater. The nodules look like potatoes with size, ranging from a millimeter to more than a metre in diameter. These nodules are widely distributed especially in the Pacific Ocean. The sediments often contain high concentrations of radioactive elements, such as radium and uranium.

Many benthic forms are detritus feeders. The plant debris is abundant on the abyssal bottom particularly in the tropical areas. Similarly the dead bodies of large animals such as whales provide food for the deep sea organisms.

3.2.2.1. *Characteristics of Abyssal Benthic Animals*

The chief characteristics of the abyssal organisms are associated with their life in the dark, calm, and changeless environment with soft bottom substrate. Most of the animals are pale in colour. Most of the isopods are small in size, and this may be related to the poor availability of food. Many mobile animals have very delicate bodies with long and slender legs for easy movement over the soft substratum. The sessile animals have very long stalks, which may be an adaptation for raising them in the water layer closer to the bottom. The abyssal organisms do not have calcareous skeletons, and this is due to the poor deposition of calcium carbonate. This is due to the reason that high carbon dioxide content in the abyssal water reduces calcium deposition.

Many benthic organisms are slow moving and the availability of prey food in the environment is little. Therefore, carnivorous organisms possess highly elastic jaws and stomachs, so that they will be able to swallow the prey of any size. Many benthic animals have enlarged eyes, adapted to collect the high amount of bioluminescent light. Bioluminescence is rare in the abyssal benthic zone when compared to the bathyal benthic zone. The rate of reproduction is slow and it does not have any seasonal variations. The abyssal animals produce only a small number of eggs at a time. The true abyssal benthic species do not produce any pelagic larvae.

3.2.2.2. *Fauna of The Abyssal Benthic Zone*

Most of the major groups of marine invertebrates are present in the abyssal benthic zone. In addition to many benthic fishes, several groups such as Coelenterata, Porifera, Polychaeta, Sipunculida, Crustacea, Pyenogonida, Mollusks, Tunicata and Echinodermata are distributed in the abyssal benthic zone. Typical abyssal benthic forms are siliceous sponges, antipatharian corals, the shrimps of the genus *Hynmenopenaeus*, hermit crab, brittle stars, starfishes, sea-urchins, isopods and amphipods. In the hermit crab, abdomen is calcified as the crab does not have a shell to protect their abdomen.

3.2.3. Hadal Zone

This is the ultra-abyssal zone which includes areas from 6000 metres to the bottom of the deepest trenches. The conditions prevailing in the hadal zone are similar to those of the abyssal zone. However, hydrostatic pressure is greater in the hadal zone than that of the abyssal zone. The pressure at a depth of 10,000 metres amounts to 1000 atmospheres. This is a limiting factor on marine organisms. Decapod crustaceans in the hadal zone are totally absent. Temperature and salinity of the hadal zone are higher than those of the abyssal. The temperature ranges between 1.2 and - 3.6°C. The bottom sediments consist of soft oozes and red clay, volcanic rocks and stones of littoral origin.

Food availability is more in many trenches than that in the abyssal bottom. As a result, there is more concentration of animals notably pogonophores in this zone. Pogonophores require more food than other animals.

3.2.3.1. Characteristics of Animals of Hadal Zone

Similar to abyssal animals, the hadal animals are adapted to total darkness and soft substratum. Their adaptations are body colour (grey and white), total blindness and long-legs (in many crustaceans). Gigantism is noted in many, especially among tanaids and isopods. The reason attributed to this is the increased hydrostatic pressure, which affects the embolic rate and retards the sexual maturity or increase the longevity of the animals. The hydrostatic pressure also has growth accelerating effect on the trench bacteria.

The species diversity and biomass in the deep seafloor reduce with increasing depth and distance from the land. There are 120 different species at a depth of 1000 metres; and, the number of species is reduced to 17 between 8000-9000 metres. Biomass is between 2 and 5 grams at depths from 4000 to 5000 metres, and the biomass reduces between 20 and 30 milligrams at 9,000-10,000 metres.

3.2.3.2. Fauna of Hadal Zone

The hadal fauna are coelenterates, polychaetes, crustaceans, mollusks, holothurians and pogonophores. The most dominant group is the holothurians, followed by the pogonophora. Many species found in the trenches appear to be endemic to the hadal zone. True endemic forms are restricted to the depth above 6,800 metres. The animal groups absent in hadal zone are decapods, bryozoans, brachiopods and turbellarians. Carnivorous animals are totally absent.

- The benthic zone of the sea extends from about 200 metres depth to the deep ocean bottom. This zone is divided into an upper 'bathyal' zone (200 to 4000 metres depth of ocean), a middle 'abyssal' zone (4000-6000 metres depth of ocean) and a lower 'hadal' zone (greater than 6000 metres).
- Most of the ocean basins are flat plains, but these plains are cut deep or elevated in some places. In most cases the trenches are troughs as continuations of the continental slope. The ocean basin elevates as oceanic rises, islands, seamounts, guyots and mid-oceanic ridges. Most earth quakes and volcanoes occur where tectonic plates collide. All along the trenches, lava comes out at temperatures of about 800°C and builds up a chain of volcanic islands, followed by earthquakes. This zone is also known as 'island arc' or 'Ring of Fire'.

- The deep sea is red clay made up of fine sediment, rich in iron and manganese. The seafloor is covered with sediments derived either from terrestrial or marine sources. The rate of sedimentation is slow. Large areas of the seafloor are covered by the 'manganese nodules'. These contain metals such as copper, nickel, cobalt, iron, manganese that are precipitated as oxides or hydroxides from seawater.
- The animals live in dark and cold conditions of the deeper part of the seafloor. They are only slow growing.
- Faunal composition of the deep-sea benthos is dominated by detritus feeders.
- Deep-sea diversity particularly in the benthos is high due to stability and extensive area of the habitat.

3.3. HYDROTHERMAL VENTS AND COLD SEEP COMMUNITIES

The earth crust is made into a number of large units called 'plates'. These plate boundaries are the sites of ridges and trench systems. They are weak spots in the earth's crust where the hottest mantle pushes up, producing cracks. Cold seawater flows into these cracks and it is heated to very high temperature. The heated water emerges out, carrying with reduced sulphur compounds, mainly hydrogen sulphide, and these structures are called 'hydrothermal vents'. These vents are the sites of unique communities of deep sea organisms present at the tectonic plate boundaries. Similar communities that are not always associated with the plate boundaries are called 'cold seep communities'. Cold seeps are of two types: hypersaline brines or hydrocarbon seeps.

One of the most significant features of deep-sea biology is the presence of hydrothermal vents and cold seeps. There are 212 hydrothermal vents, and 24 cold seeps, reportedly present in the world oceans. All the vent areas and cold seeps have an assemblage of large animals. The dominant animals of the vents are large clams, mussels and giant vent worms. Other large animals associated with vents are crabs, snails, polychaete worms, siphonophore, and fishes. The hydrothermal vents and cold seeps are known to have 443 and 211 species, respectively, and most of the species (>90%) are endemic to the vents or cold seeps. The animals are ancient and highly adapted to toxic chemicals present in the area.

The vent systems and cold seeps depend on the primary productivity of chemoautotrophic bacteria. These bacteria depend mainly on the utilization of sulphides for energy in the vent systems; and on sulphides or methane or other hydrocarbons in the cold seep areas. These reduced compounds are oxidized by the chemoautotrophic bacteria to release energy, which they use to produce organic matter from carbon dioxide. The bacteria get oxygen from the surrounding cold water. The productivity is very high to a tune of 19 µg carbon per gram per hour. These bacteria also endosymbiotically associated with the giant vent worm, *Riftia pachyptila* (red tipped tube worms) and the largest clam, *Calyptogena magnifica.* The animals get nutrition from bacteria. In general, sulphide is a toxic chemical which will interfere with oxygen uptake and respiration of animals. However, it is not happening due the presence of a protein factor in the animal's blood and this protein binds to sulphide, and releases it out of animals. In the cold seeps, mussels (*Bathymodiolus*) have symbiotic association with methane oxidizing bacteria. These bacteria oxidize the methane, and provide energy to the mussels. However, most other vent species do not have symbiotic bacteria, and they feed on bacteria by grazing or filtering.

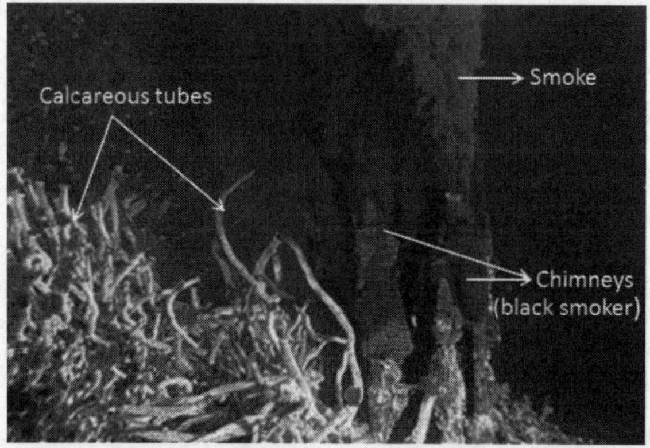

Fig. 3.10: Hydrothermal vent environment, showing the giant worm (*Riftia pachyptila*) with calcareous tubes and black smoker chimey

Hydrothermal vents release hot water at temperature over 350°C through 'black smoker" chimneys (Fig. 3.10). The animals in the vent do grow much faster than surrounding deep-sea animals away from the vents. The life span of individual vents is only in the tens of years. However, vent associated species remain in the inactive vents. To cite an example, the hydrothermal vent called 'The Rose Garden' was first discovered in 1977 in the Galapagos. A study of this first vent has revealed a rapid change in the vent community after 1977. Initially the vents were dominated by giant vent worms, clams and mussels, but that over time the worms and other animals diminished and leaving the mussels as the most persistent one. The way forward is in developing deep sea observatories for long term investigation using internet connections with fibre optic cables and global satellite communications for providing information of the deep sea environment remotely.

The deep-water hydrothermal vents are species-rich communities, while the shallow- water hydrothermal vents at depths of less than 200 metres are low in animal diversity and their ecosystems are simple- only two species of brachyuran crab, *Xenograpsus novaeinsularis* and *X. testudinatus* have been identified so far.

3.3.1. Adaptations

Life on vent has adapted to unfavourable environment. A special adaptation is in the eyes of shrimps and crabs in the vent community. In the shrimp *Rimicaris exoculata*, compound eye is replaced by an extended naked retina, and this helps in detecting bacterial food and also protecting the animal from heat of vents. A similar adaptation occurs in the adult crab, *Bythograea thermydron*. This crab has naked retinal structures in adults and normal compound eyes in larvae. Its early 'zoea' larvae live not in the vents, but as deep sea plankton and the zoea have normal compound eyes. These eyes progressively lose their optics during development and turn into the naked retina of the adult crab. The pigments in the eye that respond to light also change from blue light (477nm) in the planktonic zoea larvae to blue-green light (487nm) in the adult eye. Thus eyes of crab larvae contain a short-wave length photo-pigment in their eyes that is well matched to the dim blue-light of pelagic waters. However, in the vent environment, there are no sufficient visible light to support vision of the adult crabs. However, the adult crabs contain photo pigments most sensitive to the blue-green wave lengths. The vent environment is gifted with chemiluminescence in the visible range of 380-620 nm by sulphide oxidation.

The shallow-water vent off Kueishan Island in northeastern Taiwan is an unusual system with highly acidic condition (pH 1.75-4.0), temperatures of 65-116°C and with toxic gases (mainly carbon dioxide, nitrogen, oxygen, sulphur dioxide and hydrogen sulphide). These conditions make the habitat poor in species diversity. It is interesting that a vast crab population of *Xenograpsus testudinatus* (at an average density of 364 crabs per square metre) thrives in the extremely toxic habitat. The sulphur-dependent bacterial mats are not present as food sources for the crab in the shallow-water vent system, in contrast to deep-water vent where sulphur bacteria are predominant food source. The crab in the shallow-water vent is feeding on zooplankton. The zooplankton are killed by the vent's discharges, and they rain down like 'marine snow'. The crab in large numbers is swarming out of their crevices when water currents are weak and is consuming the marine snow on the seafloor. When the water currents increase, the dropping of dead zooplankton reduces and the crabs return to their crevices. Thus the crab, *Xenograpsus testudinatus* lives at high densities around the hydrothermal vents due to abundant availability of zooplankton as food materials.

3.3.2. Significance of hydrothermal vents

- The life was originated on the earth in the form of thermophilic bacteria in hot water environment at hydrothermal vent at the bottom of ocean, some 3.5 billion years ago.
- The hydrothermal activity represents a significant cooling process for the planet as a whole, according to the earth scientists.
- For oceanographers, hydrothermal fluxes of chemicals, notably iron, manganese, methane, hydrogen and sulphide help to buffer the chemical balance of the ocean and atmosphere.
- The physicochemical conditions of hydrothermal vents are extreme. The water emerging from the vents is hot (>350°C), anoxic, acidic (pH 3-4) and with highly toxic dissolved hydrogen sulphide and heavy metals such as cadmium, arsenic and lead. This water from the vents mixes with the normal surrounding deep water and this water is cold (2°C), oxygen-rich, with normal pH of 7.5 and without any hydrogen sulphide. As a result of the mixing of vent water and surrounding seawater, there is a steep change of physico-chemical conditions over a short distance.
- Microorganisms dominate the vent sites. A dense bacterial biomass is supported in the vent waters by the primary production of sulphide oxidizing bacteria.
- The vent sites support exceptionally productive biological communities in the deep sea, and the organisms living there range from tiny chemosynthetic bacteria to tube worms, giant clams, and crabs.
- Hydrothermal vents are short lived ecosystems with diverse endemic fauna.
- About 92% of species in and around vents are endemic.
- Deep-water hydrothermal vents are species-rich, while the shallow-water hydrothermal vents at depths of less than 200 metres are species-poor and their ecosystems are also simple.
- The food source is mainly sulphur-dependent bacteria for animals of the deep-water hydrothermal vents and it is mainly zooplankton for those of the shallow-water vents.

- One of the most significant features of deep-sea biology is the presence of hydrothermal vents and cold seeps.
- The hydrothermal vents are the sites of unique communities of deep sea organisms present at the tectonic plate boundaries and they release hot water at temperature up to 350°C through 'black smoker" chimneys. Similar communities that are not always associated with the plate boundaries are called 'cold seep communities'.

- The vent systems and cold seeps depend on the primary productivity of chemoautotrophic bacteria. The organisms depend mainly on the utilization of sulphides for energy in the vent systems; and on sulphides or methane or other hydrocarbons in the cold seep areas.
- The animals in the vent do grow much faster than surrounding deep-sea animals away from the vents.
- The life span of individual vents is only in the tens of years.
- The dominant animals of the vents are large clams, mussels and giant vent worms. About 92% of species in and around vents are endemic.
- A special adaptation is in the eyes of shrimps and crabs in the vent community. Compound eye is replaced by an extended naked retina, and this helps in detecting bacterial food and protecting the animal from heat of vents.
- The hydrothermal vents are of great ecological significance. The life was originated on the earth in the form of thermophilic bacteria at hydrothermal vent, some 3.5 billion years ago. The hydrothermal activity represents a significant cooling process for the planet. Hydrothermal fluxes of chemicals, notably iron, manganese, methane, hydrogen and sulphide help to buffer the chemical balance of the ocean and atmosphere.
- The deep-water hydrothermal vents are species-rich communities, while the shallow- water hydrothermal vents at depths of less than 200 metres are low in animal diversity and their ecosystems are simple. Major food source is the sulphur-dependent bacteria for animals of deep-water vents, and the zooplankton for shallow-water vents.

Do you know?

- In Antarctic sea, one baleen whale population consumes 190 million metric tonnes of krill (*Euphausia superba*) per year (Fig. 3.11).
- Nearly 90% of all volcanic activity is oceanic.
- Billions of fish eggs are laid in the oceans. However, only 1 in 1,000,000 fish eggs actually makes it to adulthood.
- *Taq* polymerase which is a thermostable DNA polymerase named after the thermophilic bacterium *Thermus aquaticus* was originally isolated by Thomas D. Brock in 1965 from hydrothermal vents.

Fig. 3.11: Antarctic krill (*Euphausia superba*)
(http://www.aquaria.com.ua/krill.html)

Questions for Review/Exercise*

(1) Define: Microbial loop, Guyots, Holoepipelagic fish, Photophores, Volcanoes, Mid-oceanic ridges, Ring of Fire.

(2) Classify and explain about pelagic zones of ocean with a suitable diagramme.

(3) How are pelagic organisms adapted?

(4) What are the striking differences in the seafloor of the Arctic and Antarctic oceans?

(5) Explain the sub-divisions of benthic zone of the ocean with a suitable diagramme.

(6) Explain with a diagram about the ocean basin.

(7) Hydrothermal vents and cold seeps are significant features of deep sea- Discuss.

(8) Discuss the adaptations of deep sea organisms.

* Regular field visits to the nearby landing centers, collect fishes of ocean, identify them and analyze their adaptive features.

Coastal Ecology 4

The coastal ecology deals with the interrelationships between marine organisms and their coastal environment that occurs between littoral zone ('intertidal') and continental shelf that falls up to a depth of 200 metres. The coastal ecology has two major components: (1) shallow-water subtidal; and (2) intertidal. The shallow-water subtidal component is also termed 'infralittoral zone' that lies between the lowest low tide water on the shore to the continental shelf at a depth of 200 metres. The intertidal component lies between the highest high tide and the lowest low tide. The coastal environment has two major categories: (1) water column (pelagic zone) and (2) sea bottom (benthic zone). The pelagic environment is quite uniform, while the benthic environment is varied with different habitats, such as rocky, sandy, muddy, flat, sloped, irregular and mixed bottom. The pelagic organisms do not experience extreme environmental changes, and there is no need to adapt greatly for their survival in the pelagic environment. In contrast, the benthic organisms are strongly adapted to the environmental changes. In general, marine species are diverse in the benthic environment due to structural complexity and diversity of habitats in the sea. Coastal environment is extremely important in serving as breeding, feeding and nursery grounds for fishes and other aquatic organisms, and also in providing habitats for colonization of mangroves, saltmarshes, seaweeds, seagrasses and coral reefs.

Estuarine and coastal systems are more productive than the open ocean. This is because of the fact that unlike Open Ocean, the estuarine and coastal systems receive nutrients not only from riverine discharge, land run-off and atmospheric deposition, but also from underlying sediments. Although continental shelves occupy just about 10% of the global oceanic area, 30-50% of the marine primary production occurs in this region, and it also sustains about 90% of the fisheries resource. Coastal and continental shelf sediments receive about 80% of the organic matter that reaches the sediment in the oceans and these amounts to 10-50% of the annual primary production in the overlying waters.

- The coastal ecology deals with the interrelationships between marine organisms and their coastal environment that occurs between littoral zone ('intertidal') and continental shelf that falls up to a depth of 200 metres.
- The coastal ecology has two major components: (1) shallow-water sub-tidal; and (2) intertidal. The shallow-water subtidal component is also termed 'infralittoral zone' that lies between the lowest low tide water on the shore to the continental shelf at a depth of 200 metres. The intertidal component lies between the highest high tide and the lowest low tide.
- Coastal environment is more productive than Open Ocean and it is extremely important in serving as breeding, feeding and nursery grounds for marine organisms, and also in providing habitats for colonization of mangroves, saltmarshes, seaweeds, seagrasses and coral reefs.

4.1. PELAGIC ZONE OF COAST

The pelagic zone of the coast lies over the littoral (intertidal) areas and the continental shelf to a depth of 200 metres (Fig. 4.1). This is also termed as 'inshore' or 'neritic zone'. The pelagic zone of the coast has two components: (1) epipelagic and (2) a part of mesopelagic. This zone is much influenced by land mass, and it is the most turbulent area of ocean. There is greater organic production in coastal waters than in the open ocean, since plant nutrients such as nitrates, phosphates and silicates are more readily available. The waves, tides and coastal currents have their significant effects in the coastal waters. The pelagic zone of the coast has high organic content and high dissolved oxygen levels. The coastal waters have a strong variation of environmental conditions, as compared with open sea or deep sea.

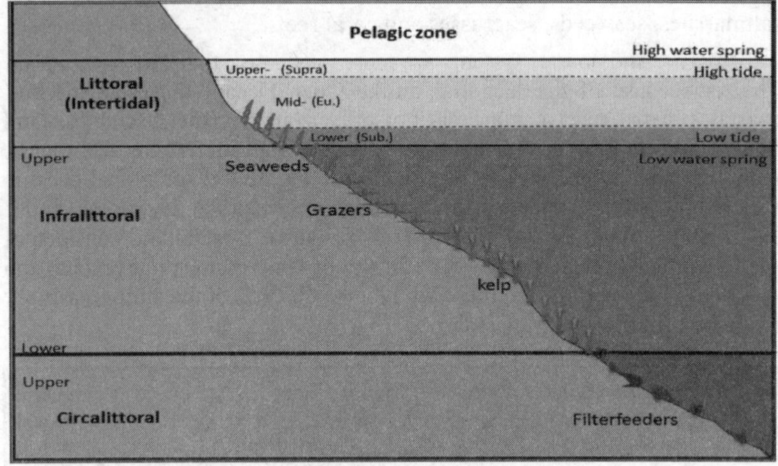

Fig. 4.1: Major divisions of coastal zone

Upper part of epipelagic zone has sufficient light to support photosynthesis. This is the 'photic zone' dominant with phytoplankters. The zooplankton are composed of holoplanktonic organisms. Phytoplankton of the epipelagic zones of temperate, tropical and polar seas are dominated by larger photosynthetic organisms (net phytoplankton, such as diatoms, dinoflagellates), whereas smaller photosynthetic organisms (nanoplankton and picoplankton) are less dominant in coastal neritic waters. The phytoplankters display much seasonal variation in biomass. The phytoplankters are grazed by copepods as the first step in the food chain of epipelagic zone of coastal waters. In contrast, the picoplankton and nanoplankton are grazed by nanozooplankton as the first step in the food chain of epipelagic zone of open ocean.

Bacteria are most abundant near the sea surface. The bacteria are associated with organic particles in the water column collectively called 'particulate organic carbon' or on various gelatinous zooplankton pieces, known as 'marine snow'. The bacteria are usually motile, gram negative, occurring in coccoid, rod and spiral forms. The dissolved organic matter in the waters is utilized by bacteria directly. The bacteria also attack both living and dead organic particles, thus releasing more dissolved organic matter. The bacterial population is higher (5×10^6 cells/ml) in nutrient-rich waters than that (10^5 cells /ml) in nutrient-poor waters. Bacteria have a very high assimilatory efficiency, converting a considerable amount of dissolved organic carbon into particulate organic carbon in the form of bacterial cells. The bacteria are lysed by viruses thereby releasing dissolved organic matter, which in turn support the bacterial growth and reproduction. The bacteria are then consumed by nanozooplankters and can channel a significant amount of energy into planktonic food webs. This is termed the 'microbial loop' (Fig. 4.2). The microbial loop is a major player governing the carbon flux in the sea. The bacteria are responsible for the regeneration of nutrients in the photic zone, helping the continued productivity of the phytoplankton even in the nutrient-poor water.

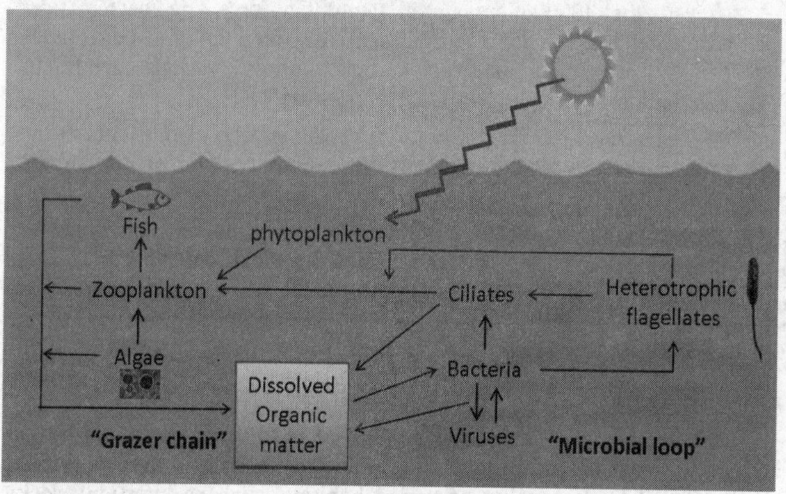

Fig. 4.2: Microbial loop and grazer chain in the epipelagic zone of coastal waters

The water overlying the continental shelf is mainly constituted by the larvae of benthic organisms. The plankton of the coastal water contain considerable numbers of larvae of sponges, echinoderms, annelids, mollusks, crustaceans and fishes as well as the medusae of coelenterates. Besides plankton, there are many swimming crustaceans, cephalopods, fishes and marine mammals such as seals and whales. The dominant fishes are herring (Clupeidae), mackerels, tunas (Scombridae), goat fishes (Mullidae) and sharks (Elasmobranchii).

- The pelagic zone of the coast lies over the littoral (intertidal) areas and the continental shelf to a depth of 200 metres. This is also termed 'inshore' or 'neritic zone'.
- The pelagic zone of the coast has two components: (1) epipelagic and (2) a part of mesopelagic.
- The coastal waters have a strong variation of environmental conditions, as compared with open sea or deep sea. The pelagic zone of the coast has high organic content and high dissolved oxygen levels.
- Phytoplankton of the epipelagic zones of seas are dominated by larger photosynthetic organisms (net phytoplankton such as diatoms, dinoflagellates), whereas smaller photosynthetic organisms (nanoplankton and picoplankton) are less dominant.
- Bacteria are most abundant near the sea surface and are equal or exceed the total biomass of phytoplankton. The bacteria are associated with organic particles in the water column collectively called 'particulate organic carbon' or on various gelatinous zooplankton pieces known as 'marine snow'.
- The bacteria are responsible for the regeneration of nutrients in the photic zone, helping the continued productivity of the phytoplankton even in the nutrient-poor water.
- The water overlying the continental shelf is mainly constituted by the larvae of benthic organisms. Besides plankton, there are many swimming crustaceans, cephalopods, fishes and marine mammals.

4.2. BENTHIC ZONE OF COAST

The benthic zone of coast can be divided in to three zones: (1) littoral zone (intertidal), (2) infralittoral zone and (3) circalittoral zone (continental shelf) (Fig. 4.1). It consists of varied habitats, such as intertidal, rocky shores, sandy shores and muddy shores. The coastal benthic zone holds high species diversity due to the habitat diversity that provides a wide variety of physical and nutritive conditions to marine organisms. The organisms live at the sea-bottom from shore to continental water have the same species diversity, but their biomass decreases. Zonation is a characteristic feature of the intertidal region, remarkably distinct on rocky shore, muddy shore, not in sandy shores.

4.2.1. Intertidal Zone (littoral zone)

The intertidal zone comprises the smallest area of the world's oceans. It is situated between high tide and low tide waters, and the area is alternatively exposed to the air or covered with water. This marine habitat is the most intensively studied due to its easy accessibility to humans. The intertidal zone has the greatest variation in environmental factors. The most prominent physical factors of this area are tide and wave. Waves are caused by winds blowing over the sea surface. Tides are the periodic rise and fall of the sea and are caused by the gravitation interactions between the Earth and the Moon (lunar waves) as well as between the Earth and the Sun (Solar waves). When sea level is at its great height, the tide is called "high tide". When sea level is at its low height, the tide is said to be "low tide". Spring tides occur when the sun and moon are aligned in a straight line, whereas neap tides occur when sun and moon are at right angles to each other. The tidal amplitude is the height of water column between high tide and low tide, and this amplitude varies from few centimetres to 15 metres with different regions. The intertidal area increases with increased tidal amplitude.

4.2.1.1. *Effect of environmental conditions*

Tides have two direct effects on organisms: (1) Duration of exposure to the air. If the duration of exposure to the air is longer, there is a chance for desiccation of animals and (2) time of exposure to the air. Exposure to air during mid-day in the tropics may lead to lethal temperatures, whereas exposure at night in the cold temperate zones in winter may lead to freezing. Intertidal organisms vary in their tolerance to either desiccation or freezing, and these results in different patterns of distribution of the organisms. Tides influence the feeding and reproduction activities of intertidal organisms. Temperature has an effect of desiccation on intertidal organisms, but death occurs only in extreme conditions. Wave actions affect the marine organisms by mechanically smashing the organisms, destroying their substrates, and dislocating the organisms to live higher than the tides. Salinity fluctuates in the intertidal area due to the flooding of the area with freshwater runoff or heavy rains.

There are three zones for the intertidal area: (1) Upper supra-littoral, (2) mid eulittoral zone, and (3) lower sub-littoral (Fig. 4.1). Supra-littoral fringe is the top margin of the intertidal zone and it is an arid zone influenced by waves only in extreme spring tides. This zone is inhabited with snails (*Littorina*), isopods (*Ligia*) and lichens. These organisms are able to withstand prolonged drying, high temperature and strong illumination. Mid-eulittoral zone is the major part of the intertidal region which is alternately exposed and submerged once or twice daily. In this zone, barnacles and mussels are dominant. Lower sub-littoral zone is the lowest part of the intertidal zone. This zone is colonized with rich population of organisms especially algae.

Organisms of the intertidal zone areas are often densely populated for several reasons: (1) The organisms are well adapted to varying conditions of wave action, temperature, light, salinity and desiccation; (2) The area is located between land and sea and hence colonized by both terrestrial and marine organisms; (3) The organisms get abundant supply of food; (4) The animals get adequate supply of oxygen released by plants during photosynthesis and dissolved in water from the atmosphere; (5) There is a lush growth of rooted plants and animals associated with them due to stable substratum with nutrients and adequate illumination; and (6) The area has plenty of nutrients brought in by rivers and distributed by tides and waves.

4.2.1.2. Adaptations of intertidal organisms

4.2.1.2.1. Adaptation to water loss:

The intertidal organisms are exposed to air during low tide, and they may lose water by evaporation. In general, intertidal organisms are tolerant to high temperature. The gastropods display some behavioural adaptations by clustering on low tide to decrease environmental temperature. The fiddler crab is attracted to light in the morning, but it avoids light during the hottest part of the day by hiding into the burrow. The organisms tolerate the water loss or develop mechanisms for avoiding water loss. Organisms avoid water loss through several methods: (1) moving into moist habitats at low tide, (2) closing off soft tissues from the environment by closing shells, and (3) secreting a mucous covering. Barnacles avoid water loss by closing their shells at low tide. The anemone and the hydroid produce mucus that reduces water loss. Crabs move from the exposed surface areas into cracks, crevices, algal mats or burrows of the intertidal area to reduce water loss. The seaweeds can not move and have no mechanism to avoid water loss, and they simply tolerate a great water loss during low tide and rehydrate during high tide.

4.2.1.2.2. Adaptation to heat balance:

Intertidal organisms are exposed to high temperature during low tide in tropics. In order to avoid high temperatures, the organisms reduce heat gain from the environment, and increase heat loss from their body. Large body size of animals takes a long time for heat gain from the environment than small body size, and hence the larger animals avoid heat as in the case *Littorina littorea* (Fig. 4.3). Many snails attach with substrates by means of mucous thread, thereby eliminating body contact with the substrate, and this mechanism also helps in avoiding heat. Shell colour also helps the animal in maintaining internal heat balance. Dark shells absorb heat, while light-coloured shells with sculptures reflect the heat.

Fig. 4.3: *Littorina littorea* (http://www.natuurlijkmooi.net/zeeland/slakken_en_keverslakken/slakken_en_keverslakken.htm)

4.2.1.2.3. Adaptation to mechanical stress:

The most common adaptations to wave stress are body shape and flexibility, attachment and migration. Most common organisms of intertidal areas are small in size with streamlined shapes that minimize exposure to the lift and drag forces of waves. The seaweeds are flexible to the wave action, and hence they are not damaged. Some organisms such as barnacles, oysters and seaweeds are fixed firmly to the substrates to avoid mechanical stress. Mobile animals such as crabs avoid wave stress by moving into sheltered areas in crevices or under rocks.

4.2.1.2.4. Adaptation to hypersalinity:

The fiddler crabs (Fig. 4.4) can even survive in hypersaline conditions on the sand flats with salinity of 90 ppt. Many other organisms can survive in low saline waters by active transport of salts from the medium to the body fluids, reduction of salt excretion from body surface and production of hypo-osmotic urine.

Fig.4.4: Fiddler crabs (*Uca* sp.), adapted to hyper-salinity

4.2.1.2.5. *Adaptation to respiration:*

Most intertidal animals are adapted to extract oxygen from seawater. In general, the respiratory surfaces of gills are thin walled extensions from the body surface, and these respiratory organs are sensitive to desiccation in air during low tide. In order to protect the organs, the intertidal organisms enclose the respiratory surfaces in a protective cavity. Mollusks have gills protected in a mantle cavity and also protected by the shell. In the case of gastropods and barnacles, there is a reduction of the gill and formation of a vascularized mantle cavity which serves as a lung for aerial respiration. To conserve oxygen and water, most animals are quiescent during low tide. Intertidal fishes can have cutaneous respiration through proliferation of blood vessels in the skin.

4.2.1.2.6. *Adaptation to reproduction:*

Most of the intertidal animals have pelagic larval stages. These are guided to the suitable place at the right time of settling to ensure better survival of organisms especially for sessile intertidal forms. Many intertidal organisms are sedentary or sessile. They depend on fertilized egg or larvae that are free floating as plankton for dispersal. Therefore, there is a high incidence of planktonic larvae among the intertidal organisms. In most intertidal animals, breeding cycles are synchronized with the tides. In the mussels such as *Mytilus edulis*, gonad matures during spring tides and spawning occurs on subsequent neap tides.

There are three types of intertidal areas based on the nature of substratum: rocky shore, sandy shore, and muddy shore and these have striking differences as given in table 4.1.

Table 4.1: Comparison among the three intertidal areas – rocky shore, sandy shore and muddy shore

Aspects	Rocky shore	Sandy shore	Muddy shore
Substrate composition	Hard rock	Sand particles of fine and course grain size	Fine grains of sand, silt and clay with organic content
Stability of substrate	Very stable	Unstable due to shifting nature of sand particles	Stable
Physical conditions (temperature and desiccation)	Highly varying	Uniform	Varying
Habitat diversity	High	Low	Moderate
Species diversity	High	Low	High
Zonation pattern	Very distinct	Not distinct	Distinct
Plant communities	Rich (seaweeds)	Poor	Very rich (mangroves, seagrass, saltmarshes)
Type of burrowing animals	Very slow burrowers	Fast burrowers	Slow burrowers
Colonization of organisms	Favourable for sedentary organisms	Unfavourable for many organisms; absence of large plants and sessile animals	Favourable for microbes, plant communities and animals

- The intertidal zone is alternatively exposed to the air or covered with water. The most prominent physical factors of this area are tide and wave. The intertidal area increases with increasing tidal amplitude.
- Organisms of the intertidal zone areas are dense because: (1) The organisms are well adapted to fluctuating conditions; (2) The area is colonized by both terrestrial and marine organisms; (3) The organisms get abundant supply of food; (4) The animals get adequate supply of oxygen; (5) Due to stable substrates, there is a lush growth of rooted plants and animal associates; and (6) The area has plenty of nutrients brought in by rivers and distributed by tides and waves.

- The most common adaptations to wave stress are body shape, flexibility, attachment and migration. Many can survive in low saline waters by active transport of salts by the body surface from the medium to the body fluids, reduction of salt excretion from the body surface and production of hypo-osmotic urine.
- Intertidal organisms are tolerant to high temperature and avoid water loss by (1) moving into moist habitats at low tide, (2) closing off soft tissues from the environment by closing shells, and (3) secreting a mucous covering.
- Most intertidal animals have respiratory surfaces that are adapted to extract oxygen from seawater. In order to protect the organs, the intertidal organisms enclose the respiratory surfaces in a protective cavity.
- Most of the intertidal animals have pelagic larval stages.
- Many intertidal organisms are sedentary or sessile.

4.2.2. Rocky shores

The rocky shores are stable and hard rocky substratum which favours many sedentary organisms for colonization. The attached forms are barnacles, mussels, oysters, and seaweeds. Some organisms such as sponge, bivalve and sea-urchin bore into rocks, and riddle the surface with holes. Of all the intertidal shores, the rocky shores are the most densely colonized by large organisms, and have the greatest diversity of species, especially in temperate regions. The rocky shores present a variety of habitats such as exposed rock, sheltered overhangs, cervices, deep or shallow pools which offer habitats to many species. The Pacific coast is dominated by rocky intertidal areas, as compared to the Atlantic coast.

4.2.2.1. Zonation and distribution of rocky shore organisms

Zonation of organisms is a striking feature of the rocky intertidal shore. The zonation is a distinct horizontal banding, and these bands succeed each other vertically on the shore. The zonation of the rocky shores depends on the tidal amplitude. There are three intertidal zones of rocky shore: (1) the upper supra-littoral fringe, (2) the mid-littoral zone and (3) the lower infra-littoral fringe (Fig. 4.5). In the supra-littoral fringe, the dominant organisms are the snails and black encrusting lichens. Blue-green algae occur at upper part of this zone. The snails living at lower part of this zone may move up to feed on the algae. The lower limit of the supra-littoral zone is colonized with barnacles. Mid-littoral zone is the middle part of the shore and is usually the widest of the three zones and it contains the typical organisms of rocky shore. Barnacles are most common in this zone, but absent in the Polar Regions. The large zone of mid-littoral zone is divided into three sub-zones: (1) Uppermost sub-zone of oysters; (2) the middle sub-zone of limpets; oysters and barnacles; and, (3) the lowermost sub-zone of seaweeds. Infra-littoral fringe is the lowermost part of the three zones of rocky shore. This is the upper boundary of the sub-littoral zone, and it is exposed only during low spring tides. Therefore, this zone is composed of organisms that can tolerate only limited exposure to air. It is often covered with the encrusting sponges (*Halichondria)*, hydroids, chitons and ascidians.

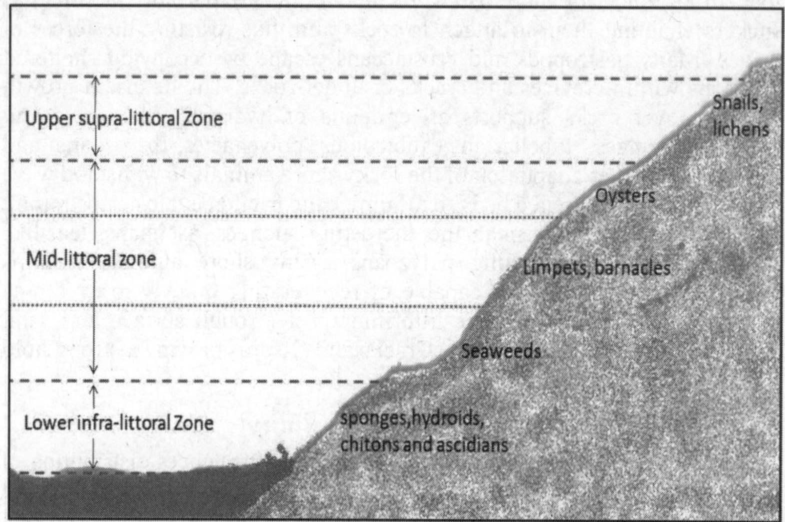

Fig. 4.5: Zonation and distribution of rocky shore organisms in upper supra-littoral, mid-littoral and lower infra-littoral zones

The zonal patterns on rocky shores are based on the variations of the tidal action and tolerances of the organisms to desiccation and temperature extremes. Physical factors such as tides, wave action, temperature and season are strongly influencing the upper limits of intertidal organisms. Biological factors such as competition, predation, grazing and larval settlement, also determine the establishment of zonal patterns. Some prey animals escape predation and contribute to zonal patterns. In the rocky intertidal area, there is an intense competition among organisms for space. The Pacific barnacle *Balanus cariosus* escapes predation by the snail *Nucella* by growing in a large area, while *Balanus glandula* escapes by settling so high in the intertidal area that the predator snail cannot reach them due to excessive desiccation. Thus, zonation of intertidal organisms is one of the functions of biological competition.

4.2.2.2. Adaptation to wave action

The animals are adapted to the wave action of the rocky shore. In the case of limpet and barnacles, the body is conical shape to offer resistance to wave action. Seaweeds are less branched with sturdy thallus to absorb the wave shock. Organisms are firmly attached to the rocky substratum by means of holdfasts in seaweeds, tube feet in starfishes, and byssus threads in mussels. Barnacles and oysters secrete a substance and this strongly cements them to the substratum. In the case of the barnacles the attachment is made by cement produced by cement glands. In gastropods, the adhesion is made by the mucus secreted by the mucous glands of the flat ventral foot. In a number of shore fishes, the pelvic fins are partially modified as suckers enabling them to attach to rocks and thus resisting the forces of waves. Many gastropods and crustaceans escape by occupying sheltered positions within crevices and cracks or under rocks. The luxuriant growth of algae over rocks supports an epifauna of hydroids, calcareous and silicious sponges, tubellarians, tubicolous polychaetes, bryozoans and tunicates. Another adaptation of the rocky shore animals to withstand wave action is to have flattened body to minimize the friction against substratum. This adaptation is present in the encrusting sponges, ascidians, leaf-like turbellarians and encrusting polyzoans. Many shore animals such as echinoderms and crabs are capable of regenerating quickly to overcome abrasive action of waves. The littorinids prefer rough surface and sand stones to withstand wave action. Cracks and fissures provide a strong hold to laminarians for protection.

4.2.2.3. *Adaptation to temperature and salinity*

Temperature is important physical factor that influences distribution of rocky shore organisms, as rocky substrates are quickly heated up. The intertidal animals have ability to regulate the high temperatures through evaporative cooling. Some behavioural adaptations such as clumped distribution patterns and diurnal phototactic rhythms. Algae have a thick cuticle and mucilaginous secretions to reduce evaporation and desiccation. The anemone is coated with a slimy secretion which helps to conserve water. The hard tightly fitting shell plates of barnacles and the opercula of gastropods protect animals from desiccation. The limpets fix the rim of its shell so close to the rock that only a narrow gap is left for gaseous exchange without allowing much evaporation. Many forms are euryhaline, exhibiting a wide range of salinity tolerance. Certain gastropods like *Littorina* species can tolerate salinities up to 50 ppt.

4.2.2.4. *Food of rocky shore animals*

The animals have different modes of feeding to utilize all the available food. They are filter feeders, algal film grazers, herbivores, several carnivores, and scavengers. The barnacles, bivalve and tunicates of the rocky shores are filter feeders. Gastropods such as littorinids and limpets are the algal film grazers. They are adapted for attachment on to the rock surface by means of broad flat muscular foot and scraping off the algal film with the help of radula with arms. Herbivores like limpets also scrap the algae, eelgrass and seaweeds. Sea urchins have well-adapted mouth parts with teeth that help to cut the larger vegetation. Carnivores and scavengers colonize in large numbers in the rocky shore due to the abundance of sedentary and slow moving animals. The carnivores are sessile like sea-anemones, or slow moving like starfishes, nudibranchs and whelks. Anemones capture and feed on large fishes and crustaceans. Hydroid colonies with very small polyps capture small organisms of zooplankton. Many crustaceans such as crabs, hermit crabs, amphipods, lobsters and shrimps are predators or scavengers or both.

- The rocky shores are stable and hard rocky substratum which favours colonization of many sedentary organisms.
- Of all the intertidal shores, the rocky shores are the most densely colonized by large organisms, and have the greatest diversity of species, especially in temperate regions.
- Zonation of organisms is a striking feature of the rocky intertidal shore. The zonal patterns are based on the variations of the tidal action and tolerances of the organisms to desiccation and temperature extremes.
- There are three intertidal zones of rocky shore: (1) the upper supra-littoral fringe, (2) the mid-littoral zone, and (3) the lower infra-littoral fringe
- Organisms are strongly attached to the rocky substratum by means of holdfasts, tube feet cementing glands and byssus threads.
- Another adaptation is to have flattened body to minimize the friction against substratum and to withstand wave action.
- The intertidal animals have ability to regulate the high temperatures through evaporative cooling and behavioural adaptations such as clumped distribution patterns and diurnal phototactic rhythms.

4.2.2.5. Tide-pools of rocky shores

A unique feature of many rocky shores is the presence of tide pools of various sizes, depths and locations. Tide pools are formed by trapping tidal water in the hollows of rocks when the tide goes out. Tidal pools are produced in the gullies between irregularly sized rocks. During high tide, all the tide pool is subjected to interchange of water with the sea. The tide pools differ from the exposed rocks and the neighbouring seawater and are subjected to different environmental conditions.

Temperature, salinity, oxygen and pH are the important environmental factors in the tidal pools. Temperature varies widely from normal temperature of the seawater to over 40°C in the pools. Salinity of the tide-pools varies from a nearly freshwater to about 50 ppt. In the tide-pools having a lush growth of algae exhibit increase of pH to 9-10 during day time, and fall of pH to 7.5 in night owing to fall in oxygen content, and subsequent increase in carbon dioxide. In general, numbers of species decrease with increasing height of the tide pool.

Tide pools provide a habitat for many organisms and they are areas of refuge from desiccation for intertidal organisms. But the tidal pools can have rapid changes in temperature, salinity and oxygen concentrations that may strongly affect the organisms. Sea anemones are the most important among the fauna of the tide pools of the rocky shore. Sponges, hydroids, polyzoans, nereid and scale worms, crabs, chitons, star-fish and brittle stars are also present in the tide-pools. All animals of rocky shores are generally not tolerant to pool conditions. For example, barnacles, a major component of the rocky shore fauna, are rarely present in the tide-pool.

- A characteristic feature of many rocky shores is the presence of tide pools. Tidal pools are produced in the gullies between irregularly sized rocks. During high tide, all the tide pool is subjected to interchange of water with the sea.
- Temperature, salinity, oxygen and pH are the important environmental variables in the tidal pools.
- In general, numbers of species decrease with increasing height of the tide pool. Tide pools are areas of refuge from desiccation for intertidal organisms.

4.2.3. Sandy shores

The sandy shore is an unstable substrate or shifting terrain, situated between land and sea. There are two types of sandy shore components: (1) sand beaches and (2) sand flats and the differences between them are given in table 4.2. Sandy beaches and sand flats are common throughout the world. They are well-known to humans for their use in recreation. Not all beaches have yellow sand: Hawaiin beaches have black sand, formed from volcanic lava, the Bermuda islands have pale pink beaches due to the presence of red shells and the Barbados have white beaches due to the presence of white shells. The sandy beach is unfavourable place for many organisms mainly because of the shifting nature and instability of the sand substrate. However, softness and porosity permit some infauna to burrow through and inhabit it. The sand particles allow exchange of gases especially oxygen which is essential for survival of aerobic infauna. Interstices among the sand grains also provide living space for a number of specialized organisms namely 'interstitial fauna'.

Table 4.2: Comparison between sand beaches and sand flats

Sl. No.	Sand beaches	Sand flats
1	Sand beaches are intertidal exposed habitat.	Sand flats are protected habitat.
2	Facing the open sea and subject to wave action	Facing a bay or lagoon protected from the open sea and has less exposure to wave action
3	Usually have a slope toward the sea	Have little or no slope
4	Consisting of grains of widely variable sizes	Consisting of fine sand grains of less variation in sizes
5	Devoid of macroscopic life and the organisms bury themselves in the substrate	Often colonized with large numbers of macroorganisms
6	Less in species diversity on surface layer due to wave action	High in epifaunal population due to low wave action

4.2.3.1. Environmental conditions

The most important physical factors that govern life on exposed sand are wave action, particle size and slope of sand beaches. Fine sand holds much water in its interstices through its capillary action, whereas coarse sand and gravel allow water to drain away rapidly. The intertidal organisms are well-protected from desiccation in a fine sand beach, but not so in a coarse gravel beach. Moreover, fine sand is more suitable for burrowing by animals, as compared to coarse gravel. The slope of a beach is the result of the interaction among particle size, wave action, and the relative importance of 'swash' and 'backwash water'. Swash is the action of water running up a beach after a wave break and this action adds sand to the beach (accretion), while backwash is the action of water flowing back down the beach, and this action removes sand from the beach (erosion). There are two beach types: dissipative beach and reflective beach. A dissipative beach is an eroded one by strong wave action and beach slope is gentle. On the other hand, a reflective beach is a well- developed one by weak wave action and beach slope is steep.

The abundance and diversity of macrofaunal organisms on open beaches are related to particle size and slope, rather than to wave action. Because of the movement of the surface layers of sediment on sand beaches, only few large organisms occupy the unstable surface. As a result, biodiversity is low in the sand beaches. This low biodiversity is due to lack of habitat diversity in the sandy shores, in contrast to the rocky shore which has diversified habitats, such as crevices, overhangs and tide pools.

Environmental factors such as temperature, desiccation and wave action are uniform at each tidal level on the beach or flat, due to their lack of habitat diversity. Sand is an excellent buffer against large temperature and salinity changes due to the fact that it holds large amounts of water in its interstices. Oxygen may become depleted in the sedimentary substrate, especially in fine-grained sediments where water exchange with the seawater is reduced.

The sandy shore changes according to season and wave action which mobilize the sand. The width of the sandy shore varies with the tide ranging from 116 metres during premonsoon period, and minimum of 66 metres during monsoon period in south-west coast of India. Extensive erosion of the beaches owing to strong winds during the monsoon period is characteristic phenomenon of the south-west coast of India. The erosion of the sand reduces the width of beach which, in turn, affects organisms living there. The profile of the beach and also the size of the particles of the substratum are largely decided by the waves.

Besides the tides and waves, the particle size of the sand on the shore also influences the profile of the sandy shore. The larger the particle size, the greater will be the slope. Thus the steepest slopes are found on coarse sand beaches. The size of the sand grains also governs the amount of water that is retained by capillary action within the minute spaces between the individual granules of sand. The water content may be about 20% of the wet weight, falling to 10% if there are much gravels or small stones.

Temperature and desiccation conditions are uniform as those on rocky shores. In tropical shores, a very sharp gradient has been observed in the top 3 to 5 centimetres, with only a very slow change below that level. Thus, in normal conditions, extremes of temperatures do not seriously affect the sandy shore organisms for making burrows.

Salinity of the interstitial water is usually similar to that of the sea. The surface layers may be high in salinity, due to evaporation and salinity may also be low as a result of freshwater input. All sands contain varying amount of organic matter. The organic matter is lighter than sand particles and therefore the organic matter can settle down in calm water. The higher concentration of organic matter may clog the interstices between the grains.

The available oxygen will be used up for microbial oxidation of organic matter at depths varying from a few millimeters to nearly a metre, depending on the organic content of sand. The anaerobic layer is marked by a black layer known as the 'sulphide layer', caused by the presence of ferrous sulphide. Most of the larger infauna and interstitial fauna are aerobic, and they are normally restricted to the upper layer. If they inhabit anaerobic layer, they have to retain a connection with the oxygenated surface layers and maintain a flow of water to and from the body. The lugworm *Arenicola*, lives in a 'U' shaped burrow through which a flow of water is maintained at high tide by means of peristaltic movements of the body (Fig. 4.6). At low tide, when the water level falls below the surface, the animal is able to expose its gills to moist air in the upper part of the tail, and so is able to make use of the atmospheric air.

Open sand beaches generally have three zones: (1) supra-littoral, (2) mid-littoral and (3) infralittoral. An upper supra-littoral fringe is colonized by amphipods or ghost crabs, a broad mid-littoral with sand crabs and isopods, and an infra-littoral fringe is inhabited with large number of species. In contrast to rocky shores, sand beaches have tidal and seasonal migrations of their animals. Both competition and predation decrease as wave exposure of the beach increases.

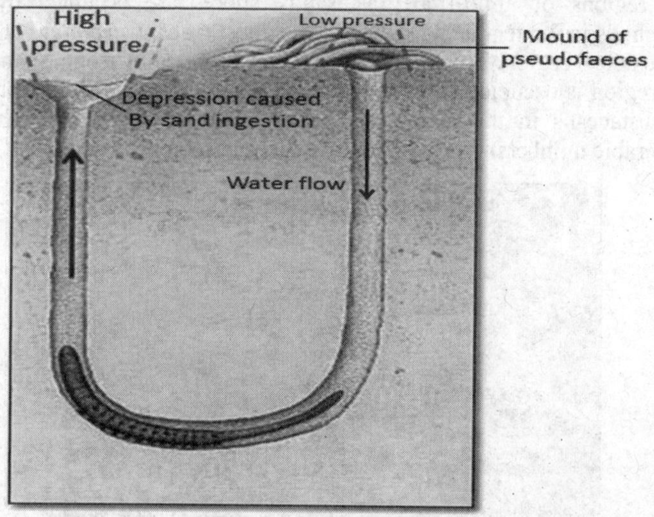

Fig. 4.6: Lugworm *Arenicola* in a 'U' shaped burrow

Thus, physical factors, mainly wave action, particle size and beach slope are the major contributors to the sandy shore environment. The protected sand flats are dominated by suspension feeders in fine sand areas, and deposit feeders in coarse sand shore. The prime biological factors that change the structure of sand flat communities are predation, competition and disturbance. The role of competition in structuring the sand flat communities is different from that observed on rocky shores. The severity of direct competition is reduced by animals, three dimensional structure of the substrate, and interactions between juveniles and adults. In contrast to rocky shores, sand flats are dominated by highly mobile predators. In higher latitudes, there may be seasonal changes in the sand flat community due to migrations of the dominant organisms.

The macrofauna of intertidal sand flats displays high variation in relation to time and space. The numbers of species may vary greatly within and between years. This variability is due to the high predation pressure and biological disturbance that control the infaunal community structure.

4.2.3.2. *Organisms of sandy shores*

The dominant animals on sand beaches and sand flats are polychaete worms, bivalve mollusks and crustaceans. Large plants and sessile animals are, in general, absent on open sand beaches. There is little primary productivity on open sand beaches, and the animals found there are mainly filter feeders, detritus feeders and scavengers. On protected sand flats, there is a substantial primary productivity and all feeding types of animals are present. Most of the sandy shore animals live buried in the sand and therefore a zonation is not evident, unlike that on the rocky shores. The upper regions of almost all the sandy shores are occupied by the amphiphods in the temperate region or the ghost crabs (*Ocypode;* Fig. 4.7) in the tropical region. The middle region is dominated by isopods, and the lower region is occupied by anomuran crabs in the tropics, and amphipod and crustaceans in the temperate regions. The isopods are present in considerable numbers, especially during the monsoon period.

Fig. 4.7: Ghost crabs (*Ocypode quadrata*)
(http://www.bukisa.com/articles/50745_the-worlds-most-colorful-crabs)

Donax is specially adapted to move through the sand very rapidly, and they move up and down the shore. The heart-urchin *Echinocardium* with the help of the spines actively dig down straight to a depth of 15 -20 centimetres. The walls of the burrow are made strong with mucus secreted, and applied by tube feet. A respiratory current is maintained inside the burrow by an extremely long tube foot which extends up to the surface. Besides these, terrestrial forms such as beetles, sand-pipers and seagulls visit this region in search of food. There are many marine animals visit the intertidal sands mainly for breeding as in the case of the king-crabs and the sea turtle. The female turtle excavates a hole in the sand and deposits the fertilized eggs and these eggs are soon covered with sand by receding tide. The female king crab lays egg which is fertilized by male on the sandy shore. After breeding, the young ones hatched out and return to the sea. The breeding migration is well correlated with the lunar cycle and that the egg laying takes place along the highest spring tide level.

4.2.3.3. Adaptations of sandy shore organisms

Burrowing is the main adaptation in the unstable sandy substrate by animals. Other adaptations are smooth shells and reduced spines to aid in burrowing, tidal migration to keep the animals in the swash zone, avoidance of diurnal predators by becoming nocturnal, and preventing of clogging of respiratory surfaces through development of meshes that filter out sand. To escape from the heavy wave action, bivalve mollusks and mole crab are able to burrow rapidly and deeply. Many organisms exhibit tidal migration, and they may move up and settle in the narrow beach. Such migration occurs in the case of the bivalve (*Donax*), the mole crab (*Emerita*) and the gastropod (*Bullia*).

4.2.3.4. Food of sandy shore animals

Many animals are plankton and suspension feeders, and they are the anomuran crabs and bivalves. When the wave recedes, the crabs extend their antennae into the water, which act as passive filters in the water current. Then each antenna is drawn alternately through the mouthparts to scrape off the plankton and other organic particles collected, and the mouthparts are so arranged to prevent the entry of sand grains. Similarly the bivalves extend their siphons to the surface of the sand to take in a current of water which brings suspended food particles; and the siphon is provided with rings of tentacles to prevent sand grains from entering the mantle cavity. The terebellid worms and the tellinid mollusks are deposit feeders. The terebellid worms spread out a network of ciliated tentacles at the opening of their burrow to collect particles or organic matter. The tellinid mollusks feed by using long inhalant siphon. The lug-worm *Arenicola* is a non-selective deposit feeder. With the help of muscular proboscis, it takes the sand and digests the organic debris contained in it. Many amphipods of the temperate and ocypode crabs of the tropical sandy shores are scavengers. They are omnivorous forms, scavenging on the upper beaches during the night, but during the day time they do not feed and live hiding themselves. Many shore fishes and some of the crabs are true carnivores. *Ocypode* crabs are predating on the bivalve like *Donax*. The isopod and polychaetes of the sandy shores are also predators in the habitat.

- The sandy beach is unfavourable place for many organisms mainly because of the shifting nature and instability of the sand substrate. However, softness and porosity permit some infauna to burrow through and inhabit it.

- Fine sand is more suitable for burrowing animals and to overcome desiccation, as compared to coarse gravel.
- The abundance and diversity of macrofaunal organisms on open beaches are related to particle size and slope, rather than to wave action. As a result, biodiversity is low in the sand beaches. This low biodiversity is due to lack of habitat diversity in the sandy shores.
- The dominant animals on sand beaches and sand flats are polychaete worms, bivalve mollusks and crustaceans.
- Large plants and sessile animals are, in general, absent on open sand beaches.
- Most of the sandy shore animals live buried in the sand and therefore a zonation is not evident.
- Most of the larger infauna and interstitial fauna are aerobic, and they are normally restricted to the upper layer. If they inhabit anaerobic layer, they have to retain a connection with the oxygenated surface layers and maintain a flow of water to and from the body.
- Burrowing is the main adaptation in the unstable sandy substrate by animals. Other adaptations are smooth shells, reduced spines, tidal migration, avoidance of diurnal predators by becoming nocturnal, and preventing of clogging of respiratory surfaces through development of meshes that filter out sand.

4.2.3.5. Interstitial fauna

Interstices provide living space among the sand grains for a number of specialized organisms namely 'interstitial fauna'. The animals live in the thin film of interstitial water. Almost all the groups of invertebrates thrive in the interstices. The temperature of the interstitial habitat depends on the amount of moisture held by the sand, attributed to the cooling effect of the surface evaporation or the continuity of the capillary water. Temperature also fluctuates diurnally and seasonally. The interstitial fauna have ability to withstand such fluctuations. The activity is much lower at low temperatures than that at high temperatures.

Salinity decreases horizontally landwards. However, the salinity of the interstitial water is normally the same as that of the seawater which may be due to the continuity of the water as a result of capillary action. Salinity of the interstitial water is reduced as result of heavy rainfall during the monsoon period, and there is a remarkable reduction in the number of animals during the low saline monsoon period. However, the salinity of the interstitial water increases during rainless summer or pre-monsoon period.

The presence of oxygen in the interstitial water is essential for the respiration of fauna. The oxygen level varies with grain size. In tidal beaches, the level of oxygen is below 0.5 milliliter in one litter of interstitial water at 5-15 centimetres depth. In areas where oxygen is lacking, a black layer is usually formed. This layer will be at a higher level if the beach is formed of fine sand grains. Species of nematodes and copepods withstand a wide range of oxygen concentrations in the interstices.

The increase in the interstitial animal population is correlated with the increase in levels of nutrients such as phosphates, nitrates, sulphates and trace elements. The pH of the interstitial water does not differ from that of the seawater, but often slightly less than those of adjacent seawater due to the higher concentration of carbon dioxide.

The sand grains are the dominant component of the substrate with grain size ranging from rounded to angular forms of varying diameters. The sand grain size determines the volume of available pore spaces and the quantity of capillary water. Both in turn, affect the rate of evaporation from the beach surface, and the rate of capillary ascent of water from the deeper levels. The distribution of sand grains varies due to factors, such as slope of the beach, amplitude of the tide, wind, wave action and local currents. The space factor is of ecological importance in the interstitial environment, since the fauna inhabiting this habitat are totally dependent on the volume of space available among the sand grains for their activities. The space among fine sand particles is small to accommodate smaller animals and those in coarse sands are large to provide microhabitat for larger animals. The total pore space ranges between 38 and 43%. The loss of water through evaporation is compensated by the capillary ascent of water from the neighbouring source. Since the water retained in the capillary space functions as the medium for the inhabitants, the supply of this water and the rate at which it is able to rise through the capillary channels are important factors. The rate of capillary ascent is slow when the sand is coarse.

The content of organic matter in the substratum is of food importance to the interstitial forms and it is more in fine sand than in coarse sand due to the fact that the fine sand retains more organic matter.

Interstitial fauna and their adaptations:

The organisms of interstitial habitat are small ranging from 1 to 3 millimetres in most cases. Another important feature of most of these organisms is that the body is vermiform in the cases of cnidarian, opisthobranch and copepod. Great contractility of the body wall and the presence of well developed cuticle with scales or spicules, provide resistance to mechanical wave abrasion. In interstitial fauna, large numbers of special epidermal glands, papillae or adhesive tubes are present for holding on to sand grains.

A majority of the interstitial forms are free living; some are semi-sessile, while a few are completely sedentary in Foraminifera, coral reefs and brachiopod. Locomotion is affected by ciliary gliding (nematodes and harpacticoid copepod) or by crawling (most crustaceans). There are predators (nematodes and turbellarians), diatom feeders (ostracods, harpacticoids, mollusks, gastrotrichs) and suspension feeders (ascidians).

The reproductive rate of a majority of these forms is very low, producing only a very small number of eggs and sperms. About 98% of the species lack pelagic larvae. Brood protection and viviparity are also important adaptation to balance the low rate of reproduction. To minimize the loss of reproductive elements, some species are hermaphrodites as in the case of gastrotrichs, and copulation occurs in harpacticoid copepods.

- Interstices provide living space among the sand grains for a number of specialized organisms namely 'interstitial fauna'. The animals live in the thin film of interstitial water. Almost all the groups of invertebrates are present in the habitat.
- The salinity of the interstitial water is normally the same as that of the sea as a result of capillary action.
- There is a remarkable reduction in the number of animals during the low saline monsoon period.
- The increase in the interstitial animal population is correlated with the increase in nutrient concentrations.
- The content of organic matter in the substratum is of food importance to the interstitial forms and it is more in fine sand than in coarse sand.
- In interstitial fauna, large numbers of special epidermal glands, papillae or adhesive tubes are present for holding on to sand grains.

4.2.3.6. Coastal sand dunes

Coastal sand dunes are one of nature's most interesting and beautiful formations. The coastal sand dunes are mounds or ridges on the shore and they are formed by sand deposited from the sub-tidal and intertidal regions. Wind is one of the most important factors, helping in formation, movement and distribution of the sand dunes. The sand dunes are colonized by vegetation beyond the highest high tide level. Thus main pre-requisite for the formation of sand dunes are wind, sand and vegetation. The sand dune vegetation is totally a different plant community with remarkable ability to tolerate adverse conditions of drought, nutrient deficiency, high winds, salt sprays and sand blast.

There are three types of sand dunes: (1) embryonic dune, (2) mid shore dune and (3) hind shore dune. The embryonic dune is distributed nearer to the sea and often unvegetated towards the sea. It is just above the high tide level with its steeper face inland. This zone is formed by sand delivered to the beach by wave action. The dominant plants growing on the embryonic dunes are *Ipomea pes-caprae, Spinifex littoreus* and few other herbaceous species (Fig. 4.8,4.9). Embryonic dune is also known as 'fore dune' which attains a maximum height of 10 feet. The mid shore dune is characterized by the presence of shrubs and is more or less stable. The common species found here are *Leucas aspera, Vitex negundo* and *Clerodendrum inerme.* Mid shore dune is moved away if the wind velocity is high and it is also termed 'wandering dune' or 'yellow sand dune' which attains a height exceeding 10 feet. The hind shore dune is rich in vegetation due to substrate stability and it has trees with long root system. The dominant plants growing on these dunes are *Vitex negundo, Clerodendrum inerme, Anacardium occidentale, Pandanus tectorius, Casuarina equisetifolia,* Palmyra, *Pongamia pinnata, Thespesia populnea, Prosopis* sp., *Azadirachta indica* and *Cocos nucifera.* If wind velocity is less, the wandering dune faces an obstacle due to another big dune and the wandering dune becomes arrested. This hind shore dune is also termed 'arrested sand dune' or 'grey sand dune'.

Fig. 4.8: *Ipomoea pes-caprae* (L.) Sweet

Fig. 4.9: *Spinifex littoreus*, also known as "Mustache of Ravana"

The dunes are very important in protecting the hinder land from the sea against the powerful physical forces of storms, waves and cyclonic surges. The dunes are also of other values: (1) dunes are typical features of coastal stability and they act as nature's line of defense, (2) sources of beach nourishment and protects coast from erosion and replenish the loss of sand dune to wave and current energies, (3) sand dunes act as 'sand banks' to maintain the sedimentary and dynamic equilibrium of the dune–beach ecosystem, and (4) serves as an ideal place for recreation.

- Coastal sand dunes formed by sand deposited from the sub-tidal and intertidal regions.
- Wind is one of the most important factors, helping in formation, movement and distribution of the sand dunes. The sand dunes are colonized by vegetation beyond the highest high tide level. Thus main pre-requisite for the formation of sand dunes are wind, sand and vegetation.
- The sand dune vegetation is totally a different plant community with remarkable ability to tolerate adverse conditions of drought, nutrient deficiency, high winds, salt sprays and sand blast.
- There are three types of sand dunes: (1) embryonic dune, (2) mid shore dune and (3) hind shore dune.
- The dunes protect the hinder land from the sea against storms, waves and cyclonic surges. The dunes also help in stabilizing coast, maintaining beach ecosystem and providing recreation.

4.2.4. Muddy shores

Muddy shores are present in intertidal areas. Muddy beaches are present in sheltered coast, estuaries and embayment along the coasts. When the water flow is not strong, the suspended mud will settle down to form extensive mud flats. Muddy shores have very fine particles that reduce the water interchange between the shore and seawater. The high bacterial population depletes oxygen in the muddy sediments below the first few centimetres of the surface. This anaerobic condition is one of the significant features of the muddy shore. Between the aerobic and anaerobic layers, there is a transition area, called the 'redox potential discontinuity (RPD) layer', characterized by a rapid change from positive to negative redox potential (Fig.4.10). Positive redox potential indicates the presence of oxygen, whereas negative redox potential refers to the low quantity of oxygen. The RPD layer of sediment is grey in colour, the oxidized layer above is brown or yellow and the anoxic layer is black. The RPD layer is biologically important for recycling of nutrients, in particular phosphate. Chemoautotrophic bacteria present in the RPD oxidize the reduced compounds, fix carbon dioxide and thus produce more organic material.

Muddy shores accumulate organic material for food supply to the resident organisms.

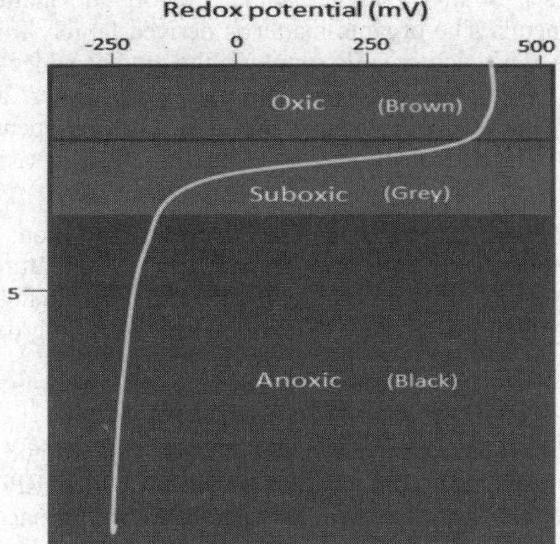

Fig. 4.10: Redox potential discontinuity (RPD) layer

Muddy shores are different from sandy shores in the intertidal areas in many ways. The muddy areas have the finest grain sizes in contrast to sand beaches with larger grain sizes. The muddy shore is much flatter than that of sand beaches. These areas are more stable than sand substrates and suitable for colonization of permanent burrowers. The water retention time is longer in muddy shore than that in sandy shore. The muddy shore has high numbers of bacteria in a narrow aerobic layer overlying a large anaerobic layer, while sandy shore contains low numbers of bacteria and less anaerobic condition. The presence of burrowers in mud flats occurs in the form of holes of different sizes and shapes, unlike the open sand beaches. The muddy shore organisms do not develop rapid burrowing, in contrast to sandy shore organisms. The muddy shores are rich in plant communities, but not so in sandy shore. The dominant macrofaunal groups on muddy shores are the same as those encountered on sand beaches – namely polychaete worms, bivalve mollusks and crustaceans. Because of the increased productivity due to both bacteria and plants, there is more food available on muddy shores than on sand beaches. Therefore, larger organisms with dense populations live on mud shores than that on sand beaches.

4.2.4.1. Physical and chemical features of muddy shores

The mud is composed of clay, sand and silt. Silt contains very fine inorganic particles. Clay is basically hydrated aluminum silicate, but normally it contains iron. Clay particles are small in size and its high percentage in the mud makes it compact and hard. This substrate is colonized with sessile fauna and even borers such as Pholas. The beach mud contains a relatively higher amount of organic matter than other marine sediments. The organic matter is derived largely from rivers and also from sea. This rich organic supply in the mud provides food for large populations of deposit feeders. The overlying water also contains considerable quantities of organic matter in a state of suspension, and this may support large populations of suspension feeders associated with muddy shores.

There is only a surface zone of aerobic decomposition; beneath this, anaerobic conditions prevail. The fluctuations in temperature and salinity in the muddy shore are not very significant and these do not significantly affect the organisms as most of them are burrowers living deep inside the mud.

4.2.4.2. Muddy shore organisms

Mud flats often develop a substantial growth of plants, and contain large numbers of bacteria. Among the bacteria in the RPD layer, the chemoautotrophic bacteria or sulphur bacteria are the primary producers of organic material using the energy of reduced sulphur compounds. The dominant animals on mud flats are polychaete worms, bivalve mollusks and crustaceans.

The dominant feeding types on mud flats are deposit feeders and suspension feeders. The major predators are fishes and birds. On mud flats, skates and rays as well as flat fish are significant predators of infaunal animals. They also excavate pits, causing widespread mortality among the other infaunal organisms. Bird predation on the intertidal areas may be intense and it plays an important role in structuring tidal flat communities.

Deposit feeders are abundant due to the availability of large amount of organic material and the large populations of bacteria in the sediments. Polychaete worms feed by burrowing through the substrate, ingest it, digest out the organic matter and pass out the undigested material through the anus. For example, *Arenicola* species form a U-shaped burrow in which one arm of the U is permanently open to the surface, while the other arm is filled with sediment that the worm ingests. The animal moves up in the open burrow to defecate at the surface making mounds (Fig. 4.6). Suspension feeders include various other species of clams, some crustaceans, and number of polychaetes. The deposit feeders are more common in fine-grained shores and the suspension feeders are more abundant in coarse sediments, where there is little organic matter.

The trophic structure of muddy shore is based on either detritus or autotrophs. Detritus along with bacteria are derived from plants and other organic sources. They are consumed directly by the large macrofaunal invertebrates. The deposit feeders ingest the detritus. The second food web is based on the microscopic diatoms as the autotrophic base. The diatoms are consumed by polychaetes, mollusks and crustaceans. These in turn, are consumed by the large predatory birds and fishes.

4.2.4.3. Adaptations of muddy shore organisms

One of the striking adaptations of mud flat organisms is the ability to burrow into the substrate. Another significant adaptation is the tolerance to live under low oxygen conditions. In order to obtain the oxygen-rich surface water and food, the muddy shore organisms form burrows, holes and tubes. Moreover, the animals produce the tubes permanently in the muddy soil, and the tubes also increase stability of the fine sediments.

4.2.4.4. Zonation of muddy shore organisms

Zonation is a characteristic feature of the intertidal zone, and is well-recognized in the muddy shores, as in the case of rocky shore. It is broadly classified in to three sub-zones: (1) supra-littoral fringe (top part of intertidal zone), (2) a mid-littoral region, and (3) a sub-littoral fringe (bottom part of intertidal zone). The supra-littoral fringe or sub-terrestrial zone is occupied usually by salt marshes in temperate regions or by mangrove forests in tropical shores. The upper area is often inhabited by burrowing crabs. The mid-littoral area is very extensive and provides habitats for most of the common species of clams and polychaetes.

The mid-littoral zone of muddy shore:

The mid-littoral zone is colonized by a number of mollusks primarily the burrowing forms. In the muddy substrates, suspension feeding bivalves are burrowing deeper with increasing age of the animals and so their foot gradually reduces and the siphons grow longer. In contrast, the bivalves inhabiting the sandy beaches have massive foot for rapid burrowing in the sand.

Sub-littoral zone of muddy shore:

The sub-littoral fringe is occupied by large numbers of polychaetes and hence this zone is designated as the polychaete zone. This zone is also inhabited by the bivalve *Mya arenaria.* It lives more than 30 centimetres below the surface of the mud. It feeds on the detritus of the seafloor and its elongated siphons help in respiration. During the period of oxygen deficiency, this species can resort to anaerobic respiration. Besides polychaetes and bivalves, some anomuran decapod crustaceans of the family Thalassinidae also occupy the sub-littoral fringe of muddy shore.

The sub-littoral region begins below the low water line and extends to the edge of the continental shelf lying below a depth of 200 metres. The physical factors of the environment are not of the extreme type and the changes of salinity and temperature are seasonal and gradual. The light is very much diminished in the deeper parts of the continental shelf. The force of tides and currents are also little, compared with the tidal zones. The particle size of the sediments and also the organic content in them are of much significance as they play a major role in determining the bottom fauna. The most variable factor of this environment is the nature of the bottom, usually composed of sand, mud or muddy clay, frequently mixed with stones and rocks. In the deeper parts of the continental shelves, on the upper parts of the slopes marine phosphorite nodules are present.

Organisms of the sub-littoral zone:

The animals occupying the continental shelf region of the ocean bed are the epifauna which live upon a hard substratum, such as rocks, coral reefs, stones, shells, algae, pilings, and over other animals and the 'infauna' which includes all others which burrow or bore into the substratum or the sediment of the seafloor. The epifauna constitute 80% of all species of benthic invertebrates in the tropical regions, and they show decrease in number of species towards the poles, where their number may be equal to that of the infauna. Infauna are closely associated with the epiflora, the permanently attached algae. The algae are abundant on all rocks, stones and other hard substrates in the photic zones. The algal mass contributes a major source of food and shelter to many benthic organisms. The underwater caves in the continental shelf zone provide a good habitat for a number of animals such as sponges, polyzoan and hydrozoan, sea anemones, and octopuses. Among the other benthic animals forming epifauna, the most important are the mussels of the genus *Mytilus* and *Perna* inhabiting the shallow waters. Infauna are dominant with bivalves *Macoma, Tellina,* and the sand dollar, *Dendraster* forming distinct communities in the continental shelf zone. Besides these macro-fauna, the bottom fauna of the continental shelf include the meiofauna and microfauna also. The meiofauna with size ranging from 50μ to 1000μ include nematodes, harpacticoid copepods, flat worms, minute species of mollusks, worms and other invertebrates such as rotifers, gastrotrichs and larger foraminiferans. Of these, the nematodes may number over one million individuals per square metre on shallow water bottoms, the harpacticoids over 4,00,000 individuals per square metre and the foraminiferans about one million per square metres. Most of the meiofauna form an important source of food for the macrofauna. Microbenthos of the sub-littoral zone includes animals below 0.5 millimetres in size, and they are composed primarily of small protozoans, especially flagellates, ciliates, amoebae and benthic diatoms and bacteria. The bottom mud of West Indies has been estimated to contain 160 million bacteria per cubic centimetre.

Nutrition and feeding mechanisms in sub littoral benthic animals:

The primary sources of food for the benthos are plankton, organic detritus, and larger algae. Benthic diatoms contribute significant amount of organic matter. Detritus feeders are the typical benthic animals. They ingest the sediment containing microbes and other organic detritus and derive their nutrients from it. Suspension feeders such as sponges, many mussels, and brachiopods, obtain their food from the detritus occurring in the overlying waters. The polychaetes are the most important deposit feeders of the sub-littoral zone. The seahare *Aplysia* and the sea-urchin, *Echinus* feed on the larger algae. Several species of fishes, starfishes, ophiuroids, Nereid polychaetes, snails, cephalopods and large crustaceans are among the important predators and scavengers of this zone. *Nautica* is a gastropod feeding on lamellibranches by drilling holes in their shells. If the desired type of lamellibranch is not available, they starve to death. Many species are known to change their mode of feeding at times. Thus some bivalves that are normally suspension feeders may feed directly on surface deposit. Certain polychaetes may function both as deposit feeders and as scavengers. The whelk *Neptunea* is feeding on live polychaetes, even though it is usually feeding on dead organisms.

- Muddy shores are present in the intertidal areas of sheltered coast and estuaries.
- When the water flow is not strong, the suspended mud will settle down to form extensive mud flats.
- The anaerobic condition is one of the striking features of the muddy shore. The muddy shore has high numbers of bacteria in a narrow aerobic layer overlying a large anaerobic layer.
- Mud shores are colonized by larger organisms with dense populations because of food availability. The muddy shores are rich in plant communities. The dominant macrofaunal groups on muddy shores are polychaete worms, mollusks and crustaceans.
- Striking adaptations of mud flat organisms are the ability to burrow into the substrate and to tolerate low oxygen conditions.
- Zonation is a characteristic feature of the intertidal zone. It is broadly classified in to three sub-zones: (1) supra-littoral fringe (top part of intertidal zone), (2) a mid-littoral region, and (3) a sub-littoral fringe (bottom part of intertidal zone).
- The supra-littoral fringe is colonized by salt marshes in temperate regions or by mangrove forests in tropical shores, and this area is often inhabited by burrowing crabs. The mid-littoral area provides habitats for clams and polychaetes. Sub-littoral fringe is colonized by large numbers of polychaetes.

Definitions

- The Neritic Zone is the shallow waters of the ocean from the littoral zone to the edge of the continental shelf.

- The part of marine environment alternately exposed and submerged by the tides is called as the intertidal zones.
- A trophic pathway in the marine microbial food web where dissolved organic carbon (DOC) is returned to higher trophic levels via the incorporation into bacterial biomass and coupled with the classic food chain formed by phytoplankton-zooplankton-nekton is known as 'microbial loop'.
- Swash is the action of water running up a beach after a wave break and this action adds sand to the beach (accretion), while backwash is the action of water flowing back down the beach, and this action removes sand from the beach (erosion).
- The animals that live in the spaces within sediment particles (interstitial spaces) are known as interstitial fauna.
- The transition area between aerobic and anaerobic layers is called as 'redox potential discontinuity (RPD) layer.

4.3. CONTINENTAL SHELF OR SUB-TIDAL ZONE OF COAST

The sub-tidal area is a shallow area of less than 200 metres depth, situated between intertidal and ocean region. This area is the continental shelf region of the world's oceans. The water overlying the continental shelf is termed the 'neritic zone'. The continental shelf area is divided in to infralittoral and circalittoral. The infralittoral area is sub-divided into upper, middle and lower infralittoral zones (Fig. 4.1). The sub-tidal zone is the major fishing ground of the world. This area is colonized with productive communities, such as coral reefs, seagrasses and large seaweeds. Most of the area in the continental shelf is composed of soft sediments, sand and mud, and lesser area of hard rocky substrate. The substratum is dominated by polychaetes, crustaceans, mollusks and echinoderms. Organisms are of different groups, based on size. Macrofauna are organisms that are greater than 0.5mm in size. Meiofauna have the size range of 0.5-0.062 mm. Microfauna have the size of less than 0.062 mm and this group is primarily composed of protozoa.

4.3.1. Environmental conditions

The continental shelf waters show more variability in environmental conditions than open ocean and deep sea. The most important physical factor that acts on the bottom communities is turbulence, created by the interaction of waves, currents, and upwelling. This turbulence keeps the waters nutrient-rich, and hence the productivity is generally high. This high productivity supports large populations of zooplankton and benthic organisms. The food supply is abundant in this area, because of the increased productivity by the plankton, and also because of the production of seaweeds and seagrasses. The soft bottom in the continental shelf area has no large topographic diversity. There are presence of ripple marks, worm tubes and fecal mounds. The substrate varies with grain size and composition, in terms of silica and calcium carbonate. The sub-tidal area has four components: (i) un-vegetated sedimentary environment; (2) rocky sub-tidal communities; (3) kelp beds and forests; and (4) seagrass beds.

4.3.2. Un-vegetated sedimentary environments

Marine organisms have patchy distribution. This distribution changes with different time periods and produces patches on different areas. In other words, the pattern of distribution of marine organisms varies with space and time. These changes result from a complex interplay of physical and biological factors. The patchiness and variability result from disturbances, created by water motion or biological activity such as predation. These factors restrict the population in a small area (patch) and this is followed by recolonization and change with time. The result is that different areas have different species abundance and composition, depending on the time and the stability of the substrate. The organisms may be categorized into functional groups such as sediment stabilizers, sediment destabilizers, detrivores, carnivores, omnivores and herbivores. Community structure changes due to physical factors or biological factors such as predation, competition or combination of both factors.

Predation is an important biological factor in determining the species structure of infaunal communities. Both invertebrates and vertebrates act as predators on infaunal organisms. There are three categories of predators: (1) surface predators (fishes and crabs), (2) burrowing predators (nemertean worms and gastropods), and (3) digging predators (blue crab and ray fish).The nemertean and polychaete predators consume about 20-50% of the standing stock of prey, but not completely. The predatory activity is responsible for clearing some macrofauna in certain small areas followed by recolonization of the cleared organisms. This results in patchy distribution of organisms on the bottom.

Competition among the organisms may be direct or indirect. Direct competition is due to inhibitory activities of one species on another. For example, aggressive movement of the mobile bivalve within the substrate disrupts the burrow of the more sedentary bivalve species leading to its replacement. Burrowing deposit feeders disturb the substratum, and make the water turbid with the suspended particles dispersed from the substratum. The suspended sediment clogs the fine filtering parts of the suspension feeders, making the feeding difficult. Thus deposit feeders maintain their own community and exclude suspension feeders. This exclusion of one trophic group through modification of the environment by another group is called "Trophic group amensalism". In general, burrowing deposit feeders are most abundant in mud, whereas suspension feeders are most abundant in sand. In the sandy substratum, the deposit feeders find difficulty in burrowing, and getting food as the organic matter is less. Tube-building organisms can stabilize the substrate with their tubes. They also reduce the space for burrowing animals. The burrowers can not burrow through the rigid tubes, and thus the tube builders indirectly limit burrowing animals. Competition for space may occur in benthic infauna both vertically and horizontally, and may affect the density of organisms. Organisms share space of other organisms, due to severe competition for space. For example, the clam lives in burrows of shrimp or worm.

- The sub-tidal area is a shallow area of less than 200 metres depth, situated between intertidal and ocean region. This area is the continental shelf region of the world's oceans. The water overlying the continental shelf is termed the 'neritic zone'.
- The sub-tidal area is the major fishing ground of the world. This area is colonized with productive communities, such as coral reefs, seagrasses and large seaweeds.
- Most of the area in the continental shelf is composed of soft sediments, sand and mud, and lesser area of hard rocky substrate.
- Macrofauna are organisms with size greater than 0.5mm, the meiofauna with size range of 0.5-0.062 mm and microfauna have the size less than 0.062mm.
- The substratum is dominated by polychaetes, crustaceans, mollusks and echinoderms.

4.3.3. Rocky sub-tidal communities

Rocky sub-tidal surfaces are common in some areas. The hard substrate communities include low growing plants, sessile or sedentary invertebrates, kelp forests and coral reefs. The prominent motile animals are usually sea urchins, crabs and gastropod mollusks. Most animals of sub-tidal rock are colonial and encrusting species and these are able to spread over surfaces by asexual means. Most of them are suspension feeders. In areas of cold or warm temperate seas, the communities are dominated by encrusting coralline algae, ascidians, or sponges that compete for space among themselves by over growth. In all cases, sea urchins are the 'keystone species', responsible for the community changes. If urchin densities are high, the communities are dominated by crustose coralline algae or mussels, both are very resistant to urchin grazing. If urchins are absent, communities of fleshy algae or kelp or both dominate.

- Rocky sub-tidal surfaces are colonized by sessile or sedentary invertebrates, kelp forests and coral reefs.

- The prominent motile animals are usually sea urchins, crabs and gastropod mollusks. Most animals of sub-tidal rock are colonial and encrusting species.

4.3.4. Kelp beds or forests

In the cold temperate regions of the world, the hard sub-tidal substrates are covered by a community dominated by very large brown algae known as 'kelps'. It is known as 'kelp beds' if the algae do not form a surface canopy while it is also called 'kelp forest' if the algae form a floating surface canopy (Fig. 4.11). Kelps are attached to the substrate by root-like holdfast. From the holdfast, arises a stem-like stipe which ends in many flat leaf-like blades. At the base of the blades is a float called 'pneumatocyst' which makes the blades to float on the surface. Kelps absorb their nutrients directly from the seawater. The major kelps that dominate the structure of the kelp forests are the genera, *Macrocystis* and *Nereocystis*, which grow a length of 20-30 metres like the trees on land. Kelp beds are dominated by the genera, *Laminaria, Pterygophora* and *Ecklonia,* which grow few centimeters in length. The extent of kelp beds and forests is dependent upon hard substrate for attachment, sufficient light, water clarity, low temperature, nutrients, and grazing. The kelp beds and forests are present wherever the water is cool but absent from warm temperate and tropical areas. They are extremely productive and fast growing. Kelps have very high productivity, and annual net productivity ranges from 800 to 2000 gram carbon in one metre square area. *Macrocystis* grows 50 centimetres in a day. Kelp plants have a life cycle that alternates between an asexual macroscopic stage termed the 'sporophyte' kelps, and a sexual microscopic stage called the 'gametophyte' kelps.

Fig. 4.11: Kelp forest

Only 10% of the net production enters the food chains of the kelp bed through grazing. The remaining 90% enters the food chains in the form of detritus or dissolved organic matter. There are many other species of algae, invertebrates and fishes, associated with the kelp forest or bed. The causes of kelp mortality are wave action, nutrient depletion, warm temperatures, and herbivore grazing. Adult plants are vulnerable to destruction by mechanical forces, mainly wave action and to nutrient depletion, mainly nitrogen. Grazing, particularly by sea urchins and fishes is another cause of kelp mortality. The urchins are controlled by the sea otter (*Enhydra lutris*), a dominant predator, and the sea otter is also known to be predated by killer whales. The large otter populations maintain low urchin densities and a lush growth of kelps. Places where otters are absent are without kelps and have enormous urchin densities. Other predators of sea urchins exist in kelp beds are starfish species.

- In the cold temperate regions of the world, the hard sub-tidal substrates are covered by very large brown algae known as 'kelps'. It is called 'kelp bed' if the algae do not form a surface canopy, while it is called 'kelp forest' if the algae form a floating surface canopy.
- The extent of kelp beds and forests is dependent upon hard substrate for attachment, sufficient light, water clarity, low temperature, nutrients, and grazing.
- The kelp bed and kelp forest occur wherever the water is cool but absent from warm temperate and tropical areas.
- The kelps are extremely productive and fast growing.

4.3.5. Seagrass communities

Seagrasses are aquatic grasses growing throughout the world's ocean. They are the only flowering plants living under the sea throughout life cycle. Like the grass on land, seagrasses grow in vast meadows under water. The seagrass meadows are one of the most beautiful sights in the world. They are abundantly found in shallow sub-tidal areas to depths of 50-60 metres, and they do occur in mid-intertidal region of soft-bottomed marine coasts and estuaries. They are anchored in sand or mud bottoms. They are particularly found in areas where there is no intense wave action due to low tide. Seagrasses are the ancient species of 100 million years old. They originated on land and later adapted to underwater life. The seagrasses maintain the physical, chemical and biological integrity of marine ecosystems in particular mangroves and coral reefs. Seagrasses alter the flow of water, recycle the nutrients, and impact the food chain of marine life systems. They also absorb nutrients from coastal run-off and stabilize sediment, helping to keep the water clear. The seagrass habitats provide nursery and feeding grounds for fish, shellfish and wildlife including several endangered and numerous economically valuable species. The seagrasses are often confused with land grasses and seaweeds. They are not at all related to land grasses and they are different from marine macro-algae as given in table 4.3.

Table 4.3: Comparison between seagrass and seaweeds

Seagrass	Seaweeds (Marine macroalgae)
Flowering plants	Non-flowering plants
Complex root structure to fix plant in the sediment, and extract nutrients	Simple holdfast to fix the algae to hard substrates, such as rocks or shells
Photosynthesis is restricted to leaf cells.	Photosynthesis is undertaken by all cells.
Reproduction through flowers, fruits and seeds	Reproduction through spores

Seagrasses are present in most parts of the world. There are about 60 species world wide. In temperate areas, the species diversity is low, dominated with a few species such as the eelgrass *Zostera marina*, whereas in tropical areas, the species are more diverse with many genera such as *Cymodocea, Enhalus, Halodule, Halophila, Syringodium, Thalassia* and *Thalassodendron*.

Seagrasses have long, thin, strap-like leaves that arise from a horizontal creeping stem called 'rhizome' (Fig. 4.12).

The dense intertwined rhizomes and roots of the grasses form a strong mat, and this stabilizes the plant and its substrate against water movement. Seagrass can reproduce through sexual or asexual methods. In sexual reproduction, the plants produce flowers and transfer pollens from the male flower to the ovary of the female flower. Most seagrass species produce flowers of a single sex on each individual, so there are separate male and female plants.They undergo pollination and complete their entire life cycle underwater.

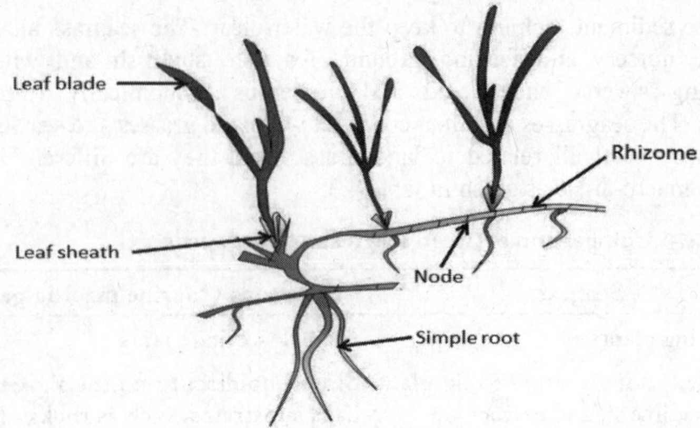

Fig. 4.12: Seagrass and its plant parts

Environmental conditions

Seagrass beds are most strongly influenced by water currents and wave action. The reduction of wave action and currents will make seagrass beds to accumulate more sediment. Temperate grass beds (*Zostera*) accumulate sediments from sources outside the beds, whereas tropical seagrass beds derive most of their sediments from within. Seagrass beds accumulate fine-grained sediments with high organic content. The microbes in the sediments are active due to high rate of metabolism and anaerobic processes. The microbes are important for regenerating and recycling of nutrients and carbon. Other physical factors that affect the seagrass beds are light, temperature and desiccation. Water depth and turbidity control the amount of light available to the seagrasses and the depth of seagrass development. Most seagrasses are tolerant to changes in salinity and can also be found in brackish waters.

Environmental services

Seagrasses are 'ecosystem engineers', because they create their own habitat, their leaves slow down water-currents, and their roots and underground stem stabilize the seabed by increasing sedimentation. The seagrasses are important for associated species in providing shelter. The seagrasses have higher primary production rates, which are closely linked to the higher production rates of associated fisheries, and thus the seagrass communities make significant contributions to the coastal productivity. Seagrasses provide coastal zones with ecosystem services such as fishing grounds, wave protection, oxygen production and protection against coastal erosion.

The seagrasses support other marine life. The habitat complexity within seagrass meadows enhances the diversity and abundance of animals. These plants support numerous herbivores and detritivores, and are considered very productive pastures of the sea. Seagrass/algae beds are rated the third most valuable ecosystem globally, next to estuaries and wetlands. Seagrass meadows produce a greater amount of organic matter, and serve as a good substratum for a variety of marine organisms, for example juvenile and adult fish, epiphytic and free-living macroalgae and microalgae, mollusks, bristle worms, and nematodes. The seagrass herbivory is a highly important link in the food chain, with many species feeding on seagrasses worldwide. Seagrass meadows provide food directly to animals such as green turtles, dugongs (sea cow or sea dog), urchins, parrot fish, gastropods and others. There are about 340 animals including sea cows and green turtles which directly feed on the seagrasses and their epiphytes. Without the seagrasses, these animals cannot survive.

SEA TURTLES

Of seven world species, five are found in India and they are Olive Ridley, Green Turtle, Hawsbill, Laggerhead (all are endangered) and Leatherback (critically endangered) (Fig. 7.14).

Olive Ridley Green Turtle Hawsbill

Odisha is the world's largest mass nesting of Olive Ridley turtles. Up to 0.6 million number of Olive Ridley turtles, over a short stretch of beach within fortnight in Gahirmatha coast. Sea turtles visit the shore in huge numbers to lay their eggs. This congregation for nesting is known as 'arribada" (Mexican word).

Laggerhead Leatherback

Fig. 4.13: Five species of sea turtles in coastal India

Temperature determines the sex of hatchlings of sea turtles. Males are produced when it is 29°C and females are produced when it is 31°C. On hatching after 70 days, the hatchlings enter into the water and making a 356° turn. Reason for this is not clear.

Behavioural patterns of seagrass animals are interesting. Animals such as urchins and spiny lobsters feed on seagrass meadows in night times, whereas some fishes, parrot fish and surgeon fish feed on the seagrass in day times. During decay the plant material is colonized by microorganisms which convert a portion of the seagrasses into protein-rich microbial biomass, and mineralize the remaining fractions. This detritus food is utilized by large organisms such as polychaetes, amphipods, gastropods, mullets and prawns. About 80% of the detritus food formed within the seagrass beds is transported to other coastal systems through water currents and supplying food for other marine organisms. Seagrass meadows provide a substrate for attachment of less mobile or sessile animals which spend their whole life in the meadows. The seagrass meadows are nursery ground for young fishes, lobsters, and prawns which spend only part of their life in the meadows. Seagrass meadows enhance the biodiversity and habitat diversity of coastal waters. It has been estimated that over 153 species of microalgae (mostly diatoms), 359 species of macroalgae, and 178 species of invertebrates are present on the seagrass blades as epiphytes and epizoics.

Seagrass meadows are capable of recycling nutrients. The meadows account for 15% of the ocean's total carbon storage. The seagrasses are efficient in carbon absorption and fixation. It is estimated that seagrasses per square meter are capable of fixing about 1000 grams of carbon, every year. This ability makes them 3-5 times more effective than near shore phytoplankton. The average global value of seagrasses for their nutrient cycling services has been estimated as US$ 19,004 from one hectare area in a year. In the Mediterranean, seagrass meadows of *Posidonia oceanica* account for an annual input of 57×10^{10} gram nitrogen in the ecosystem.

With flexible leaves, seagrass can withstand water movements and reduce the velocity of water currents, and thus provide a safe habitat. Extensive root system of the seagrass meadows trap sediments, stabilize substrates, enhance clarity of seawater and also functions as barriers to coastal erosion. Some of the species form reef-like structures close to the water surface that dissipate the wave energy.

Tropical seagrasses are important in their interactions with mangroves and coral reefs. When the water movement is slow, suspended sediment settles down, and the seagrasses trap the sediment. This trapping of sediment by providing clean water with reduced sediment loads which benefit the adjoining coral reefs. Sediments accumulated by seagrasses may also form substrate for colonization of mangroves. Both mangroves and seagrasses export nutrients as dissolved organic matter (DOM) or particulate organic matter (POM) which, in turn, nourishes the reef organisms. Mangrove runoff and water flowing out of seagrass beds have DOM which on decomposition releases inorganic nutrients. Likewise, large amount of POM also enters into sea as partially decomposed detritus from mangroves and seagrasses. Many organisms act on this detritus, and convert it into small fragments, which become a food source for larger marine organisms. Thus, a flow of dissolved nutrients from mangroves enhances primary productivity of seagrasses, whereas seagrass meadows and mangroves enhance the secondary productivity of the coral reefs by providing alternative feeding sites. Thus, coral reefs, mangroves and seagrass systems are interconnected, wherever they occur together.

- Seagrasses are the only flowering plants living under the sea. They are abundantly found in shallow sub-tidal areas. They do occur in mid-intertidal region of soft-bottomed marine coasts and estuaries.
- There are about 60 species of seagrasses in the world. In temperate areas, the species diversity is low, whereas in tropical areas, the species are more diverse.
- Seagrass beds are most strongly influenced by water currents and wave action. The reduction of wave action and currents will make seagrass beds to accumulate more sediment. Other physical factors that affect the seagrass beds are light, temperature and desiccation.
- With flexible leaves, seagrass can withstand water movements and reduce the velocity of water currents, and thus provide a safe habitat. Extensive root system of the seagrass meadows trap sediments, stabilize substrates, enhance clarity of seawater and also functions as barriers to coastal erosion.
- The seagrasses have high primary productivity.
- Seagrass meadows provide food directly to green turtle, dugongs, urchins, parrot fish, gastropods. They also supply detritus food to detritivorous fishes.
- Seagrasses provide fishing grounds, wave protection, oxygen production and protection against coastal erosion.

Do you know?

- Barnacles are monoecious, with both sexes are found in a single individual. This enables them to self-fertilize if there are no other barnacles nearby.
- Cling fish pelvic fins are modified into suckers which allows them to attach firmly to solid surfaces, ultimately preventing them to be washed away by the strong currents.

- The fiddler crabs can survive in hyper-saline conditions on the sand flats with salinity of 90 ppt.
- Black Sea Hare of California are among the world's largest gastropods, from the phylum molluska, weighing up to 35 pounds and growing up to 35 inches long.
- Bull kelp can grow up to 80 feet long, and the most amazing thing about the bull kelp is that all of this growth occurs in just one single summer season.
- The Shark Bay Marine Park of Western Australia has the world's largest meadows of seagrass and the largest number of seagrass species growing together in one place in the world.

Questions for Review/Exercise*

(1) Define: Neritic zone, Tide pools of rocky shore, Kelp beds and forests, Trophic group amensalism

(2) Explain the characteristics of coastal ecology.

(3) Explain the features of pelagic zone of coast.

(4) How are the intertidal organisms adapted? Explain.

(5) Explain with diagram about the microbial loop and grazer chain in the epipelagic zone of coastal waters.

(6) Differentiate between sand beaches and sand flats.

(7) Write a short note on the divisions of benthic zone of coast.

(8) Tabulate the differences between the rocky shore, sandy shore and muddy shore.

(9) Explain the zonation patterns of rocky shore and muddy shore.

(10) Write an account of adaptations of organisms in sandy shore, muddy shore and rocky shore.

(11) Characteristics of interstitial fauna- Explain.

(12) Write a short note on coastal sand dunes.

(13) Explain the concept of redox potential discontinuity (RPD) layer with a suitable diagramme.

(14) What are the characteristics of continental shelf? Explain.

(15) How are seagrasses different from seaweeds?

(16) Give an account of seagrass communities.

* Visits to sandy shore, rocky shore and muddy shore, observe the zonation patterns of organisms, identify the organisms present in the shores and study their adaptations. Analyze the seagrass communities and coastal sand dunes for their physical and biological features.

Ecology of Coral Reef Ecosystem 5

Coral reefs are structurally complex, biologically diverse and most beautiful underwater ecosystem (Fig. 5.1). They occupy about 600,000 square miles of the earth, which is about 0.17% of the total area of the earth planet. They occur in tropical sub-tidal shallow waters. Coral reefs exist in the coastal waters that are clear and shallow coastal waters with low nutrients at temperatures ranging from 18°C to 33°C. Coral reefs prefer mostly the natural substrates, such as rocks, coral stones, molluskan shells to settle and grow, but rarely prefer the man-made structures. Coral reefs are rich in biodiversity with about 91,000 species, which is about 4% of all marine species. Covering only 0.2% of the ocean's floor, the coral reefs contain 25% of marine species, and they are often termed the 'tropical rainforests of the oceans'. Coral reefs are valuable ecosystem, providing a source of food, coastal protection and income for millions of people.

Fig. 5.1: Coral reef ecosystem

The coral reefs are among the most productive systems in the marine environment. The gross primary productivity is 1500-5000 gram carbon in an area of one square metre for a year (gC/m^2/year) in the coral reef system, in contrast to a gross primary productivity of 18-50 gC/m^2/year in the open tropical oceans. Coral reefs are highly productive due to the presence of large amount of bacteria and plants and to the rapid cycling of nutrients by microbes. Coral reefs are biologically diverse due to structural complexity and protective nature from stresses. They form a complex structure which, in turn, provides habitats for other reef associated organisms. However, the coral reefs occur in nutrient-poor waters. This high species diversity and productivity of coral reefs even in the nutrient-poor waters is the unique feature.

There are two types of corals: hermatypic and ahermatypic. Hermatypic corals produce reefs in association with algal plant cells called 'zooxanthellae' (Fig. 5.2). In contrast, ahermatypic corals do not form reefs and are mostly not in association with algae (Fig. 5.3). Ahermatypic corals are distributed throughout the oceans of the world, but hermatypic corals (coral reefs) are restricted to the tropical and subtropical seas. The ahermatypes are generally leading solitary life and occur in deep waters down to depths of 2000 metres, whereas the hermatypes lead colonial life and occur in shallow coastal waters.

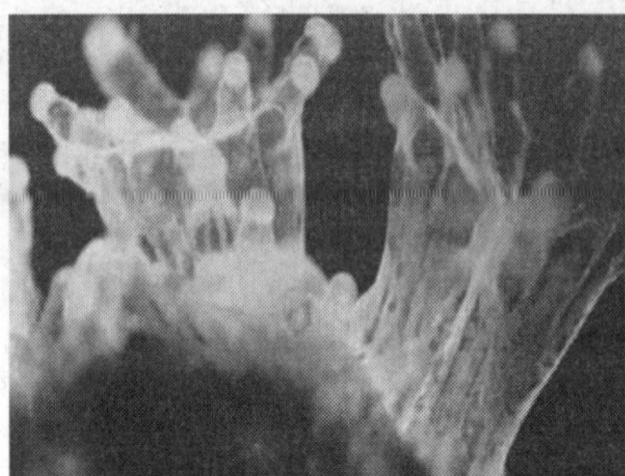

Fig. 5.2: Hermatypic coral Polyp with zooxanthellae (http://serc.carleton.edu/eslabs/corals/2b.html)

Fig. 5.3: Ahermatypic Sun coral without zooxanthellae (http://www.flickr.com/photos/klicking/3301171667/)

Coral reefs are unique among marine communities in that they are built up entirely by biological activity. Coral reefs are globally significant in the calcium cycle. The coral reefs remove about 700 billion kilograms of carbon every year. Coral reefs are formed by precipitation of calcium carbonate by converting 50% of the calcium that enters the world's oceans each year into calcium carbonate. The carbon dioxide dissolves in seawater and produces carbonic acid. This acid dissociates to form bicarbonate which in turn, reacts with calcium in the seawater, forming calcium bicarbonate. During algal photosynthesis, carbon dioxide is removed from the calcium bicarbonate (water soluble) and this leads to the formation of calcium carbonate (water insoluble) which precipitates to form coral reefs. Calcium carbonate also deposits on the surface of some algae and foraminiferans (Fig. 5.4).

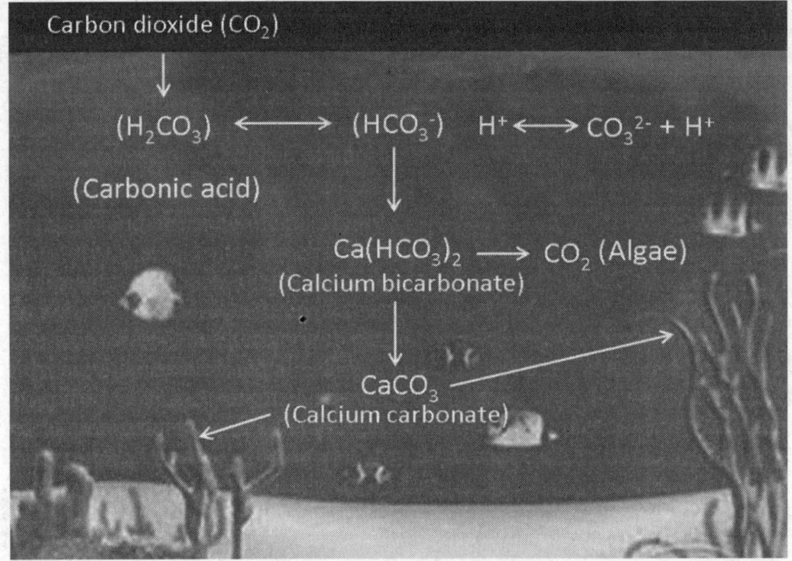

Fig. 5.4: The reactions of calcium that lead to the formation of coral reefs in seawater

5.1. IMPORTANCE OF CORAL

The coral reefs are the hotspots of marine biodiversity, providing habitat for a variety of marine life. The reefs play a key role in coastal protection from erosion and cyclone. They are ancient ecosystems of 35 million years existence on the earth planet, and the existing coral reefs had overcome many catastrophic events.

Coral reef areas have a global coverage of 62 million hectares. They have a total economic value of about 170 billion US$ annually and an annual value of about 2750 US$ for one hectare. Healthy coral reefs may support fish harvests up to 30 tonnes annually from an area of one square kilometre. Besides the economic value, there are significant ecological services, in terms of coastal protection, religious beliefs and traditions. The coral reefs are the source of protein, medicine, cultural products and souvenirs. The sand beaches around the coral reefs are used as tourist spots that in turn, help to earn foreign exchange (Fig. 5.5).

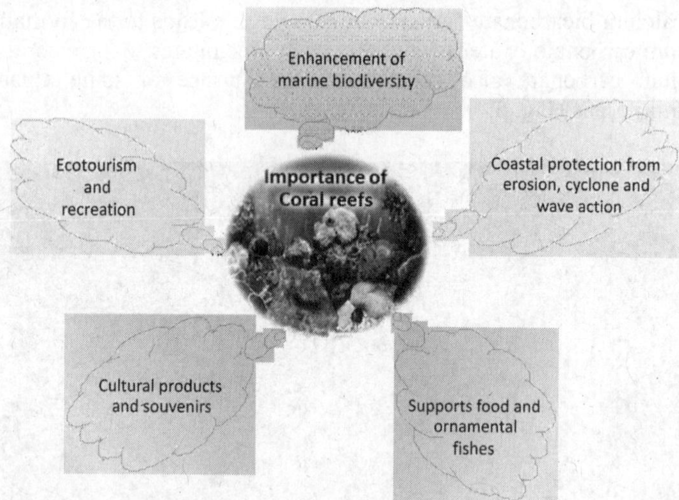

Fig. 5.5: Ecological and economic significance of coral reef ecosystems

5.2. CORAL DISTRIBUTION

The biodiversity of coral reefs is the highest in the Indo-Pacific region that includes the Philippines, Indonesia, New Guinea and northern Australia. There are many differences between Atlantic and Pacific coral reefs. The physical conditions under which reefs grow in both the Atlantic and Pacific are similar with respect to temperature, salinity and turbidity, but the biodiversity of coral reef ecosystem is different between the two areas. The Indo-Pacific regions have more coral genera and species than the tropical Atlantic reefs. For example, a dominant reef-building genus *Acropora* has as many as 200 species in the Indo-Pacific, but only three in the Atlantic (Fig. 5.6).

Fig. 5.6: *Acropora* sp. (http://urbansage.com/frags.html)

The species richness of Atlantic reefs is much less than that of Pacific reefs. Some important reef building genera of the Pacific are *Pocillopora, Pavona* and *Goniopora* and these are absent from the Atlantic. Such difference between Pacific and Atlantic is also found with most of the reef associated species. There are 5000 species of mollusks in Indo Pacific reefs, but there are only 1200 species on the Atlantic reefs. Similarly, there are 2000 species of Indo-Pacific reef fish, compared with 600 Atlantic reef fish. Atlantic reefs commonly have large numbers of sea fans and whips (Gorgonians; Fig. 5.7) whereas the sea fans are much less in Indo-Pacific reefs. Soft corals are also present in large numbers in Indo Pacific reefs, whereas these are rare in the Atlantic. The hydrocoral *Millepora* are often called 'fire coral' a significant member of Atlantic reefs. Atlantic reefs lack the giant clams and octocorals. There are no anemone fish (*Amphiprion*) in the Atlantic, though the fishes that associate with anemones do occur in Atlantic. Similarly, crabs and shrimp that guard certain corals against predators are absent from the Atlantic. Most Atlantic corals are nocturnal, but in the Pacific, many are diurnal. There are also differences in the life histories of organisms. The coral recruitment in the Caribbean is primarily asexual through fragmentation. In the Pacific, recruitment is by sexually reproduced planula larvae. Herbivory and predation are more intense on Pacific reefs than on Atlantic reefs.

Fig.5.7: Seafans (Gorgonian)
(http://www.atlasomega.com/2010/09/marvels-of-the-darkness-cold-water-corals)

5.3. TYPES OF CORAL REEFS

There are three major types of coral reef systems: (1) fringing reefs, (2) barrier reefs and (3) atoll (Fig. 5.8). There are also other minor types: (i) patch reef (smaller in size than a fringing reef), (ii) ridge reef (a long ridge of corals) and (iii) table reef (like a table on the seafloor). Fringing reefs are the early system, formed by settlement of corals, and growing near the coastline and bordering the land. As the corals propagate towards the seaward direction, a gap is formed in the form of lagoon between the shoreline and reef, and this stage of reef development is called a 'barrier reef'. Atoll is a ring like elevated reef structure of very deep water far from land, and it encloses a lagoon. This lagoon may contain lagoon reefs or patch reef. In general, massive and branching forms of coral reefs are common in the shallow waters, while platform type of coral reefs predominate in the deep waters. The atoll is found only in the Indo-Pacific area.

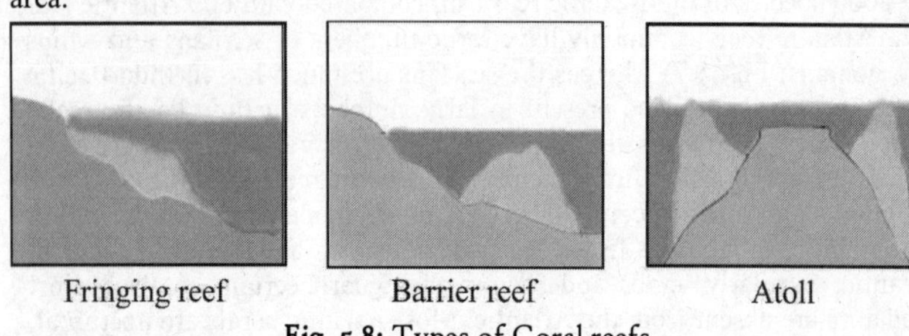

Fig. 5.8: Types of Coral reefs

The largest barrier reef is the Great Barrier Reef of Australia, which stretches for almost 2000 km and covers an area of 2,00,000 square kilometers along the eastern coast of Australia. The reef is so huge it can be even seen from the Moon. The Great Barrier Reef has taken at least 15 million years to get its present size.

5.4. ECOLOGY OF CORAL REEFS

Physical factors that influence coral reef development are temperature, depth, light, salinity and sedimentation (Fig.5.9). The coral reefs require warm temperatures greater than 20°C. Some of them do occur in low temperature below 10°C, but do not form reefs. Some coral reefs can tolerate temperatures up to about 40°C. Calcification occurs mostly in warm waters. Hence, the coral reefs are distributed in the tropical latitudes and decreased towards subtropical latitudes. The coral reefs are generally lying between 25° north and 25° south latitudes.

The coral reefs are limited by depth. Coral reefs do not develop in water that is deeper than about 50-70 metres. Most reefs grow in depths of 25 metres or less. This depth restriction is due to the light requirement of hermatypic corals for associated algae. Light is one of the physical factors limiting the coral reefs. The light is required sufficiently for allowing photosynthesis of the symbiotic zooxanthellae in the coral tissues and for calcium deposition in coral reef formation. The diversity of hermatypic corals decreases with depth with decreasing light intensity. Coral growth and calcification depend upon sufficient light for the zooxanthellae. The zooxanthellae increase the rates of calcification and growth of the coral colonies. Active coral growth cannot occur below 15-20% of surface light incidence.

Salinity restricts coral reef development. Hermatypic corals are true marine organisms and are intolerant of salinities deviating from that of normal seawater. The coral reefs require seawater with salinity of 32-34 ppt. Any reduction down to 30ppt interferes with growth of coral reefs. Reefs are absent where there is a discharge of fresh water. The coral reefs do occur in the Persian Gulf regions of elevated salinity of 42ppt. The coral reefs are not well-developed in estuaries and very hypersaline conditions.

Sediment has an adverse effect on the coral reefs. Many corals can remove sediment by trapping it in mucus and carrying it off by ciliary action. Species with large polyps can more easily remove suspended mater than small polyp species. Massive and slow-growing hermatypic corals produce large amounts of mucus when sedimentation is high, while rapidly growing branching species, such as *Acropora* species produce less mucus that can not remove particulate matter effectively. Most hermatypic corals cannot withstand heavy sedimentation as it also clogs feeding structures of coral reefs. The sediment in the water causes turbidity of water, and reduces the light essential for photosynthesis by the zooxanthellae in the coral tissue. As a result, coral reef development is reduced in areas of high turbidity.

Coral reef development is greater in the areas which are subjected to moderate wave action. Massive skeletons of calcium carbonate in the coral reef provide resistance to damage by wave action. The wave action provides fresh oxygenated and plankton-rich seawater and also prevents sediment from settling on the colony.

Exposure to air limits the growth of coral reefs. Most corals are killed by long exposure to air, and hence their upward growth is limited to the level of the lowest tides. The duration of exposure to air, affects the coral reefs, and hence they secrete mucus to prevent dehydration for a short time.

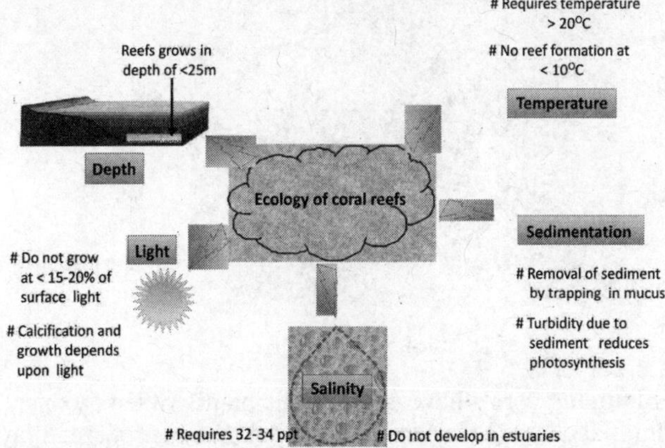

Fig. 5.9: Ecological factors that determine colonization of coral reef ecosystem

Do you know?

- The Great Barrier Reef is the only natural structure that can be seen from space.
- Reefs remove about 700 billion kilograms of carbon every year.
- Sabah (Malaysia) is at the epicenter of the coral-triangle and has at least 550 different species.

5.5. BIOLOGY OF CORAL REEFS

Corals are members of the phylum Cnidaria and class Anthozoa. Reefs are solid limestone structures created mainly by corals on the shallow sea floor. Coral animal is called a 'polyp', a hollow cylinder of tissues with tentacles and a central mouth opening at the top (Fig. 5.10). The tentacles contain microscopic stinging capsules called nematocysts. A nematocyst is a bulbous double-walled structure containing a spirally folded, venom-filled thread with a minute barb at its tip. A tiny sensor projects outside the nematocyst. When the sensor is stimulated physically or chemically, the capsule explodes and ejects the thread. The barb penetrates the victim's skin and injects venom. The size of the polyp ranges from 0.5 to 20mm in diameter. The external surface of the polyp has some special cells called 'calcioblasts' which construct a solid skeleton by precipitating the dissolved calcium from seawater. The polyps occupy little cups or corallites in the massive skeleton. Each corallite has a series of sharp blade-like sclerosepta rising from the base. The pattern of these septa is used a taxonomic tool for identification.

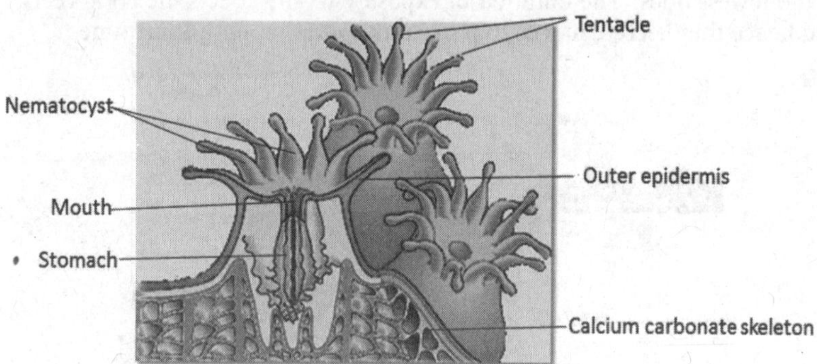

Fig. 5.10: Structure of polyp

All reef-building corals have microscopic plants called 'zooxanthellae'; they are dinoflagellates living inside their cells of gastrodermal layer. The colour of the corals is due to the presence of algal pigments. In the coral reefs, the coral animals and algae live together, helping each other symbiotically. The algae produce organic matter by photosynthesis and supply a part of the organic matter for coral nourishment. The corals excrete nitrogen, phosphorus and carbon-dioxide, essential for photosynthesis by the algae. The removal of carbon dioxide by algal photosynthesis changes the acidic condition in to alkaline, which is favourable for calcium deposition by the corals. The coral growth is greatly enhanced by the presence of the zooxanthellae. Organic phosphate inhibits calcification, and so the removal of phosphates by algae during photosynthesis increases the rate of calcium deposition. The algae may secrete some protein that can code for calcium deposition on coral skeleton. The coral reefs break-down lipid of their diet into acetate, and this acetate is converted into fatty acids by zooxanthellae. Unsaturated fatty acids are more in coral reefs of shallow waters than that in coral reefs of deep waters. Thus, zooxanthellae play vital role in coral ecology in providing photosynthetic products for nutrition of coral reefs, accelerating the process of calcification of coral reefs and also synthesizing lipid.

There are four types of coral reefs depending upon size and shape: massive (stone like), foliaceous (leaf like), branching (plant like), and encrusting (coating like) types. The rate of growth varies with coral types: a stony brain coral takes about 20 years to the size of a human brain, while a branching coral grows faster as much as 18-20 centimetres in a year(Fig. 5.11).

The corals are mostly carnivores, feeding on zooplankton, and they do feed on decomposing organic matter (detritus) and dissolved organic matter directly from seawater. All corals are nocturnal, feeding actively during night time. Many corals are dormant during the daylight hours, feeding in the dark when plankton rise into the waters over the reef at dusk and spend the daylight on the bottom.

(a) Massive brain Coral (*Favites* spp.) (http://www.aquariumdomain.com/forums/showthread.php?161-Closed-Brain-Coral-%28Favites-spp.%29-Discussion)

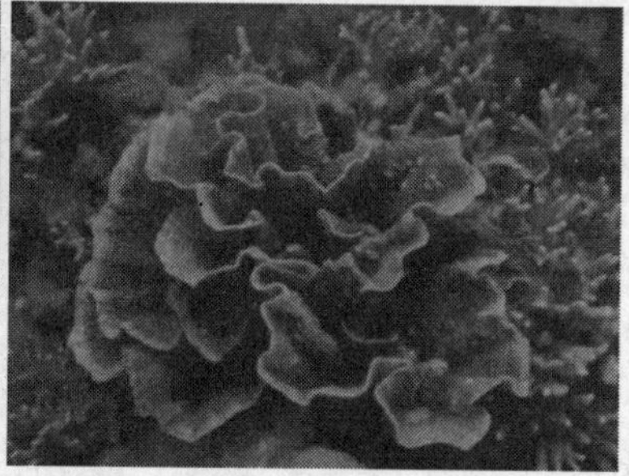

(b) Foliaceous coral (*Pachyseris* sp.) (http://adiklupet.blogspot.in/2009/02/on-coral-taxonomy.html)

(c) Branching staghorn coral (*Acropora cervicornis*) (http://storiented.blogspot.in 2010_06_01_archive.html)

(d) Encrusting coral (http://oceanservice.noaa.gov/ education/kits/corals/media/ supp_coral03g.html)

Fig. 5.11: Types of coral reefs based on size and shape

Corals reproduce both sexually and asexually. Asexual reproduction is usually by budding off new polyps from the parent and increases the size of the colony. Fragmentation of the parent colony can also lead to new colonies. Each polyp is a hermaphrodite and can produce both male and female gametes. At certain phases of the moon, the corals release the gametes simultaneously into the surrounding seawater and fertilization takes place. The larva comes out of the fertilized egg called 'planula', settles at hard substrates, and grows by repeated budding processes. The rate of recruitment of new coral colonies to the reef system varies with time and space.

Can we measure growth and age of coral reefs?

The growth rate of coral reef species can be measured using x-ray method. A coral skeleton appears as dark and light bands. The dark band refers to high calcium deposition which occurs with fast growth of coral reefs in summer months. Whereas the light band refers to low calcium deposition which is associated with slow growth of coral reefs in monsoon months. Based on this principle, the age of the coral is determined. Other method of measuring age of coral reefs is carbon dating using dead corals.

In order to study the coral reefs and their growth over a period of time, underwater camera is installed. In addition, the data buoy is fitted with two sensors, one above the water and the other under the water. The one above the water collect and send data regarding humidity, wind speed, air temperature besides latitude and longitude, as it moves slowly in the water. The underwater sensor provides data on turbidity, salinity levels, water current speed and direction.

5.6. ORGANISMS ASSOCIATED WITH REEFS

The corals are the major organisms that form the basic reef structure. However, many other organisms of all phyla and classes get associated with the reefs. This makes coral environment most rich in biodiversity. Coral reefs are complex associations of organisms that have a high diversity of habitat types, which is one reason for the high diversity of species. The coral reef system has diverse habitats such as coral reefs, sand, caves, crevices, algal bed, shallow and deep waters. Other organisms that contribute strongly to the calcium carbonate budget of reefs are the coralline algae, green algae (genus *Halimeda*), hydrocoral (*Millepora*) and mollusks including the giant clams. Other prominent groups on coral reefs are the soft corals, gorgonians, echinoderms (sea urchins, sea cucumbers, and starfish), sponges, crustaceans, polychaete worms and fishes. Coral reefs are richest environment for fish on earth. Bacteria are also abundant on reefs and are responsible for high productivity of the reef and cycling of organic matter.

In addition to the stony corals, other cnidarians contribute significantly to the reef. They are relatives of the corals called 'gorgonians' (sea fans and sea whips). Soft corals (order Alcyonacea) lack a true skeleton but have spiny material (spicules) in their tissues which give them a leathery and tough texture. The soft corals are profusely present in most coral reef areas.

The coralline algae construct and maintain reefs. They are encrusting and spreading in thin layers in the form of algal ridge over the reef. This is the most rapidly calcifying zone on the reef. The algae are cementing various pieces together. Some green algae, such as *Halimeda* are non-encrusting, non-coralline, grow erect and secrete calcium carbonate. These provide sand in coral reef systems after breakage. Thus coral reefs, coralline red algae, and non-coralline green algae are dominant calcium carbonate secreting organisms of the reef system. In addition to algae, the giant clams, such as *Tridacna* and *Hippopus* contribute to the deposition of calcium carbonate on reefs. These giant clams are absent in the Atlantic reef.

Tridacna is the giant clam (Fig. 5.12) having endosymbiotic association with unicellular alga zooxanthellae. This unicellular alga produces 80% of the clam's food. A 90 kg giant clam releases one billion eggs per spawning.

Fig. 5.12: *Tridacna* sp., the giant calm

Algae are important primary producers of the reef ecosystem. They are able to cement calcium carbonate pieces together, production of coral sand and also in erosion of reefs. The algae also create habitats in and around themselves and provide shade from the hot sun for associated organisms. In addition to algae, large number of bacteria exists in lagoon and sediment of the coral reef ecosystems. The bacterial production depends on the dissolved organic carbon (DOC) from the water flowing over the reef and the particulate organic carbon (POC) from breakdown of reef organisms. The bacterial biomass and production are higher in the sediments than in the water column, and the bacterial production is more or less equal to the total reef primary production. The bacteria represent a significant energy source for the reef community and suggest the presence of a microbial loop system similar to that found in the plankton community.

5.7. SPECIES INTERACTIONS AND ECOLOGY OF REEFS

Competition for space and light occurs among coral colonies. Fast growing corals give shading to other slow-growing colonies, a process that is a type of 'exploitative competition'. In such competition, upright, branching coral colonies reduce the light required for encrusting or massive corals. Such exploitative competition makes the dominance of *Pocillopora* (cauliflower coral) at the shallow depths in the eastern Pacific and *Acropora* in the Caribbean. The slow growing species can kill the tissues of adjacent, competing coral species by means of their nematocyst-loaded digestive filaments, a process that is a type of 'interference competition'. Another group of organisms in which space competition occurs is the cryptofauna, the smaller colonial encrusting invertebrates, such as sponges, bryozoans, polychaetes, bivalves and tunicates that occupy the undersides of coral reefs.

Predation

Live corals are fed by predatory animals. Two taxa of predators are capable of destroying whole coral colonies and modifying the structure of reefs. The first one is the star fish *Acanthaster planci* which can consume whole reefs, when it is present in large numbers. The second one is the corallivorous fishes that consume the coral polyps and the predators are puffer fish, butterfly fish, file fish, trigger fish, surgeon fish and parrot fish (Fig. 5.13). The parrot fish make a special jelly-like bubble around themselves for protection at night.

Fig. 5.13: Predators of coral reefs:
(a) Parrot fish; (b) Butterfly fish; (c) Crown-of-thorns star fish (*Acanthaster planci*); (d) Surgeon fish; (e) Puffer fish

Grazing

Algae are the major space competitors with coral reefs. As the algae grow faster than corals and they can out-compete the corals. It does not occur when the algae are grazed by urchins (*Diadema*) and herbivorous fishes (rabbit fish, damselfish). Grazing can also produce shifts in the abundance of different groups of algae. The grazers prefer fleshy and filamentous algae over coralline algae and therefore the coralline algae increase at the expense of the fleshy forms due to grazing pressure. The coralline algae are less palatable to the grazers and this poor palatability is attributed to high content of calcium carbonate and production of noxious chemicals in the algae. More over, some algae can settle in to the habitats inaccessible to grazers. The grazing is high in the shallow reef areas and decreases markedly in the deeper reef area and in sandy lagoon. When the grazing intensity is less, fleshy filamentous algae grow faster and may eliminate the slow-growing coralline algae. Thus the density of algal species is determined by the intensity of grazing activity. In general, moderate grazing prevents dominance of any algae and increases diversity. Another important grazing fish is Damselfish. This fish will graze on algae within their territory and they also protect the territory against other grazing and corallivorous fishes. The damselfish even control the algal species composition in its territory by selective feeding. This may permit some algae to overgrow shade and kill slow growing massive coral colonies, and also to enhance the growth of fast growing branching corals to dominate in the territory.

Bioinvasion of *Kappaphycus alvarezii* on coral reefs

Kappaphycus alvarezii (Doty) Doty ex. P. Silva (Rhodophyta: Solieriaceae) (Fig. 5.14) is a Philippine-derived red seaweed. It was first introduced in to Gulf of Mannar, India in 2002 for the purposes of mariculture as this species is a source for sulphated polysaccharides, carrageenans, and is used widely for food products and pharmaceuticals. It was reported earlier that the species has successfully invaded and established on coral reefs in Hawaii islands, where it was initially introduced for mariculture. Studies on invasion of *K. alvarezii* on coral reefs at Hawaii islands revealed that it had spread from the initial site of introduction to other reefs at a rate of 250 metres per year. The seaweed is an invasive species due to many unique features such as vegetative propagation, adaptation to low and high-wave energy environments, extraordinary phenotypic plasticity, high growth rate and chemical defense against herbivores.

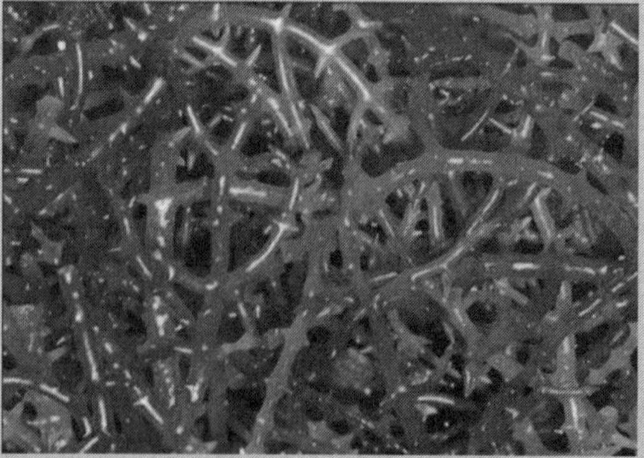

Fig. 5.14: *Kappaphycus alvarezii*, an invasive species

5.8. ECOLOGY OF REEF FISHES

Coral reefs harbour more species and diverse fish communities of attractive colours than any other environment on the earth. There are four hypotheses to explain the high local fish diversity of reefs. These are the competition model, the lottery hypothesis, the predator disturbance model, and the recruitment limitation model. According to competition model, the high diversity is the result of strong competitive interactions following recruitment. According to the lottery hypothesis, competition is not important and recruitment dominates due to the fact that most coral reef fishes produce large numbers of larvae that are dispersed in the plankton. According to the predator-disturbance model; predation, catastrophe and unpredictable recruitment keep the fish population less and there is no competitive exclusion. According to the recruitment limitation model, the adult population indicates variation in the larval recruitment and not post-recruitment events.

A striking feature of reef fishes is their attractive colour and patterns. This colour and pattern may be important in visual communication of fishes and also serve the fish in sex recognition, courtship and mating behaviours. The colour pattern also camouflages the species and also advertises the toxic nature of the fish to predators. Another striking feature is the marked difference in the activity of fish species on the coral reef between day and night. Most of the fish species are visible during the day and some nocturnal fishes are seen only at night. This permits a greater number of species to exist on the reef without competing directly. The largest group of fishes on reefs are predators (carnivores); constituting 50-70% of the species. The second largest trophic group of fishes is the herbivores (15% of species), while the remaining fishes are generally classed as omnivores with a few plankton feeders. Cleaning behaviour is a specialized form of predation, in which small fishes or shrimps remove ectoparasites from larger fishes. This occurs on all reefs. The fish being cleaned are vulnerable to predation.

A large number of fishes have toxic secretions on their body surfaces and they are parrot fish, wrasses and surgeon fish. The presence of toxic secretion in the fishes is to deter predation by carnivores. Ciguatera is a human disease caused by eating certain fishes whose flesh contains toxins produced by the dinoflagellate, *Gambierdiscus toxicus*. Venomous fishes are rare on reefs and they are stone fishes and scorpion fishes.

The cryptofauna refers to those invertebrates that inhabit the coral substrates, boring into calcium carbonate, living in cracks, crevices, caves and in coral heads and rock. They are bivalve mollusks, crustaceans and some polychaete worms. Colonical cryptofaunal organisms are sponges, bryozoans and ascidians. Their ability to reproduce asexually allows them to overgrow and compete with other species. They are important bio-erosion forces that break down coral reefs.

- The coral reefs are among the most productive systems in the marine environment. The main factor that sustains high productivity of coral reefs is the presence of large amount of plants and bacteria along with the rapid cycling of nutrients by microbes.
- The coral reefs are biologically diverse and they are often termed the 'tropical rainforests of the oceans'.
- Coral reefs are built up entirely by biological activity. Coral reefs are formed by calcium carbonate by converting 50% of the calcium that enters the world's oceans each year into calcium carbonate.
- Coral reefs are of two types: hermatypic and ahermatypic. Hermatypic corals produce reefs in association with algal plant cells called 'zooxanthellae'. In contrast, ahermatypic corals do not form reefs and are not in association with algae. Ahermatypic corals are distributed throughout the oceans of the world, but hermatypic corals are restricted to the tropical and subtropical seas.
- There are three major types of coral reef systems: (1) fringing reefs, (2) barrier reefs and (3) atoll. There are also other minor types: (i) patch reef (smaller in size than a fringing reef), (ii) ridge reef (a long ridge of corals) and (iii) table reef (like a table on the seafloor).
- Based on size and shape, the coral reefs are of four types: massive (stone like), foliaceous (leaf like), branching (plant like), and encrusting (coating like) types.
- The corals are mostly carnivores, feeding on zooplankton and they do feed on decomposing organic matter (detritus) and dissolved organic matter directly from seawater. All corals are nocturnal, feeding actively during night time.
- Corals reproduce both sexually and asexually.
- Algae are the major space competitors on coral reefs. The algae grow faster than corals and may kill the corals. It does not occur when the algae are grazed by urchins and herbivorous fishes.
- A striking feature of reef fishes is their attractive colour and patterns serve the fish in visual communication of fishes, sex recognition, courtship and mating behaviour.
- Most of the fish species are visible during the day and some nocturnal fishes are seen only at night. This permits a greater number of species to exist on the reef without competing directly.
- A large number of fishes have toxic secretions on their body surfaces and the presence of toxic secretion in the fishes is to deter predation by carnivores.

Questions for Review/Exercise*

(1) Define: Zooxanthellae, Planula, Fire coral, Cryptofauna, Crown-of-thorns

(2) "Coral reefs are tropical rainforests of the oceans" – Justify.

(3) How are the coral reefs highly productive in nutrient-poor waters?

(4) Differentiate between hermatypic and ahermatypic corals.

(5) Write a short note on the structure of polyp with a well labeled diagram. Add a note on its biology.

(6) Write a short note on ecological and economic significance of coral reef ecosystems.

(7) Why the species richness of Atlantic reefs is much less than that of Pacific reefs?

(8) What are the major types of coral reef systems? Explain their formations.

(9) Explain how the coral reefs are formed? Add a note on favourable conditions for the coral reefs.

(10) Write an account of organisms associated with coral reef systems.

(11) Discuss the species interactions in the coral reef system.

(12) What are the characteristics of coral reef fishes? Explain.

(13) What are the predators of coral reefs? Explain.

* Field visits to coral reef areas for observation on the coral reef communities. Analyze the reef fishes in the landing centres for their colour pattern and adaptations.

Ecology of Estuary 6

An estuary is a place where freshwater and seawater meets and mix. It is a partially enclosed water body. The estuary is a system through which the river passes to the sea and the seawater enters the river mouth. An estuary has a free connection of the sea with the freshwater source, at least during a part of the year (Fig. 6.1).

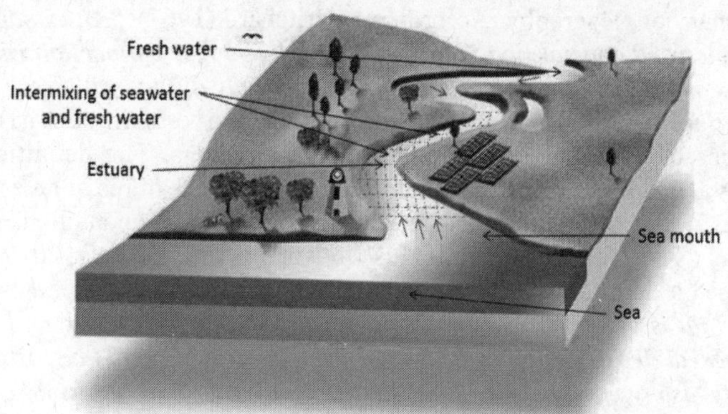

Fig. 6.1: An estuarine system

Estuaries are the focal point of human activities. Many of the major cities of the world are established along estuaries. Estuaries act as buffer zones capable of receiving heavy river discharges during monsoons. They act as a nursery and breeding grounds for many marine and estuarine organisms. Majority of fishes found in estuaries are the juveniles. There are marine species which spend part of their life cycle especially juvenile stage in the estuary. Examples are croakers, anchovies, flat-fishes and shoals. There are species which spend their entire life cycle in the estuary. These estuarine residents include gobids, sygnathids, ambassids, atherinids, mullets and some clupeids. Estuaries serve as highways for migrations of anadromous fish species, such as *Hilsa* and Salmon from the sea to river and for migrations of catadromous species, such as eel fishes from river to sea for spawning process. If the estuarine system is absent, life cycle of many fish species will be disrupted. The estuaries and their fishery resources provide livelihood for fisherman living around these areas and help in supplying food fishes. Estuaries are of recreational importance, since they provide water front along with sea, thick greeneries of mangroves and thus they promote tourism. Estuaries are of great ecological significance especially in the water cycle. As the sun heats up the sea, millions of litres of water rise up into the air in the form of water vapour. The vapours cool to form clouds. The clouds drift over the land it rains or snows. Rivers collect the water and carry it back into the sea. In this process, estuarine systems play a vital role.

6.1. DEFINITION OF ESTUARY

Estuary is defined as '*a wide tidal mouth of a river*' in the Oxford English reference Dictionary, and as '*the mouth of a river where tidal effects are evident and where freshwater and seawater mix*" in the

Dictionary of Geography. According to Pritchard (1967), "*An estuary is a semi-enclosed coastal body of water which has a free connection with the open sea and within which seawater is measurably diluted with fresh water derived from land drainage*". All the definitions of estuaries are based on the characteristics of estuaries in temperate regions. The definition for estuary has not taken into account of the important features of an estuary, such as periodic closure of their mouths and hypersaline conditions of dry seasons. Therefore, Potter and his co-workers in the year 2010 have proposed a definition: "*An estuary is a partially enclosed coastal body of water that is either permanently or periodically open to the sea and which receives at least periodic discharge from a river(s) and thus, while its salinity is typically less than that of natural seawater and varies temporally and along its length, it can become hypersaline in regions when evaporative water loss is high, and freshwater and tidal inputs are negligible*".

- An estuary is a system through which the river passes to the sea and the seawater enters the river mouth. An estuary has a free connection of the sea with the freshwater source, at least during a part of the year.
- Estuaries act as buffer zones capable of receiving heavy river discharges during monsoons.
- They act as a nursery and breeding grounds for many marine and estuarine organisms. There are marine species which spend part of their life cycle especially juvenile stage in the estuary.

6.2. TYPES OF ESTUARIES

Based on geology and geomorphology, there are four types of estuaries: (1) Coastal plain, (2) Tectonic, (3) Semi-enclosed bay or lagoon, and (4) Fjord.

6.2.1. Coastal plain estuaries

They were formed at the end of the last ice age when the rising sea level invaded low-lying coastal river valleys. These are also called as "Drowned river valley". Mandovi and Zuari estuaries in India and the Chesapeake Bay are the best examples for this type of estuary.

6.2.2. Tectonic estuaries

They were formed similar to the coastal plain estuaries, but due to subsidence of the land, but not as result of a rising sea level. The San Francisco Bay is an example.

6.2.3. Semi-enclosed bay or lagoon

Sand bars are built-up parallel to the coastline and partially cut off the water behind them from the sea. This creates a shallow lagoon behind the sandbars, which collects the freshwater discharge from the land. The water in the lagoons varies in salinity depending on the climatic conditions. These are common in North Carolina, along the Texas and Florida Gulf coasts, in North Western Europe and parts of Australia.

6.2.4. Fjord-type estuary

This type of estuary is a valley, originally cut by glacial action and then invaded by the sea. This type of estuaries are deep estuaries that have a moderately high river input, but little tidal mixing. In this type of estuary the river water remains at the surface and moves seaward with little mixing of the underlying saltwater. The salinity of the bottom layer will not vary significantly. The Fjord-type estuaries often have a shallow bar of rock at the mouths, and this situation restricts water interchange between surface water and deeper fjord waters. Examples for this type of estuary are present in Norway, Chile, Scotland, Alaska, New Zealand, and British Columbia.

Chilka lake: A success story

Chilka Lake is the largest brackish water lagoon in Asia, located in Orissa state of India. It faced a serious problem of siltation until the year 2000. An attempt was made to open the mouth of the lagoon. As a result, there has been a drastic change in fish resources in the lake. Six fish species which were thought to have extinct got reappeared gradually during post restoration period. There was about 22 fold increase of Hilsha fish landing from 7 tonnes per annum prior to opening of the mouth to 153 tonnes per annum in nine years. Thus the hydrological intervention for restoration of the Chilka lake is a success story resulted in swelling of fish resources.

6.3. CLASSIFICATION OF ESTUARIES

The salinity of estuarine water is between 0.5ppt and 35ppt. This salinity variation is due to dilution of seawater by fresh water. This dilution varies with estuaries depending on the volume of freshwater, the range of tidal amplitude, and the extent of evaporation from the water within the estuary. Based on salinity gradient, there are three types of estuaries: (1) Positive or salt wedge, (2) Negative or evaporate and (3) Neutral.

6.3.1. Positive estuary or salt wedge estuary

It is a river-dominated estuary. In this, the evaporation is less than the volume of freshwater entering the estuary (Fig. 6.2). In such positive estuaries the outgoing freshwater floats on top of the saline water and water gradually mixes vertically from bottom to the top. Examples for this type of estuaries are the Hooghly, Mahanadi, Godavari and Krishna estuaries in India.

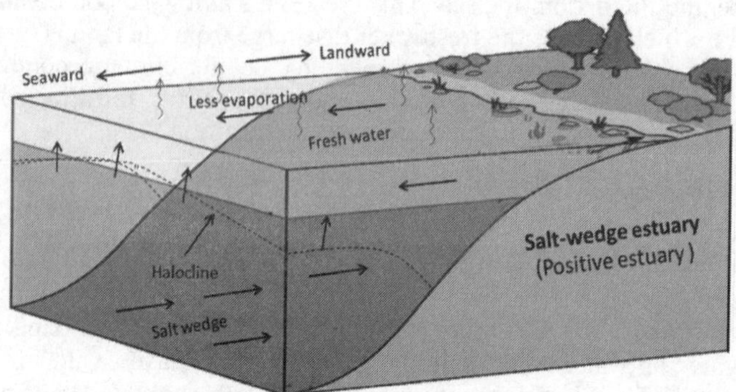

Fig. 6.2: Positive estuary or salt wedge estuary

6.3.2. Negative estuary or evaporate estuary

In the negative estuary, the evaporation from the surface exceeds the freshwater run-off entering the estuary (Fig. 6.3). This type of estuaries is mostly found in tropical regions and it can also occur in temperate regions. In this type of estuary the freshwater input is limited. In negative estuaries evaporation causes the surface salinity to increase. This saltier surface water is denser than water underneath and therefore sinks. Example for this type of estuary is the Coovum estuary in India.

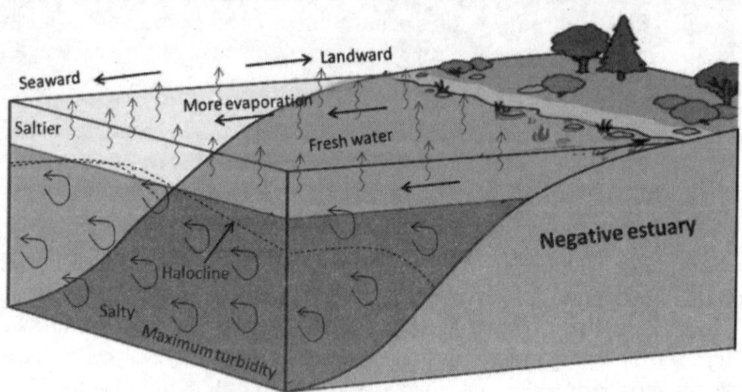

Fig. 6.3: Negative estuary or evaporate estuary

6.3.3. Neutral estuary

The freshwater input to the estuary exactly equals the evaporation, and in this situation, a static salinity regime is established (Fig. 6.4). Such an estuary is termed 'neutral' but it is a rare type, as evaporation and freshwater inflow are almost never equal.

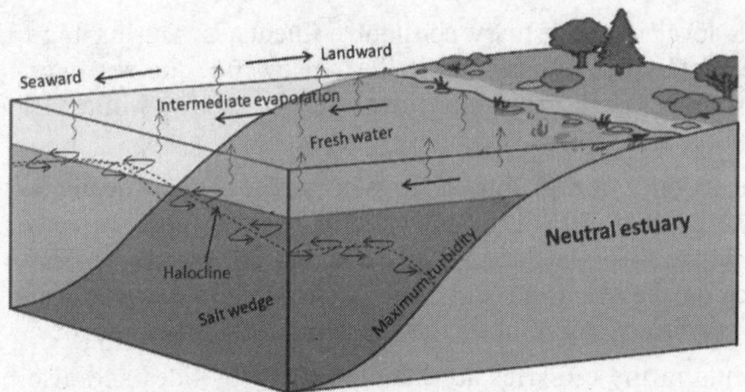

Fig. 6.4: Neutral estuary

- Estuaries are of four types: (1) Coastal plain, (2) Tectonic, (3) Semi-enclosed bay or lagoon and (4) Fjord.
- Based on salinity gradient, there are three types of estuaries: (1) Positive or salt wedge, (2) Negative or evaporate and (3) Neutral.

6.4. ENVIRONMENTAL FEATURES OF ESTUARIES

The estuarine ecology is influenced by both physical and chemical environmental factors (Fig. 6.5).

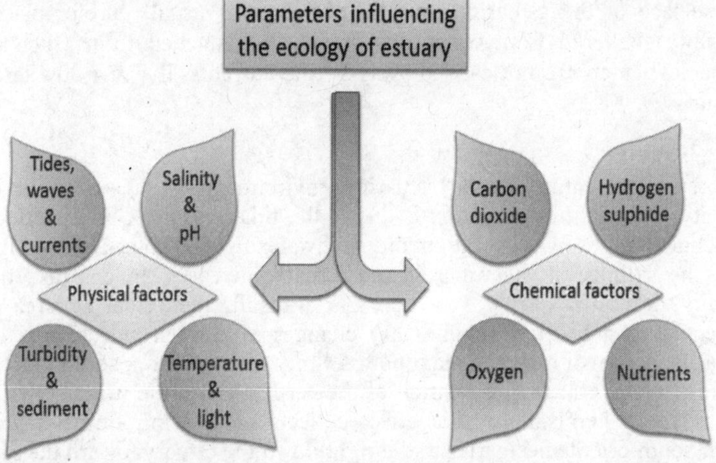

Fig. 6.5: Factors influencing ecology of estuary

6.4.1. Physical environmental features

6.4.1.1. Tides, waves and currents

Water level in the estuary constantly fluctuates. During high tide the seawater enters the estuary, while during low tide the seawater recedes. During summer, less freshwater flows into the estuary while during rainy season, more freshwater enters the estuary.

Wave action is minimal in estuaries due to the shallow depth. Within the estuary, it is often difficult for the wind to generate large waves because of the shallowness of the area. Waves in an estuary do not have much influence on the estuarine organisms. Estuaries are places of calm water. Reduced wave action promotes the deposition of fine sediments.

Currents in the estuaries are mainly caused by tides and also by river flow which in turn, depends on the size and volume of the estuary. If the estuary is broad, the river current will be slow, whereas if the channel is narrow the speed of the current will naturally increase. If the river opens in to a shallow open estuary or sand bar, the tidal currents will be reduced. The currents in an estuary are stronger at the low tide than those at the high tide, and the currents are largely confined to the deeper channels. The tidal action helps to mix the fresh water and saltwater. When there is no tidal action, the lighter freshwater will simply flow over the heavier seawater. The time required for a given mass of fresh water to be discharged from the estuary is called as 'flushing time', and it is a measure of the stability of the system. Long flushing times are important to the maintenance of estuarine plankton communities. Swifter currents combined with a steeper slope are not favourable for burrowing organisms and those that crawl on the surface of the estuarine mud. The organisms usually prefer places specially sheltered from water movement. But attached forms such as barnacles which are not carried away by the currents, live at a low level habitat.

6.4.1.2. Salinity

A striking feature of the estuarine environment is the variation in salinity. This salinity variation is due to the tidal regime, Coriolis effect, seasonal changes in evaporation and freshwater flow or both. During high tide, the salinity of the water in the estuary increases. In contrast, the salinity decreases during low tide. As a result, the estuarine area is subjected to a salinity regime that changes with each tide. Thus the estuarine area will have the maximum salinity fluctuation. A second factor is the Coriolis effect. The rotation of the earth deflects the water flow. In the northern hemisphere, this effect deflects out-flowing freshwater in north-south oriented estuaries to the right down the estuary toward the sea. Seawater flowing into the estuary from the ocean also displaced to the right from the sea toward the estuary. The opposite trend occurs in the Southern

Hemisphere. As a result, salinity variation occurs on opposite sides of the estuary.

Salinity is changed seasonally in the estuary. During summer, salinity is high because of the less inflow of freshwater as well high evaporation, whereas during rainy seasons, salinity is low because of more inflow of freshwater. The salinity is relatively less at the surface, but increases with depth. Salinity changes in the substrate are less than in the water column. Therefore, organisms living within the substrate are subject to less drastic salinity changes than organisms in the water column. This effect is most effective in the lower intertidal zone, but not in the high intertidal zone.

Estuarine animals encounter different salinities. They are adapted either to avoid unfavourable salinities or to withstand such conditions by physiological adaptations. To avoid unfavourable salinity conditions, the barnacles close their valves, the mussels close their shells, some organisms burrow and other organisms migrate up or down the estuary. Most of the estuarine animals are euryhaline, capable of tolerating wide fluctuations in salinity. They can passively oppose reduced salinities by means of membranes having a reduced permeability to water and salts. And they can also react functionally by actively pumping out the excess water which tends to flow into their body and by absorbing salts from the surrounding medium. Most of the estuarine animals have effective osmoregulatory adaptations. Some marine species such as shore crab *Carcinus* living in the estuary regulate their salt content to levels higher than the surrounding when the salinity of the medium is low. This is 'hyper-osmotic regulation'. In some marine invertebrates, such as the mangrove crab *Heloecius cordiformis*, the body fluid is maintained at a lower salt concentration than the seawater, this type of regulation is known as 'hypo-osmotic regulation'. Since these animals are normal inhabitants of brackish waters or estuaries, this is of much adaptive value. Both mechanisms do occur in the shore crabs as well as the teleost fishes and they are normally hypotonic to seawater and hypertonic to brackish water.

Crabs

Crabs have popular names: blue crab (*Portunus pelagicus*), shiva or three spotted crab (*Portunus sanguinolentus*), Christ crab (*Charybdis* sp.), Ghost crab (*Ocypoda*) and mud crab (*Scylla serrata*) and spider crab (*Macrocheira kempferi*, the largest crab in the world with 3m across).

Sex of crabs can be easily identified from abdomen. The male has narrow, triangular flap in contrast to the female which has a round watch-glass shaped abdomen, occupying most of the space between the legs. The females have four pairs of feather appendages which help to carry the tens of 1000s of yellow eggs.

The Portuguese people are eating fiddler crab meat in a sustainable method. They donot kill the crab, but break off the large claw and release the crab. In due course, the crab will grow another claw. This way the population of fiddler crab does not get depleted.

Portunus sanguinolentus
(Shiva crab)

Charybdis sp.
(Christ crab)

6.4.1.3. *Temperature*

Water temperature in estuaries is more varied than in other coastal waters. This can be attributed to a smaller volume of water, and a larger surface area for evaporation in the estuaries. Estuarine waters are colder in winter and warmer in summer than the surrounding coastal waters. Temperature also varies vertically, with the temperature at the surface showing a greater range than deeper waters. Estuarine temperature fluctuates daily and seasonally. The upper layer of water is warm, while the lower layer is cold. Sometimes, the variations depend on the clarity of the water. Thus there is wide temperature fluctuation in the waters of an estuary. This favours only eurythermal species eliminating stenothermic animals in the estuarine areas. Further, large seasonal variations in the flora and fauna of the estuaries are attributed to the temperature variation.

- A striking feature of the estuaries is the variation in salinity. This salinity variation is due to the tidal regime, Coriolis effect, seasonal changes in evaporation and freshwater flow or both. During high tide, the salinity of the water in the estuary increases. In contrast, the salinity decreases during low tide. During summer, salinity is high because of the less inflow of freshwater and high evaporation while during rainy seasons salinity is low because of more inflow of freshwater. The salinity is relatively less at the surface, but increases with depth. Salinity changes in the substrate are less than in the water column.
- Organisms living within the substrate are subject to less drastic salinity changes than organisms in the water column.

- In addition to salinity, there is wide temperature fluctuation in the waters of an estuary. This favours only eurythermal species eliminating stenothermic animals in the estuarine areas.

6.4.1.4. *Estuarine sediments*

Most estuaries have soft and muddy substrates. These are derived from sediments carried into the estuary by both seawater and fresh water. The silt particles are brought to estuary by fresh water of rivers, and these silts mix with seawater resulting in flocculation of silt particles in the presence of ions in the seawater. The particles are heavier and settle down at the bottom. The seawater brings suspended material and it settles down to form mud or sand substrate when the water velocity reduces. The deposition of the sediments is controlled by the speed of currents and the particle size of the sediments. Large particles settle out faster than small particles. Strong currents keep particles in suspension longer than slow currents. In the places of strong currents, the substrate will be coarse sand as only large particles can settle out. While in the calm places of weak currents, fine silt will settle out. Thus seawater drops its coarse sediments at mouth and fresh water drops its coarse sediments in the upper reaches of the river. The mixing of seawater and fresh water results in deposition of fine silt mud. In the river dominated estuary, the sediments carried by the river, accumulate at the river mouths. This accumulation leads to formation of delta. One or more deltas may appear. They divide the river into channels and shorten the estuary. The substrate accumulates organic material and this serves as a food source of estuarine organisms.

The deposition of particles is also controlled by currents and the size of the particles. A tidal current entering the estuary deposits gravel first, then sand and finally silt which accumulates as mud. Similarly a river current deposits gravel and coarse sand first, and then silt, when the current is halted by the inflowing saltwater. Thus an estuary with a calm central area is dominated by mudflats with sand dominating at the mouth. Clean sandy beaches are rarely seen along estuaries. Mud of varying texture is the most abundant estuarine substratum. These muddy substrates are colonized by the salt marshes in the temperate zone and mangroves in the tropical areas. These vegetations make the estuarine system very productive and also provide ideal habitats for a number of organisms.

- Most estuaries have soft and muddy substrates.
- The coarse sediments are dropped by seawater at the river mouth and by fresh water at the upper reaches of the river, whereas fine silt mud is deposited by the mixing of seawater and fresh water.
- The silts of fresh water mix with seawater resulting in flocculation of silt particles in the presence of ions in the seawater. The particles are heavier and settle down at the bottom.

- In the river-dominated estuary, the sediments carried by the river, accumulate at the river mouths. This accumulation leads to formation of delta.

6.4.1.5. *Turbidity and light*

The major cause of estuarine turbidity is the river-derived silt, reaching the maximum during the rainy season. Maximum turbidity occurs at the mouth and decreases with distance from the mouth. Another source of sediments that make the water turbid is due to tidal currents. Thus in tidal flats the water becomes turbid at low tide.

Estuary has a high silt content. Because of the great number of particles in suspension, turbidity is high in estuaries. It reduces the penetration of light and also the primary productivity. The production of organic matter is therefore, dependent largely on the activities of shore plants such as salt marshes like *Spartina* and *Zostera* and mangrove forests. The silt enriched with organic matter in estuarine water provides food for many animals and nutrients to plants. The silt also affects many animals by clogging their gills and covering the hard substrates thereby preventing the settlement of larvae.

- Estuaries that are colonized with salt marshes in the temperate zone and mangroves in the tropical areas are very productive and also providing special habitats for a number of organisms.
- Estuarine water is high in turbidity because of the great number of silt particles present in suspension. It reduces the penetration of light and the primary productivity. Maximum turbidity occurs at the mouth and decreases with distance from the mouth.

6.4.2. Chemical environmental features

6.4.2.1. *Oxygen*

The solubility of oxygen is related to the salinity and temperature in the estuarine water. The level of oxygen varies seasonally and also diurnally within estuaries. Seasonally the variations in oxygen levels are influenced mainly by variations in river discharge, tides, day length and biological effects. In the shallow estuaries with clear water, there will be a lush growth of algae. In such estuaries, there is marked diurnal variation in the oxygen levels, having the maximum during the day and the minimum at night. During day time, the oxygen level is greatly increased by the photosynthetic activities of plants. At night, these plants do not perform photosynthesis, but continue to respire, thereby making the oxygen levels very low. However, in turbid estuaries there is no diurnal change in oxygen levels. The lower limit of dissolved oxygen required for normal respiration of many marine animals is above 1 or 2 ml per litre. Burrowing animals can survive even in anoxic conditions for prolonged periods of time. The high organic content and high bacterial populations of the sediment reduce oxygen level in the interstitial water. Oxygen is in sufficient supply in the water column but is severely reduced in the substrate.

- The level of oxygen varies seasonally and diurnally within estuaries.
- In the shallow estuaries with clear water, there will be a lush growth of algae. In such estuaries, there is marked diurnal variation in the oxygen levels, having the maximum during the day and the minimum at night. However, in turbid estuaries there is no diurnal change in oxygen levels.

6.4.2.2. Carbon dioxide and pH

The solubility of carbon dioxide is determined by the amount of seawater and the temperature in estuarine water. The concentration of carbon dioxide remains more stable in the estuary and similarly the pH of an estuary does not significantly change, since the buffered seawater would overcome the acidity or alkalinity of the inflowing freshwater. In those regions of the estuary, where the influence of freshwater is greater, carbon dioxide concentration and hydrogen ion concentration vary greatly. When the carbon dioxide concentration is higher, the carbonic acid concentration goes up and the pH becomes lower. Thus, during the day, the pH goes up because of the utilization of carbon dioxide for photosynthesis while during night, pH depletes. The maximum level of carbon dioxide corresponds to minimum levels of pH and oxygen. In turbid estuaries and in bottom water, the carbon dioxide concentration is higher and the pH is lower. The amount of carbon dioxide present is lesser, in the mouth of the estuary, where the amount of seawater is greater owing to the tidal influx.

The transport of carbon by the World Rivers is a significant component of the global carbon cycle estimated to 0.9 gigatons carbon per year, of which 60% is inorganic and 40% is organic. In general, estuaries are net heterotrophic ecosystems, where total respiration exceeds gross primary production. This respiration is particularly intense in the upper part of estuaries, which coincides with the maximum turbidity zone, where oxygen depletes, CO_2 supersaturates and thereby CO_2 fluxes to the atmosphere are highest. Most of the riverine organic carbon respired in estuaries enters in the form of particulate organic carbon (POC) and originates from soils, freshwater phytoplankton and sewage. On the contrary, riverine dissolved organic carbon (DOC) appears more susceptible to bacterial degradation. CO_2 exchange with the atmosphere along the nutrient rich river-estuary is buffered by carbonate precipitation in the autotrophic river, while CO_2 dissolves in the heterotrophic estuary.

Phytoplankton blooms are frequently reported in marine coastal eutrophic areas such as bays, lagoons and fjords during summer. Some of these are often monospecific blooms caused by either dinoflagellates or diatoms and this elevates the pH up to 10 in the surrounding water. During such periods of high pH, the biomass of grazers gets increased and the algal blooms are naturally controlled.

Harmful Algal Blooms

Harmful Algal Blooms (HABs) are global phenomenon, reported from over 30 countries including India. In recent years, there have been increased numbers of HABs, economic losses from them, the types of resources affected, and number of toxins and toxic species. Among around 5000 species of marine phytoplankton, around 300 species including diatoms, dinoflagellate, raphidophytes, prymnesiophytes, cyanophytes, and silicoflagellates can cause algal blooms. Only a few dozen of these species have the ability to produce potent toxins, which result in loss of marine life and biodiversity, impacting on fishing activities, tourism, human health and the environment. However, most of the algal blooms are caused by non-toxic organisms. A sudden explosion of population of a single microalgal species up to 1 million cells per llitre of seawater, leading to discolouration of seawater. This event is mainly triggered by nutrient enrichment of seawater, possibly caused by discharge of sewage, fertilizer residue or waste materials.

The local people's belief is interesting. They equate the period of algal blooms as that the mother ocean is at menstrual period and they avoid fishing in the sea.

Harmful algal bloom with red colouration
(http://oceanservice.noaa.gov/hazards/hab/)

- The concentration of carbon dioxide remains more stable in the estuary and similarly the pH of an estuary does not significantly change.
- In turbid estuaries and in bottom water, the carbon dioxide concentration is higher and the pH is lower.

6.4.2.3. Hydrogen sulphide

Estuarine sediments have a higher content of organic matter derived from the detritus brought in by the rivers. This higher content of organic matter, in association with poor oxygen supply in the sediment will result in the formation of hydrogen sulphide and hence black layer of sediment is formed. This is also dependent on the activity of anaerobic sulphate reducing bacteria, such as *Desulphovibrio aestuarii*. Iron salts available in the nearby sand and mud are converted to ferrous sulphide. The excess hydrogen sulphide is toxic, making the area unfavourable to aerobic organisms.

- Estuarine sediments have a higher content of organic matter and this in association with poor oxygen supply in the sediment results in the formation of hydrogen sulphide. The excess hydrogen sulphide is toxic, making the area unfavourable to aerobic organisms.

6.4.2.4. Nutrients

Estuaries are usually rich in phosphates, silicates and nitrates which support large phytoplankton populations. This fertility of estuaries is dependent upon the flow of nutrients from the river water and tidal currents from the sea. In general, the nutrients are higher in estuaries than those present in the surface water of the open sea. The level of nutrient salts in the estuarine waters fluctuates seasonally. This is mainly due to variation in land drainage and as river runoff. The nutrient level is generally higher during monsoon and post monsoon and lower in summer and premonsoon.

- The nutrients are higher in estuaries than those present in the surface water of the open sea. The nutrient level is also higher during monsoon and post monsoon and lower in summer and premonsoon.

6.5. ESTUARINE ORGANISMS

Estuarine habitats are lower in species richness than the nearby freshwater or marine habitats. This is attributed to fluctuating environmental conditions, mainly salinity. Estuarine plankton are also reduced in number of species. Phytoplankton are dominated with diatoms and zooplankton such as copepods. However, bacteria are extremely abundant both in the water column and in the substrate of estuaries because of the abundant availability of organic matter for microbial decomposition. The deeper substrates of estuaries are often barren of plant life. The shallow and intertidal substrates are occupied by seagrass beds, a number of macroalgae, and an abundance of diatoms.

ESTUARINE FLORA: (1) phytoplankton which are lesser in species than in the seas, (2) benthic diatoms occurring in sediments, (3) bacteria and fungi, and (4) larger attached plants, such as seaweeds, seagrasses, salt marshes and mangroves. Estuarine systems are sediment-rich fringes favourable for colonization of large flowering plants such as mangroves (*Avicennia* and *Rhizophora*) and salt marshes (*Suaeda*, *Salicornia* and *Spartina*; Fig. 6.6).

Salicornia brachiata Roxb *Suaeda maritima* (L.) Dumort.

Fig.6.6: Predominant salt marsh plant species in tropical regions, bioindicators of salinity

ESTUARINE FAUNA: There are three types of fauna in the estuaries: (1) marine, (2) freshwater, and (3) brackish water or estuarine. Based on the salinity tolerance, the organisms are classified into five groups: (1) Oligohaline, (2) True estuarine, (3) Euryhaline, (4) Stenohaline, and (5) Migrants.

OLIGOHALINE ORGANISMS: Oligohaline species such as *Channa striata* are the freshwater species. These cannot tolerate salinities above 5 ppt, and are restricted to the upper reaches of the estuary.

TRUE ESTUARINE ORGANISMS: True estuarine organisms are present in the middle reaches of the estuary where the salinity is between 5 and 18 ppt. These species are restricted to brackish water of the estuaries, and do not occur in the sea or in freshwater. The term "true estuarine organisms" refers to all species living in or migrating into estuaries. Examples for true estuarine organisms are *Etroplus suratensis, Liza* spp., *Mugil* spp. and *Gerres* spp.

EURYHALINE ORGANISMS: Some species extend from the sea into the upper reaches of the estuary. They tolerate a wide range of salinity from 5ppt to 30ppt. Euryhaline species constitute the majority of total estuarine taxa such as *Lates calcarifer, Ambassis* spp., and *Elops machnata.*

STENOHALINE ORGANISMS: These live on open seashores and at the mouths of estuaries. They do not penetrate into estuaries below salinities of 25ppt in the lower reaches. Examples for stenohaline organisms are *Lutjanus* spp., and *Sardinella* spp.

MIGRANTS: Migrants are the transitional fauna. This group includes those animals that spend only part of their life cycle in the estuary. Examples for migrants are Salmon, Eel and *Hilsa* spp. Shrimps of the family Penaeidae spend their juvenile stage in the estuary, and adults in the sea. The transitional fauna include predators such as birds and fishes that enter the estuary for food.

- Estuarine habitats are lower in species richness than the nearby freshwater or marine habitats owing to fluctuating environmental conditions.
- Phytoplankton are dominated with diatoms and zooplankton such as copepods. However, bacteria are extremely abundant both in the water column and in the substrate of estuaries because of the abundant availability of organic matter for microbial decomposition.

6.6. ADAPTATION OF ESTUARINE ORGANISMS

Estuary is a unique habitat with fluctuating environmental conditions, and the estuarine organisms are able to adapt to the environmental conditions.

6.6.1. Morphological adaptations

Estuarine animals display morphological adaptations for living, under conditions of fluctuating temperature and salinity. Mud dwelling organisms have hair or setae which protect the respiratory organs from clogging by silt particles, as in the case of estuarine crabs and many bivalve mollusks. Marine species in an estuary have smaller body size, lower reproductive rate and lowered fecundity.

Among plants, morphological adaptations are aerenchyma tissue, salt glands, high lignin and carbohydrate, abundant stomata, thin cuticles and reduced leaf area. Rooted plants maintain high concentrations of osmotically active substances to counter the water-retaining capacity of the surrounding medium. Most plants remove excess salts, through mechanisms such as salt-secreting glands, deposition of salt on external surfaces and shedding of salty leaves. The plants have air-filled tissues called 'aerenchyma' that supplies oxygen to the roots in anoxic mud and also provides oxygen indirectly to animals in the mud. Many plants have high lignin content which may be an adaptation to add strength to plant to cope with high internal salinity. The plants also produce phenolic compounds to deter the herbivory.

6.6.2. Physiological adaptations

Many estuarine organisms are osmoregulators or osmoconformers, adapting to fluctuating internal salt concentrations. The osmoregulators can control the salt content of their internal fluids, while the osmoconfomers can not control their internal salt content. When marine animals move from seawater to brackish water, the internal salt concentration of marine species will be higher than that of the estuarine water, and therefore, water moves across their membranes into their body to equalize the concentration. When freshwater animals move from freshwater environment to brackish water of estuaries, the reverse effect will occur.

The osmoregulation occurs in some estuarine animals, primarily polychaete worms, mollusks and crustaceans. The polychaete worms are slow responding to salinity changes, whereas bivalve mollusks are fast responding to drastic salinity decreases by closing up in their shells to prevent excessive dilution of internal fluids. Crabs are highly adapted to live in estuarine conditions and have well developed osmoregulation by several means: (1) reducing the body permeability through exoskeleton, (2) regulating the ion concentrations of internal fluids, and (3) accumulating the organic solutes such as amino acids in the cells. The osmoregulatory activity varies with developmental stages of animals. Only adult estuarine crabs are able to live successfully in low salinities due to their osmoregulatory abilities, but their eggs and young ones do not have osmoregulation. Therefore, breeding of the crabs occurs in the seawater, more than in low saline water.

6.6.3. Behavioural adaptations

A striking behavioural adaptation in estuarine organisms is burrowing into the substrate. This burrowing behaviour helps the species to avoid predatory problem by birds, fishes or crabs and it also helps survival of the species with low osmoregulatory potential. In some polychaetes, the trochophore larva lives in the mud, instead of leading a pelagic life. This behaviour protects the larva from water current. The oxygen concentration in the lower part of the water column in estuaries is less than in the surface water. As a result the estuarine organisms have the adaptation to survive at low oxygen concentrations. Burrowing animals can tolerate anoxic conditions. Hence, most of the estuarine animals are adapted to detritus and ciliary feeding. Many species such as mullet fish and shrimp enter the estuary as juveniles and migrate back to the sea after maturation. Thus many marine organisms use the estuary as a nursery ground for the great amount of food available and less of predatory problem for the young ones.

- Marine species in an estuary have smaller body size, low reproductive rate and fecundity.
- Many estuarine organisms are osmoregulators or osmoconformers adapting to fluctuating internal salt concentrations.
- A striking behavioural adaptation in estuarine organisms is burrowing into the substrate.
- Mud dwelling organisms have hair or setae which protect the respiratory organs from clogging by silt particles.

6.7. ESTUARINE PRODUCTIVITY

Estuaries are highly productive areas enriched with nutrients. The estuaries act as sinks for organic material brought in from the rivers and also from the sea. Phytoplankton productivity is high in estuaries of low turbidity. On the other hand, the phytoplankton productivity is low in turbid estuaries, even though there are sufficient nutrients and lack of sufficient amount of light. In many estuaries, the phytoplankton that are produced are not fully utilized, owing to a lack of sufficient amount of zooplankton. In the Cochin backwaters in India, the estimated annual consumption of primary production by the zooplankton herbivores is approximately 25% of the total production. The unconsumed food material sinks to the bottom as detritus. Besides phytoplankton, benthic diatoms, seagrasses and algal mats also contribute to the productivity of the estuaries. However, the major source of primary productivity in estuaries is dependant on tropical mangroves and salt marshes. The saltmarsh plant *Spartina alternifera* produces annually 34,580 kilocalories in an area of one square metre. Nearly 50% of its production is available to the estuaries in the form of decomposing organic matter (detritus). There is an abundant supply of detritus in the estuary derived from land drainage, seagrasses, and also from unconsumed phytoplankton. Thus, most of the estuarine food chains are based on detritus.

- Estuaries are highly productive areas enriched with nutrients.
- Organic material is derived from primary productivity in the estuary, in addition to that is brought in from the rivers and from the sea.
- Phytoplankton productivity is high in estuaries with low turbidity and it is low in turbid estuaries.
- In many estuaries, the phytoplankton are less consumed by zooplankton.

6.8. ESTUARINE FOOD WEBS

The estuarine food web has a number of energy pathways. These are generated primarily from a detrital base. The organic matters produced in the system are decomposed by detritivorous bacteria and other microorganisms and finally detritus is produced. The detritus is fed by organisms such as clams, polychaete worms, amphipods and gastropods. The large numbers of deposit feeders and suspension feeders are consumed by a number of predators, both vertebrates and invertebrates. These include crabs, shrimps, polychaetes and predatory gastropod mollusks. The dominant predators are fishes and birds. Feeding habits of fishes change with age. Much of the productivity of estuaries that ends up in fishes is lost to the estuarine system when the fishes move offshore as adults. Estuarine birds take 4-20% of the fish populations as food. Incomplete use of energy in estuarine system allows export to other areas, primarily by fishes and birds. The incidence of omnivory is generally higher in estuaries than in other systems. It is the abundance of food and lower numbers of predators that has permitted estuaries to be nurseries and feeding grounds for the juveniles of many animals.

- The estuarine food web is primarily detritus-based.
- The incidence of omnivory is generally higher in estuaries than in other systems.
- Estuaries are ideal nursery and feeding ground for juveniles and migratory adults of many animals because of abundance of food and relative lack of predators in the estuaries.

Do you know?

- Estuaries are among the most productive natural systems on earth due to the mixing of nutrients from land and sea, producing more food per acre than the richest farmland.
- Many species of fish and wildlife rely on the sheltered waters of estuaries as protected spawning places; hence estuaries are often called as "nurseries of the sea."
- Saint Lawrence River is the largest (1,197 km long) estuary in the world, which connects the Great Lakes to the Atlantic Ocean.

6.9. SALT MARSHES

A salt marsh is a coastal wetland, rich in marine life. It is also known as 'tidal marsh'. It is located in the upper coastal intertidal zone between land and brackish water, dominated by dense stands of salt-tolerant plants such as herbs, grasses or shrubs. These plants are rooted in soils which are alternately submerged and exposed by tidal action. A distinctive feature of salt marshes is the colour - the plants have various shades of grey, brown and green. Salt marsh plants cannot grow where waves are strong, but they thrive along low-energy coasts, mostly in temperate and high latitudes and they do occur in tropical regions. The salt marsh plants are terrestrial in origin and they stabilize the salt marsh areas in trapping and binding sediments. Salt marshes are located on different physical and geomorphological settings such as deltas, estuaries, open coast, embayment and drowned-valley.

Two major groups of marshes are found in North America. The Atlantic and Gulf coast marshes are extensive, occupying gently sloping substrates, and are dominated by *Spartina* species (Fig. 6.7). Pacific coast marshes have a narrow seaward fringe of *Spartina* followed by a broad expanse of *Salicornia* occupying most of the mid-intertidal zone.

Fig.6.7: A predominant saltmarsh plant (*Spartina alternifera*)

The salt marshes are colonized world wide by *Spartina* and *Salicornia*. Marine animals in the salt marshes are crabs (*Uca, Sesarma*), mussels, snails (*Littorina, Cerithidea, Melampus*) and smaller crustaceans such as amphipods and shrimps. The major terrestrial animal components of marshes are insects and the marshes are invaded at high tide by marine and estuarine animals.

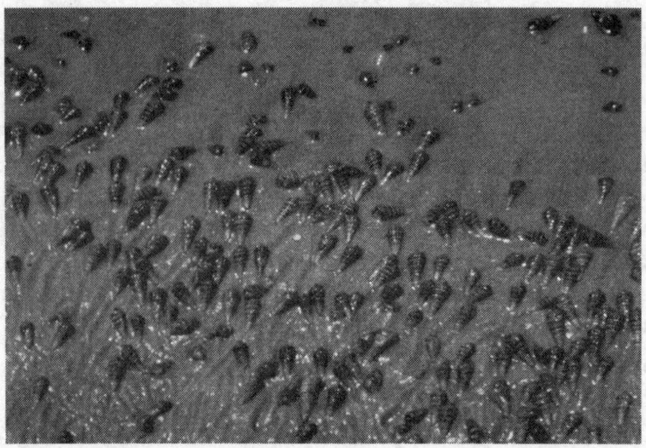

Fig.6.8: *Cerithidea cingulata*: a dominant gastropod in salt marshes

6.9.1. Environmental characteristics

The salt marsh is a dynamic environment with wide variations in salinity, temperature and substrate. The salt marshes tolerate the salinity variation ranging from 20ppt to 40ppt. Salt marshes are found in intertidal area and hence, the salinity variation may be more sudden and more extreme than that in the waters of the estuary. The marshes tolerate the temperature variation of freezing in winter and above 30°C in summer. Mud surface temperature may vary up to 10°C during a single day. Salt marsh soils are muddy and often anaerobic. The sedimentation in the marsh is higher in the lower part of the marsh than in the upper part.

6.9.2. Development of salt marshes

Salt marshes start life as mudflats. In areas of sheltered water, the sediment settles out and builds up. The salt marsh plants produce roots in the mudflats, bind the mud particles together, and make the mudflats more stable. As the mudflats build up, different types of plants can grow and live there creating a salt marsh habitat made up of vegetations with narrow channels between them. The development of mudflats and salt marsh over time is known as 'succession'.

Plant species diversity is low in salt marshes due to extreme environmental conditions such as salinity, complete or partial submersion and anoxic mud substrate. The most common salt marsh plants are glassworts (*Salicornia* spp.) and the cordgrass (*Spartina* spp.), which have worldwide distribution. They are often the first plants to take hold in a mudflat and begin ecological succession into a salt marsh. Their shoots lift the main flow of the tide above the mud surface while their roots spread into the substrate which stabilizes the mud and carries oxygen into it so that other plants can establish themselves. Plants such as sea lavenders

(*Limonium* spp.), plantains (*Plantago* spp.) and varied sedges and rushes grow once the mud is vegetated by the pioneer species.

6.9.3. Zonation of salt marshes

The salt marshes exhibit zonation pattern (Fig. 6.9). At the lowest level the pioneer *Salicornia* species can withstand frequent immersion, whereas species of the upper marsh can only withstand occasional inundation. The lower marsh has the fewest species, while the upper marsh a much more diverse community. Coastal salt marshes can be distinguished from terrestrial habitats by the daily tidal flow of water. This is an important process in supplying sediments, nutrients and water supply to the marshes. At higher elevations in the upper marsh zone, there is much less tidal inflow, resulting in lower salinity levels. Soil salinity in the lower marsh zone is constant due to regular tidal flow. However, in the upper marsh, variability in salinity occurs as a result of less frequent flooding and climate variations. Rainfall can reduce salinity, while evapo-transpiration can increase salinity during dry periods. As a result, there are microhabitats colonized by different species of flora and fauna depending on their salinity tolerance. Vegetation found at the water must be able to survive high salt concentrations, periodical submersion and water movement, while plants inland in the marsh may experience dry and low-nutrient conditions. The upper marsh zones have less species through competition, while lower marsh zones have species which are tolerant to salinity, water submergence and low oxygen levels.

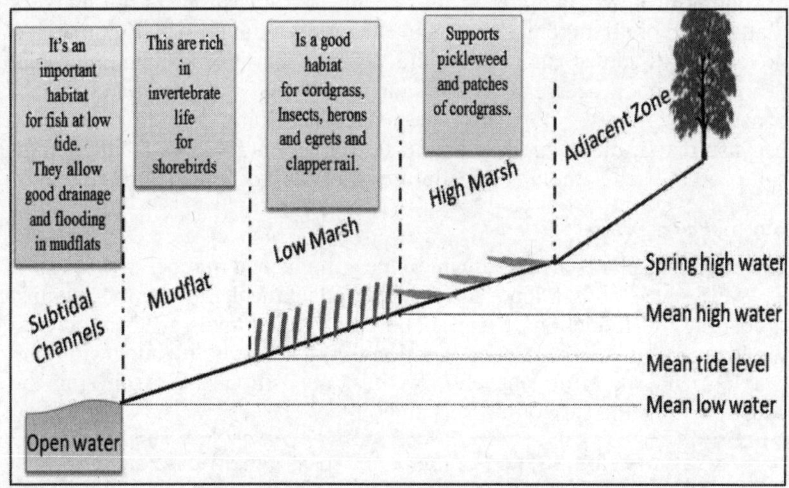

Fig. 6.9: Zonation of salt marshes

6.9.4. Ecological services

Salt marshes are photosynthetically active and are extremely productive habitats. They serve as sink for a large amount of organic matter, which feeds organisms, starting from bacteria to fish, mammals and birds. Salt marshes play a large role in the aquatic food web and in export of nutrients to coastal waters. They provide support to terrestrial animals such as migrating birds, and also provide coastal protection. In the marsh vegetation, animals can hide from predators, because the shallow brackish area physically excludes larger fish. Young fish get diets such as organics of mud substratum, marsh plant and smaller organisms from the salt marshes. The salt marsh plants produce organic detritus, an important food source for many marsh dwellers. Tidal waters distribute detritus throughout the estuary through their movement up into the marsh and then retreat. Algae are also an important food source in salt marshes. They provide nursery areas for fish, food for birds and farm animals and also nesting sites for seabirds. Farm animals may graze on the upper parts of the saltmarsh. Salt marshes may help with defense against the sea as they can reduce the strength of the waves.

6.9.5. Productivity of salt marshes

The primary producers of the marsh are the marsh plants and various microalgae. Vascular plant productivity in marshes is high, but varies with geographic locality, while algal productivity is less in all regions. Marsh productivity increases with availability of nitrogen. The reduced productivity of vascular plants and enhanced algal productivity, in the case of southern California marshes, may be due to open canopy of salt marshes that allow more light to penetrate and encourage algal growth. Salt marshes do export dissolved and particulate organic matters through microbial decomposition processes. Most of the productivity of the marsh is not consumed directly. Only few herbivores consume marsh vegetation, because marsh plants are not palatable to many herbivores for their high salt content, large amounts of cellulose, silica and chemical substances.

6.9.6. Food webs

Salt marsh plants are beneficial to the animals in a number of ways such as substrate stabilization, surface for attachment, protection against predation and reduction of waves and currents. Animals are beneficial to the saltmarsh habitats through aeration of the substrate by burrowing and grazing, incorporating nutrients by fecal deposition and stabilizing the substrate as in the case of mussels and oysters.

- A salt marsh is located in the upper coastal intertidal zone between land and brackish water, dominated by dense stands of salt-tolerant herbs, grasses, or shrubs. These plants are rooted in soils which are alternately submerged and exposed by tidal action.

- Salt marsh plants thrive along low-energy coasts, mostly in temperate and high latitudes and they do occur in tropical regions.
- The salt marsh plants stabilize the salt marsh areas in trapping and binding sediments. The sedimentation in the marsh is higher in the lower part of the marsh than in the upper part. Salt marshes are located on deltas, estuaries, open coast, embayment and drowned-valley.
- Dominant plants of salt marshes are species of *Spartina* and *Salicornia* and the animals are crabs (*Uca, Sesarma*), mussels, snails (*Littorina, Cerithidea, Melampus*) and smaller crustaceans such as amphipods and shrimps. The major terrestrial animal components of marshes are insects. The marshes are invaded at high tide by marine and estuarine animals.
- The salt marshes tolerate the salinity variation ranging from 20-40 ppt. The marshes tolerate the temperature variation of freezing in winter and above 30°C in summer. Salt marsh soils are muddy and often anaerobic.
- The salt marshes exhibit zonation pattern. The upper marsh zones have less species through competition, while lower marsh zones have much more species which are tolerant to salinity, water submergence and low oxygen levels.
- Salt marshes are extremely productive habitats. They play a major role in the aquatic food web and in export of nutrients to coastal waters. They provide coastal protection as they can reduce the waves. They provide nursery areas for fish, food for birds and farm animals, and also nesting sites for seabirds.
- Salt marsh plants are beneficial to the animals through substrate stabilization, surface for attachment, protection against predation, and reduction of wave and current.
- Animals are beneficial to the salt marsh habitats through aeration of the substrate by burrowing and grazing, incorporating nutrients by fecal deposition and stabilizing the substrate by mussels and oysters.

Questions for Review/Exercise*

(1) Define 'estuary'

(2) What are the types of estuary based on geology and geomorphology? Explain.

(3) Explain the Classification of estuary based on salinity gradient.

(4) Discuss the environmental characteristics of estuaries.

(5) Explain the flora and fauna of estuaries.

(6) How do the estuarine animals adapt themselves to fluctuating conditions of estuaries?

(7) Justify the statement "Estuaries are highly productive areas enriched with nutrients".

(8) Write a short note on the development of salt marshes.

(9) Explain the zonation pattern exhibited by salt marshes.

(10) Explain the ecological services of salt marshes.

(11) Write a note of fish migratory pattern.

(12) Explain the types of fishes based on salinity tolerance.

* Field visits to an estuary, analyse the physical and chemical parameters in different points for wet and dry seasons and make a record of the biological species present in the estuary.

Ecology of Mangroves 7

7.1. Distribution of mangroves
7.2. Types of Coastal Settings
7.3. Types of mangrove forests
7.4. Mangrove ecology
    7.4.1. Tidal amplitude and topography
    7.4.2. Temperature
    7.4.3. Salinity
    7.4.4. Rainfall and supply of freshwater
    7.4.5. Substrate characteristics
7.5. Mangrove eco-biology
    7.5.1. Morphological adaptations
    7.5.2. Anatomical adaptations
    7.5.3. Physiological adaptations
    7.5.4. Adaptation for Dispersal and Establishment
7.6. Mangrove ecological functions
    7.6.1. Screening the solar UV-B radiation
    7.6.2. Reducing the 'green house gas effects'
    7.6.3. Mitigating the fury of cyclones
    7.6.4. Mitigating the fury of giant waves
    7.6.5. Controlling the flood
    7.6.6. Prevention of the coastal erosion
    7.6.7. Trapping the sediments
    7.6.8. Deepening the creeks
    7.6.9. Trapping and recycling of nutrients
    7.6.10. Supporting the fishes and wildlife populations
    7.6.11. Supporting Food Web in Other Coastal Ecosystems
    7.6.12. Protecting other Coastal Ecosystems
7.7. Uses of Mangroves
    7.7.1. Timber and firewood
    7.7.2. Logs for multiple uses
    7.7.3. Wood-chip industry
    7.7.4. Foliage as animal feed
    7.7.5. Decoction of bark for leather industry
    7.7.6. Natural beehives and extraction of honey
    7.7.7. Fisheries and livelihood
    7.7.8. Pharmaceuticals
    7.7.9. Novel values of mangroves

Mangroves are the only woody plants, growing in the interface between land and sea in tropical and subtropical regions of the world (Fig. 7.1). The mangroves and their associated organisms constitute the 'mangrove forest community' or 'mangal'. The mangal and its associated environmental factors constitute the 'mangrove ecosystem' (Fig. 7.2). The mangrove forests are often called as 'tidal forests', 'coastal woodlands' or 'oceanic rainforests'. The mangroves constitute the only tall tree forest on the earth lying at the boundary between land and tropical sea. The mangrove habitat is the one place on earth where land, freshwater and ocean mix. They are found in intertidal areas along sheltered shores and in estuaries.

Fig. 7.1: A splendid view of a mangrove forest

Mangrove forests are among the world's most productive ecosystems on the earth. The 'standing crop' of mangroves is greater than any other aquatic ecosystems. The mangroves have diversified habitats such as water bodies, forests, litter forest floor, mudflats and adjacent coral reefs and seagrass meadows. These diversified habitats support a wide variety of organisms, and make the mangrove system profusely rich in biodiversity. Mangroves are remarkable ecosystems, providing many goods and services to humans, particularly fisheries, forest products, pollution removal and coastal protection.

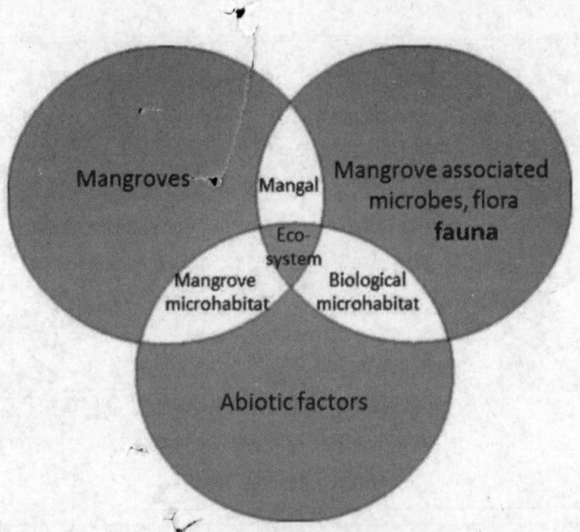

Fig. 7.2: Physical and biological components of mangrove ecosystems (From Kathiresan and Bingham, 2001)

- Mangroves are the only woody plants, growing in the interface between land and sea in tropical and subtropical regions of the world. The mangroves and their associated organisms constitute the 'mangrove forest community' or 'mangal'. The mangrove forests are often called as 'tidal forests', 'coastal woodlands' or 'oceanic rainforests'.
- Mangrove forests are among the world's most productive ecosystems on the earth.
- The 'standing crop' of mangroves is greater than any other aquatic ecosystems.
- The mangroves have diversified habitats which support a wide variety of organisms.

7.1. DISTRIBUTION OF MANGROVES

Mangroves are a rare forest type in the world. The mangroves are restricted to tropical and subtropical regions of the world and to the latitudes between 30° N and 30° S. Mangroves grow better in wet equatorial climates than they do in seasonally monsoonal or arid climates. The growth and biomass production of mangroves decrease with increasing latitudes and they are highest around the equator region. Mangroves have broader ranges along the warmer eastern coastlines of the Americas and Africa than along the cooler western coastlines (Fig. 7.3). This difference in distribution is due to the presence of warm and cold oceanic currents.

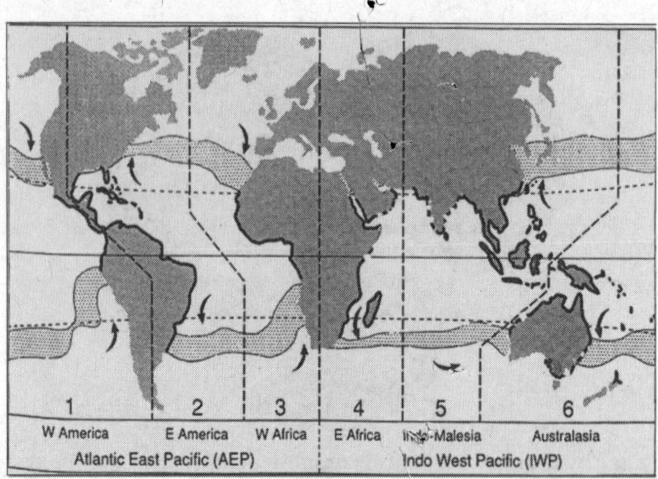

Fig. 7.3: Global distribution of mangroves with six geographic regions (Duke, 1992)

Mangroves are found in 123 countries and territories and they occupy a total of 15.2 million hectares under five regions: Asia, Africa, North and Central America, South America, and Oceania (Table 7.1). Asia has the largest extent of mangroves. Five countries - Indonesia, Australia, Brazil, Nigeria and Mexico - account for 48 % of the total global area. The large single blocks of mangrove forests are found in four areas of the world: (1) Sundarbans, (2) Niger Delta, (2) northern Brazil and (4) southern Papua New Guinea.

Table 7.1: Areal coverage of mangrove forests (FAO, 2007)

Region	**Area (sq. km)**	**Per cent**
Asia	60477.98	38.5
Africa	32427.54	20.6
North & Central America	23581.05	15.0
South America	20377.64	13.0
Oceania	20185.37	12.9

7.1.1. Old and new world mangroves

There are two main centres of mangroves: the Eastern hemisphere and the Western hemisphere (Fig. 7.3). The Eastern hemisphere is Indo-West Pacific region that includes East Africa, Indo-Malaysia and Australia. The Western hemisphere is Atlantic East Pacific region that includes West America, East America and West Africa. The Eastern hemisphere is the place of origin for mangroves, and hence the region is called as the 'Old world mangroves' whereas the Western hemisphere is the place of relatively recent origin for mangroves, and hence the region is called as the 'New world mangroves'. Eastern hemisphere occupies 57% of global mangrove area, while Western hemisphere covers 43%. The Eastern hemisphere has 62 mangrove species, while the Western hemisphere has only 11 species. In the Indo-West Pacific, the diversity is highest around South East Asia. Some mangrove genera are specific to the regions: *Peliciera*, *Conocarpus*, and *Laguncularia* are present only in the new world, whereas *Osbornia* and *Camptostemon* exist only in the old world. Only the mangrove fern *Acrostichum aureum* is common to both the new and old worlds. In the world, there are two places, richest in mangrove genetic diversity, and they are the Bhitarkanika (Odisha) in India and the Baimuru in Papua New Guinea.

7.2. TYPES OF COASTAL SETTINGS

Mangroves are present in a variety of tropical coastal settings such as the deltas, estuaries, lagoons, and coastal fringes (Fig.7.4). The deltas are formed by sedimentation at the river mouth, and this land is colonized by mangroves. Estuaries are the sites of active sediment deposition and the sites are colonized by mangroves. Coastal lagoons are often formed behind sand spits and barrier islands and these sites have less wave action and colonized by mangroves. Mangroves are also present along the coastal fringes where wave energy is low and suitable sediments occur.

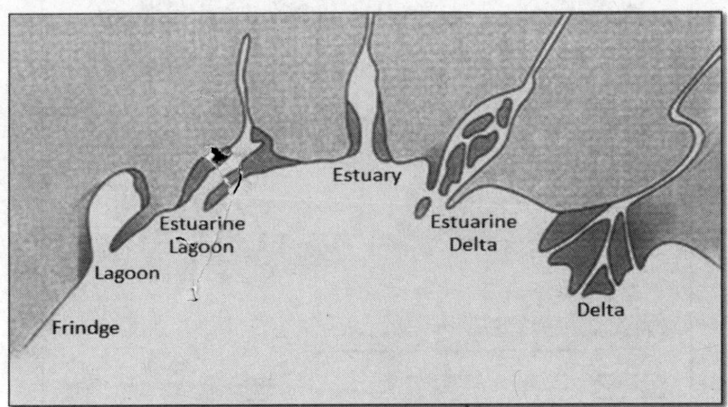

Fig. 7.4: Tropical coastal settings in which mangroves thrive (From Kjerfve, 1990)

7.3. TYPES OF MANGROVE FORESTS

There are six functional types of mangrove forests: (1) fringe, (2) riverine, (3) basin, (4) overwash, (5) scrub (dwarf) and (6) hammock forests (Fig. 7.5). The last three types are the modified forms of the first three types. Fringing mangrove forests occur along the borders of protected shorelines and islands, influenced by daily tidal range. The fringing forests are sensitive to erosion and long exposure to turbulent waves and tides. Riverine mangrove forests are luxuriant patches of mangroves that exist along rivers and creeks, which get flooded daily by the tides. Such forests are influenced by large amounts of freshwater and nutrients and thus making the system highly productive with trees growing taller. Overwash mangrove forests are small mangrove islands, frequently influenced by tidal washings. Scrub mangrove forests are dwarf mangroves present along flat coastal fringes. Basin mangrove forests are stunted mangroves located along the interior side of the swamps and in drainage depressions. Hammock mangrove forests are similar to the basin type except that they occur in more elevated sites.

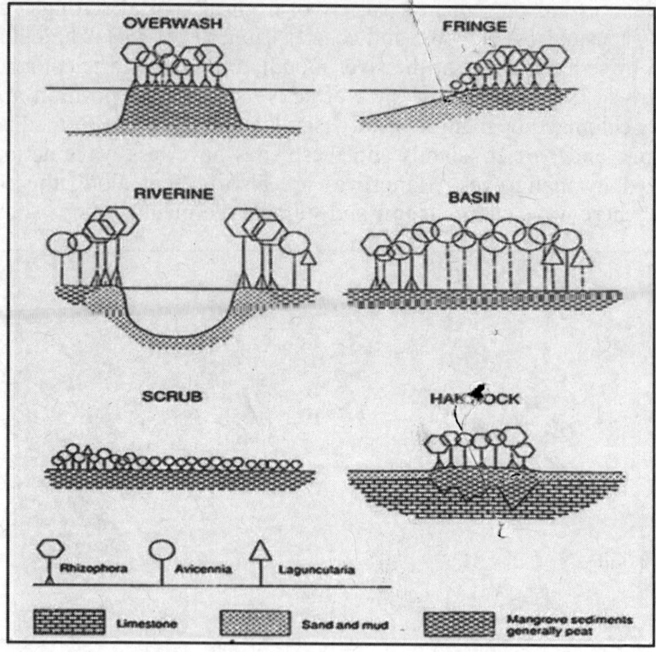

Fig. 7.5: Six types of mangrove forests of common occurrence (From Lugo & Snedaker, 1974; Woodroffe, 1992)

The type of classification noted above does not provide information on the physical processes that take place in mangrove forests. There is another classification that includes three types of mangrove forests: (1) river-dominated, (2) tide-dominated and (3) interior mangrove forests (that have less influence of river/or tides) (Fig. 7.6).

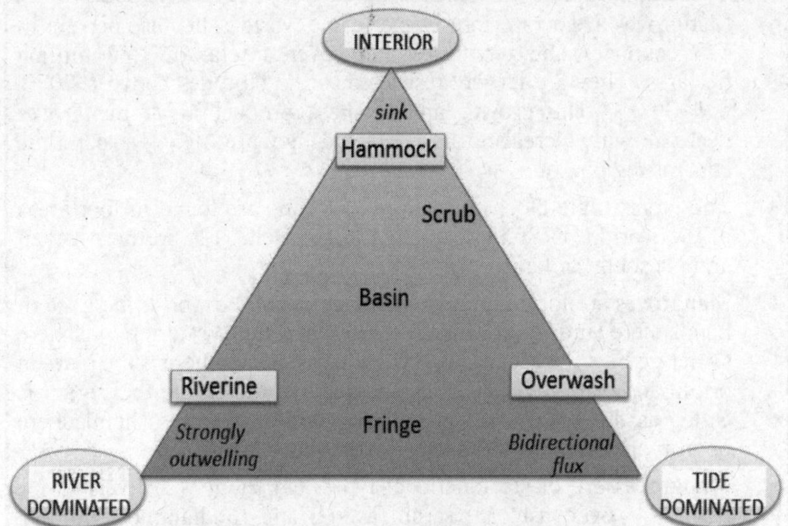

Fig. 7.6: Relationship between the types of mangrove forests and two dominant physical processes (Tide and river). Strong outwelling is a characteristic of river- dominated mangroves. Bidirectional flux is a unique feature of tide dominated mangroves. Mangroves located in the interior typically form sinks for sediment nutrients.

Based on the substrate, tidal range and sedimentation, there are six more broad classes of mangrove settings:

1. Large deltaic systems (occurring in low tidal range and substrate with very fine sediments) (*e.g.* mangroves of Borneo, Sundarbans)
2. Tidal plains (where alluvial sediments are reworked by the tides; and there is the presence of large mudflats for the growth of mangroves)
3. Composite plains (influenced by both tidal and alluvial conditions. *e.g.* lagoons formed behind wave-built barriers where mangroves grow)
4. Fringing barriers with lagoons (high wave energy conditions with sediments of fine sand and mud) (*e.g.* mangroves of the Philippines)
5. Drowned bedrock valleys (*e.g.* mangroves of Northern Vietnam or Eastern Malaysia)
6. Coral coasts (mangroves growing at the bottom of coral sand or in platform reefs) (*e.g.* mangroves of India, Indonesia and Singapore)

- Mangroves are a rare forest type in the world. They are present in 123 countries and territories and cover a total of 15.2 million hectares. They are largely restricted to the latitudes between 30° N and 30° S. The growth and biomass production of mangroves decrease with increasing latitudes and they are highest around the equator region.
- The large single blocks of mangrove forests are found in four areas of the world: (1) Sundarbans, (2) Niger Delta, (2) northern Brazil and (4) southern Papua.
- Mangroves are located in two main centres of the world: the Eastern hemisphere (Indo-West Pacific region) and the Western hemisphere (Atlantic East Pacific region). The Eastern hemisphere is the place of origin for mangroves rich in biodiversity, and hence the region is called as the 'Old world mangroves' and the Western hemisphere poor in biodiversity as the 'New world mangroves'.
- Mangroves are of six functional types: (1) fringe, (2) riverine, (3) basin, (4) overwash, (5) scrub (dwarf) and (6) hammock forests. There is another classification that includes three types of mangrove forests: (1) river-dominated, (2) tide-dominated and (3) interior mangrove forests (that have less influence of river/or tides).

7.4. MANGROVE ECOLOGY

Mangroves are extensive on the stable and sheltered coasts. The plants are able to adapt themselves to practically all types of adverse conditions except frost. Therefore, mangroves are distributed mostly in the tropical regions. Two conditions are favourable for colonization of mangroves in the muddy intertidal areas and they are smooth topography and large tidal amplitude. The mangroves find it difficult to colonize the coastal zone with waves of high energy. They grow poorly in stagnant waters. Mangroves grow better in the muddy soil regularly flooded and exposed due to tidal waters. The mangroves can also be present in substrates other than muddy soil, such as coastal coral reef areas and oceanic islands. In such areas, the mangrove plants grow on peat, which is derived from decayed vegetation. Many ecological factors that strongly influence the mangroves are geographical latitudes, wave action, rainfall, freshwater runoff, erosion/ sedimentation rates, aridity, salinity, nutrient inputs and soil quality.

7.4.1. Tidal amplitude and topography

In general, coasts with a greater tidal range can help in colonization of extensive mangrove forests. A greater tidal range increases the intertidal area which depends upon its slope. The mangrove coverage increases with tidal amplitude and smooth slope of the intertidal area, while it decreases with low tidal amplitude and higher slope of the intertidal area. In the Sundarbans, the tidal water penetrates at an average distance of 110 km inland from the shoreline, due to gentle slope of the coast and large tidal amplitude (6 metres).

7.4.2. Temperature

Mangroves attain climax growth only under tropical conditions where atmospheric temperature in the coldest months is greater than 20°C and the seasonal fluctuation does not exceed 5°C. Mangroves grow in latitudes where the average sea surface temperature is 24°C. Very high temperatures are not favourable for the mangroves. The photosynthetic activity falls to zero at leaf temperatures of 38-40°C, as against the optimum leaf temperature for photosynthesis which is 28-32°C. A small increase in temperature may not adversely affect the flowering, but may change their reproductive cycle and thus may alter the duration between flowering and the fall of ripe seeds. The arid climate that prevails in the Arabian Peninsular region is largely mono-specific with *Avicennia marina* as this species is tolerant to high temperature.

7.4.3. Salinity

Salinity plays a vital role in the distribution of species, their productivity and growth of mangrove forests. Changes in salinity are controlled by climate, hydrology, rainfall, topography and tidal flooding. Mangroves tolerate higher salinity than do non-mangrove plants, but tolerance varies among the mangroves. Mangroves, such as *Avicennia marina* and *Ceriops tagal* are highly tolerant to salinity. Some mangroves prefer low salinity and they are *Bruguiera sexangula, Heritiera fomes, Nypa fruticans* and *Sonneratia caseolaris*.

In general, mangrove vegetation is more luxuriant in lower salinities. Experimental evidences indicate that at high salinity, mangroves spend more energy to maintain water balance and ion concentration. High salinity reduces the biomass production in *Bruguiera gymnorrhiza*, and causes denaturation of terminal buds in *Rhizophora mangle*. Saline interstitial water reduces the leaf area, increases the osmotic pressure in leaf sap, increases the leaf area/weight ratio, and decreases the total N, K, and P minerals. Simple salinity fluctuations also have significant negative effects on the photosynthesis and growth of plants. The Arabian Peninsula has high water salinity (>50 ppt). However, mangroves grow relatively well in east of the Peninsula (the Persian Gulf), but not in west of the Peninsula (the Red Sea). This is due to the fact that the Red Sea has very low input of river water through very short perennial rivers, whereas the Persian Gulf receives nutrient-rich waters from river discharge.

7.4.4. Rainfall and supply of freshwater

The freshwater plays a significant role in determining the plant species, their biomass and forest structure, especially in the river-dominated mangrove forests. The rainfall has significant influence on the distribution and species composition of mangroves, and this is because the rainfall regulates salt concentration in soil and in plants. Freshwater supply is indicated by the ratio of rainfall to evapotranspiration. Where the ratio exceeds the value of 1, the mangroves grow luxuriantly. In arid climates, on the other hand, where the ratio falls below the value of 1, the mangroves exhibit stunted growth. High rainfall in humid areas leaches out residual salts from the mangrove soil and thus encourages the growth of mangroves. For example, along the Panama coast and southern coast of Costa Rica, where the annual rainfall ranges between 2100 and 6400 mm, mangroves exceed 35 metres in height and have a biomass of 280 tonnes per hectare area. Indonesia has the highest biodiversity and luxuriant growth of mangroves and is mainly attributed to high rainfall of 4000 mm and high humidity. Similarly, Northern Australia has greater mangrove biodiversity in estuaries and in areas with abundant rainfall.

In Egypt, Sudan and Ethiopia bordering the Red Sea, the mangroves are mostly monospecific with *Avicennia marina*, becoming bushy in nature due to high salinity, low precipitation and low nutrients. In Pakistan coasts, mangrove biodiversity is low and it is attributed to the arid climate with high temperature (41°-42°C), low humidity and low rainfall (~ 220mm). In the arid regions, barren salt flats often develop along the landward side of the mangroves, as a result of poor leaching of salts from the soil due to low rainfall, and the high soil salinity leads to poor growth of mangroves.

Is freshwater required for mangroves?

Conservation of coastal mangrove forests relies heavily on the management of water resources. In many places, the river water is diverted by dam constructions mainly for irrigation purposes in upstream areas. Monsoon failure is frequently occurring in many places, leading to freshwater scarcity. Thus freshwater supply that feeds coastal mangroves is reduced. This may lead to high salinity which will affect the growth of mangrove species and animal life. The reduction of freshwater flow to coastal systems affect the fishes for their growth, migration and breeding, leading to a loss in fish resources.

On the river Ganges, the Farakka barrage was constructed that has reduced flow of freshwater, consequently reduced the ability of the river to dilute and disperse pollution. The freshwater is required for germination, and sprouting of seeds and seedlings of mangroves. Due to lack of adequate freshwater, several wildlife species got extinct in Sundarbans and they were Javan Rhino, water buffalo, swamp deer, barking deer, and sweet water turtle. In fact, the mangrove species composition in Sundarbans is changed due to reduction in freshwater inputs, the freshwater loving species such as *Nypa fruticans* and *Heritiera fomes* get reduced in population density. These species are also getting replaced with the salt tolerant species such as *Ceriops* species belonging to the family-Rhizophoraceae.

A poor supply of freshwater has resulted in disappearance of species in Pichavaram mangrove forests situated in southeast coast of India. Species such as *Kandelia candel, Bruguiera gymnorrhiza* and *Sonneratia apetala* were once present in large numbers in that area. However, species such as *K. candel* and *B. gymnorrhiza* were extinct. *Sonneratia apetala* was noticed until late 1980s, but only very few individuals are present now in that area.

In Bhitarkanika, Orissa State in India the mangroves occupy the seaward edge of the estuary and lining the creeks and rivers for 35 kilometres inland depending on the penetration of tidal water up in the river Brahmani. The key requirement for sustaining Bhitarkanika mangroves is (i) abundant freshwater and silt that the river carries mostly during the monsoon and (ii) fresh ground water in the estuarine stretch in the non-monsoon season. With construction of a multi-purpose dam on Brahmani at Rengali at 220 kilometres upstream of the delta head, availability of fresh water and silt in the estuarine region are getting reduced due to abstraction for irrigation of 210,000 ha of arable land in Orissa.

A detailed hydrological investigation of the present and future water utilization in the Brahmani basin was undertaken by Regional Research Laboratory and Spatial Planning and Analysis Research Centre, Bhubaneswar, Orissa, India. This study shows that the flow pattern in the lower delta is significantly altered by operation of Rengali dam. The current annual flow at delta head around 17,000 million cubic metres (Mm^3) will reduce to 15,000 Mm^3 in normal years, but to 10,000 Mm^3 in dry years. This reduction of water flow will cause salinity change and also affect ground water. As 70% of the mangrove region is now experiencing salinity up to 15ppt, which even if increase to 20 ppt will not cause any threat. But the progressive reduction of sedimentation from 15.3 to 6.1 million grams due to dam construction can pose a threat to new colonization of mangroves.

7.4.5. Substrate characteristics

One of the important factors in the mangrove habitat is nutrient concentrations. Nutrient fluxes are closely dependent upon plant assimilation and microbial mineralisation. Two major elements (N and P) are of great significance for the growth of mangroves. The concentration of dissolved inorganic nitrogen is generally low in tropical mangrove waters.

Some microorganisms incorporate nitrogen (N_2) from the atmosphere and convert the nitrogen present in the soil into ammonium, which makes nitrogen available for the use of plants. Ammonia is the main form of inorganic nitrogen in mangrove soil. It combines with oxygen to form nitrite or nitrate by the process of nitrification, which is used by the plants. This process normally occurs in the root zone that releases oxygen. Like nitrogen, the concentration of inorganic phosphorus is also generally low in mangrove waters. It is available as a soluble salt that can readily be assimilated by the plants. Phosphate is efficiently adsorbed by the fine sediments of muddy areas rather than the coarse-grained sediments. This is the reason for mangroves that grow luxuriantly in the muddy environments. Microorganisms play vital role in supplying the soluble form of nutrients to mangroves, especially around root zone ('rhizosphere').

The nutrients influence the growth and development of mangrove forests. For example, *Rhizophora apiculata* seedlings do significantly better in plantation sites enriched with potassium. In general, the growth of mangroves is limited in carbonate soils with low-nutrients like phosphorus. This happens because whatever phosphorus may be present gets bound with calcium and is effectively held within the sediments. In an experimental study with *Rhizophora mangle* seedlings, phosphorus enrichment has produced a nearly 7-fold increase in stem elongation rates and a 3-fold increase in the leaf area. Similarly, low phosphorus availability limits the growth of the mangroves and promotes the development of hard, long-lived leaves called 'sclerophylls'. The sclerophylls are an adaptation to conserve the nutrients in these oligotrophic habitats. Mangroves may develop other types of mechanisms to conserve valuable nutrients. For example, mature young leaves have much higher nitrogen and potassium concentrations than senescent old leaves. This is a result of nutrients being translocated from the old leaves into other parts of the plant before the leaves fall. Damage to mangroves declines the levels of nutrients. Severely damaged mangrove sites exhibit significant loss of both nitrogen and phosphorus from the soil. This may be indirectly related to decline in crab population as indicated by a decrease in the density of burrows.

The presence of sulphides is a characteristic feature of mangrove sediments and it influences the distribution of mangroves. Tolerance to sulphides varies with mangrove species. For example, *Rhizophora mangle* (with its numerous aerial stilt roots) has moderately reducing soil with low sulphide levels. In contrast, the zone of *Avicennia germinans* (with its numerous aerial pneumatophores) has strongly reducing soils with high sulphides. In general, high sulphide levels can damage, reduce growth and cause high mortality of mangroves. Sulphide is formed due to reduction of sulphate and it is generally slower in young forests as compared to the older forests. This is because of the occurrence of high nutrient level and low sulphide toxicity in the young forests.

Disturbance can increase the rate of sulphide reduction. Clearing of mangrove forests or simple formation of canopy gaps can change the physical and chemical characteristics of the underlying soil, leading to anaerobic condition and an increase in the sulphide activity in the sediment. Heavy organic input can also increase sulphide production. The H_2S accumulation can kill the mangroves if their pneumatophores get covered by silt and are unable to transport oxygen to the root zone.

- The plants are able to adapt themselves to practically all types of adverse conditions except frost. Therefore, mangroves are distributed mostly in the tropical regions.
- The mangrove coverage increases with tidal amplitude and smooth slope of the intertidal area, while it decreases with low tidal amplitude and higher slope of the intertidal area.
- Mangroves are extensive on the stable and sheltered coasts. The mangroves find it difficult to colonize the coastal zone with waves of high energy.
- Mangroves grow better in the muddy soil regularly flooded and exposed due to tidal waters. The mangroves can also be present in coastal coral reef areas and oceanic islands and in such areas, the mangrove plants grow on peat, which is derived from decayed vegetation
- Mangroves grow in latitudes where the average sea surface temperature is 24°C. The arid climate that prevails in the Arabian Peninsular region is largely mono-specific with *Avicennia marina* as this species is tolerant to high temperature.
- In general, mangroves grow luxuriant in places with high rainfall and lower salinities.

7.5. MANGROVE ECO-BIOLOGY

7.5.1. Morphological adaptations

Mangroves are highly adapted to the tropical coastal environment. There are no other groups of woody plants in the entire plant kingdom with such adaptations. The mangroves are structurally and functionally unique to have breathing roots, extensive support roots, salt-excreting leaves, water-dispersed propagules, rapid rate of canopy production, frequent absence of understory, absence of growth rings, wood with narrow and densely distributed vessels, highly efficient nutrient retention mechanisms, and ability to tolerate salinity and to maintain water content. These adaptations vary among mangrove species and with the physico-chemical nature of the habitat.

Mangroves have modified root systems. The most remarkable modifications of the roots are: (1) stilt (support) roots of *Rhizophora* (Fig. 7.7); (2) pneumatophores (respiratory roots) of *Avicennia* (Fig. 7.8), *Sonneratia* and *Lumnitzera*; (3) knee roots of *Bruguiera, Ceriops* and *Xylocarpus*; (4) plank (snake-like) roots of *Excoecaria*; and (5) buttress roots of *Xylocarpus* and *Heritiera*. All these aerial roots are present above the soil (Fig. 7.9) and they provide mechanical support and also to transport oxygen to underground roots in anaerobic soils.

Fig.7.7: Stilt (support) roots of *Rhizophora mucronata*

Species such as *Rhizophora, Bruguiera* and *Ceriops* have stilt roots and these are looping branch-like supports that arch-out from the tree to penetrate the soil. These aerial roots are lignified to provide strength in support. When the aerial roots enter the soil or water, they become fibrous. Such roots are termed 'nutritive roots' to help the plants in the absorption of nutrients and water. Trees of *Sonneratia, Avicennia, Phoenix* and *Xylocarpus* have no tap roots, but develop very extensive cable roots which lie horizontally some 20-50 centimeres below the soil surface. The cable roots further give-off downward roots for absorption of nutrients and also aerial roots upwards for gas exchange. These aerial roots are called 'pneumatophores'. They may be narrow pencil-like pneumatophores of less than 30 centimetres in *Avicennia* or stout -conical structures of 3 metres long in *Sonneratia*. Knee roots are rounded knob-like projections which extend upwards from the sub-surface roots, as in the case of *Bruguiera* and *Ceriops*. Buttress roots are flange-like extensions of the trunk on the soil surface and they serve an important structural support as in the case of *Xylocarpus granatum* and *Heritiera* species.

Fig.7.8: Pneumatophores of *Avicennia marina*

The mangroves are not deep-rooted, but shallow-rooted plants and this is advantageous in oxygen-deficient conditions of soil substrates. The aerial roots are provided with minute openings called 'lenticels'. These are loose, air-breathing aggregations of cells, through which oxygen passively diffuses into the plant up to its underground root system. The lenticels close during high tide and the oxygen is used by the plant and this results in reduced pressure within the roots. When the tide recedes, the lenticels get opened and air is sucked in to the roots. Pneumatophores are normally unbranched, but branched in oil polluted areas. Following the 1991 Gulf War, mangroves in the Arabian Gulf began developing branched pneumatophores with adventitious roots. The pneumatophores are abundant in anaerobic and oil polluted areas.

Mangroves have xerophytic adaptations to live in saline environment. The cuticle is thick and smooth with small hairs, giving the plant a glossy appearance. In most species, a horn or beak-like cuticular outgrowth covers the stomatal pores. In general, stomata are sunken in mangrove leaves. Water storage tissue is present inside the leaves and it helps in filtering off heat rays (Fig.7.9). The leaves present at the top of canopy are vertical in position, while those at the lower canopy are horizontal in position. The leaves are arranged in a modified decussate pattern with each pair at an angle less than 180° to the preceding pair. This arrangement reduces self-shading and produces branch systems that fill space to facilitate efficient photosynthetic activity. Mangrove leaves are leathery with obscure leaf veins. All these adaptations prevent excessive exposure to sunlight, reduce transpiration and ultimately help in the conservation of water. The leaves accumulate flavonoids which act as UV-screen compounds, thereby preventing the adverse effects of solar ultraviolet radiation. In *Nypa*, there is neither erect stem nor pneumatophores, but the lower portion of each leaf is largely swollen due to the presence of well developed aerenchyma tissue.

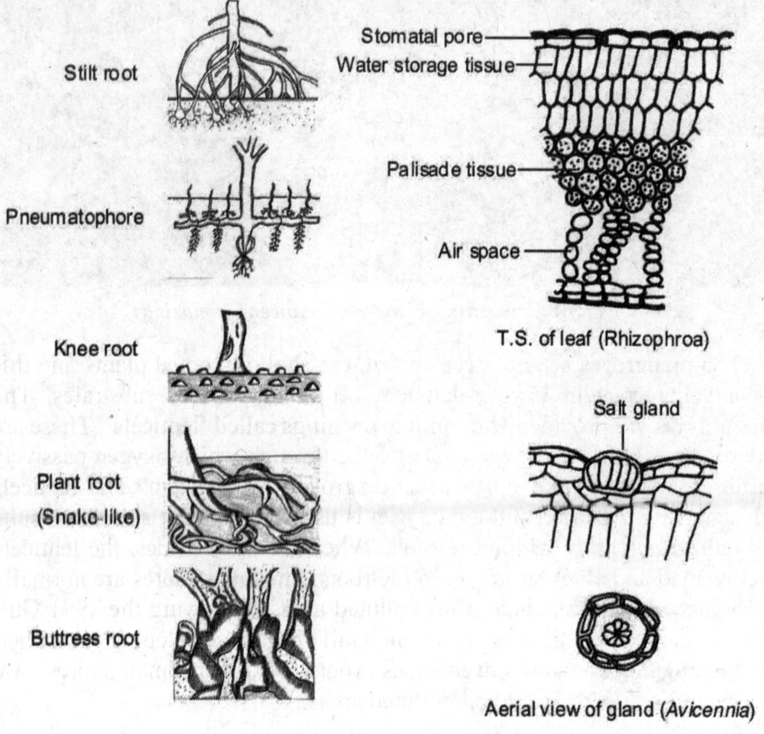

Fig. 7.9: Aerial roots and anatomy of leaves in mangroves

Desalination by mangroves

Water is prerequisite for human life. Only less than 1% is available in freshwater bodies for drinking purpose. Two per cent of water is present in ice Mountains and the remaining 97% of water is present in the sea and ocean. Technologies are available to convert the seawater into potable water, but they are costly. There is a need for developing desalination technologies using biological systems. In this context, Japan has successfully established a desalination industry by developing a filter, based on the biological mechanism of mangroves. By virtue of complex root structure, the mangroves prevent the entry of seawater inland and protect the underground-water systems in the coastal environment. Fresh water can be drunk from the cut end of aerial roots of mangroves!

7.5.2. Anatomical adaptations

7.5.2.1. Root anatomy

Anatomically, the mangrove roots are similar to most other vascular plants. There are two important differences: (1) great development of aerenchyma in pneumatophores of mangroves and; (2) high content of lignin in buttress roots and plank roots in mangroves. The mangrove root has a root cap, lateral roots arising endogenously, exarch protoxylem, and alternating strands of primary phloem and xylem. Many also have an enlarged polyarch stele with a wide parenchymatous medulla. Aerial roots are modified for life above-ground. Compared to the underground roots, the aerial roots have an exaggerated zone of elongation behind the apical meristem. They also have significant secondary thickening (similar to the stems). When the aerial roots reach the ground, they shift to having a short elongation zone. They also become thinner and to form "capillary rootlets" with a simple diarch stele and a narrow cortex. Like aquatic plants, true mangroves lack root hairs. Hence, the endodermis is an effective absorbing layer.

7.5.2.2. Wood anatomy

Two unique features of mangrove wood anatomy are: (1) absence of growth rings and (2) wood with narrow, densely distributed vessels. Water conduction through wood is strongly influenced by size and distribution of the vessels, present in the wood. Water moves most quickly through ring-porous woods in which the largest vessels are in the outermost growth layer. Conduction is much slower in diffuse-porous woods where vessels are more uniform in size and distribution. The wood of most mangroves is diffuse-porous, but *Aegialitis rotundifolia* has ring-porous wood.

7.5.2.3. Seed and seedling anatomy

The seedlings have special anatomical features that allow them to tolerate flooding. Air spaces in the ground tissue conserve oxygen that enables the mangrove seedlings to maintain aerobic metabolism during the periods of flooding. Tannin cells are present in the aerenchymatous tissues and they are helpful to deter predation. Epidermis lacks stomata, but numerous lenticels help in gas exchange.

7.5.3. Physiological adaptations

SALT REGULATION: Mangroves are tolerant to high salinity. They do require salts in their tissues mainly for the purpose of water transport. Water tends to flow from a region of low concentration of salts to a region of high concentration. Because of this, salt concentration in the root tissue has to be higher than that in the soil around the roots, for the water to flow from the soil to the roots. Similarly, the shoot tissue needs to have higher concentration of salts than the root tissue, for the water to flow from root to shoot.

The salt regulation at the cellular level is a highly remarkable adaptation. In the leaves of mangroves, cell vacuoles occupy about 80% of the cell volume and the remaining 20% by the cytoplasm. The cell vacuoles are major sites for salt accumulation and these have higher concentrations of salt than the cytoplasm. Under this situation, the water is supposed to flow from cytoplasm to cell vacuole and if that happens, the cytoplasm will get dehydrated and the cell will be damaged. To avoid this and to maintain osmotic balance, organic compounds of low molecular weights are synthesized and accumulated in the cytoplasm.

Extra salts may damage the plant tissues. The extra salts are reduced by transferring them into senescent leaves or by storing them in the bark or in the wooden parts of the trees. There are three mechanisms by which the mangroves regulate salt concentration in their tissues: (1) salt exclusion, (2) salt excretion (Fig. 7.10), and (3) salt accumulation. Mangrove species - *Rhizophora, Bruguiera* and *Ceriops* - have the mechanism of salt exclusion, in which extra salts get excluded through the ultra-filters present in the root system. Mangrove species - *Avicennia, Aegiceras* and *Acanthus* - have the mechanism of salt excretion, in which extra salts are excreted through specialized glands present in the leaves. Salt excretion is an active process, as evidenced by ATPase activity in the plasmalemma of the excretory cells. Mangrove species - *Sonneratia, Lumnitzera* and *Excoecaria* - have the mechanism of salt accumulation, in which extra salts are stored in cell vacuoles and hence the leaves of plants are succulent. Mangroves are very conservative in using water. They can regulate the loss of water vapour from the leaves. In the wet season, the fine root biomass increases in response to decreased salinity of the surface waters, directly enhancing the uptake of low-saline water.

Fig.7.10: Salt excretion by *Avicennia* through salt glands on leaf surface

Mangrove gene prospecting

M.S. Swaminathan Research Foundation (MSSRF), India has been working with the vision of transferring the gene responsible for salt tolerance from mangroves to crop plants, thereby cultivating the crop plants along the coast using saltwater irrigation. MSSRF has isolated salt-tolerant genes (AM 244) from the mangrove species *Avicennia marina,* and introduced it into a paddy crop (Pusa Basmati and IR64) via *Agrobacterium.*

7.5.4. Adaptation for dispersal and establishment

Establishment of new plants is a challenge to mangroves in unstable mangrove substrates. The mangroves disperse their offspring through flushing of the tides. Vivipary is an adaptive strategy in which the parental plants release growing plants rather than seeds or fruits. The members belonging to the family of rhizophoraceae grow embryo out of the seed and fruit while still attached to the parental plant (Fig. 7.11). The plant actually releases seedling rather than seed. In other species such as *Aegiceras, Avicennia, Nypa,* cryptovivipary exists in which the embryo grows out of the seed, but remains within the fruit and then dispersed. The viviparous germination allows seedlings to develop salinity tolerance before being released from the parent tree. The vivipary also provides a store of nutrients before the seedling falls from the parent plant, and it assists quick rooting in the muddy environment. It also helps to develop buoyancy for distribution of the seedling and structural stability to protect the seedling from damage by pests and parasites. In the case of non-viviparous mangroves – *Sonneratia* and *Xylocarpus* - seeds or fruits are larger which assist flotation.

Fig. 7.11: Viviparous seedlings hanging down from shoot branches of *Rhizophora apiculata*

Mangrove propagules have an obligate dispersal phase of few days or several weeks before the radicle extends for root development. If the propagules do not contact the sediment, they remain viable in seawater for several months. Dispersal of propagules depends on the buoyancy and longevity of the propagules and on the activity of tides and currents. The propagules of *Kandelia candel* are sensitive to light; high levels of light inhibit rooting. This is an adaptation as it keeps the floating propagules alive during dispersal periods. Most *Avicennia marina* propagules become stranded and established close to their parents. The large propagules are able to strike root and establish quickly, as an adaptation to avoid disturbances caused by water movement.

- The mangroves are structurally and functionally unique to have breathing roots, extensive support roots, salt-excreting leaves, water-dispersed propagules, rapid rate of canopy production, frequent absence of understory, absence of growth rings, wood with narrow and densely distributed vessels, highly efficient nutrient retention mechanism and ability to cope with salt and to maintain water content. These adaptations vary among mangrove species and with the physico-chemical nature of the habitat.

- Mangroves have modified root systems: (1) stilt (support) roots; (2) pneumatophores (respiratory roots); (3) knee roots; (4) plank (snake-like) roots; and (5) buttress roots. All these aerial roots are present above the soil and they provide mechanical support and also to transport oxygen to underground roots in anaerobic soils. The mangroves are not deep-rooted, but shallow-rooted plants and this is advantageous in oxygen-deficient conditions of soil substrates.
- Anatomically, the mangrove roots are similar to most other vascular plants with two striking differences: (1) great development of aerenchyma in pneumatophores of mangroves (which help in floatation) and; (2) high content of lignin (which help to prevent damage by herbivores).
- Two unique features of mangrove wood anatomy are: (1) absence of growth rings and (2) wood with narrow, densely distributed vessels.
- The seedlings have air spaces to conserve oxygen that enables the mangrove seedlings to maintain aerobic metabolism during the periods of flooding. The seedlings also have tannins, helpful to deter predation.
- Mangroves are tolerant to high salinity. They accumulate salts in their tissues mainly for the purpose of water transport from root to leaves. The leaf cells accumulate organic solutes to maintain osmotic balance between salt accumulated cell vacuole and cytoplasm.
- Mangrove propagules have an obligate dispersal phase of few days or several weeks before the radicle extends for root development. Dispersal of propagules depends on the buoyancy and longevity of the propagules and on the activity of tides and currents.

7.6. MANGROVE ECOLOGICAL FUNCTIONS

Mangroves have many ecological functions, remarkably in protecting the coast from solar UV-B radiation, 'green house' gases, cyclones, floods, sea level rise, wave action and coastal erosion. Mangrove swamps act as 'traps' for the sediments, and as 'sink' for the nutrients. The root systems of the plants contribute to a stability of the coast. The ecosystem provides a source of food, breeding grounds and nurseries for many food fishes and shellfishes, and they do very often attract other kinds of wildlife. They further help in offering protection to other associated marine systems including the islands, coral reefs and seagrasses.

7.6.1. Screening the solar uv-b radiation

Mangroves possess mechanisms to deal with intense sunlight rays and solar ultraviolet (UV) radiation. For example, *Avicennia* species grow in the areas having high sunlight, hot and dry conditions and the species are well adapted to arid zones. *Rhizophora* species exhibit greater solar UV tolerance than do other mangrove species. The mangrove foliage produces phenols and flavonoids that serve as UV-screen compounds. This ability of mangroves makes the habitat free from the deleterious effects of UV radiation.

7.6.2. Reducing the 'green house gas' effects

Mangroves are among the most carbon-rich forests in the tropics. Mangroves remove CO_2 from the atmosphere through photosynthesis. This perhaps reduces the problems that go with the 'green house gases' and global warming. They fix greater amounts of CO_2 than what the phytoplankton do in the tropical oceans. Mangroves respond well to high CO_2 by exhibiting greater accumulation of biomass with increasing CO_2 level. About 55-85% of the above-ground biomass is held in the woody trunk and branches. In contrast 15-17% is held in aerial roots. The below-ground biomass (root) is high as compared to other forest types. The ratio value between above-ground biomass and below-ground biomass is 2-3, while it is 4 in upland forests. This high root biomass helps to ensure stability in the soft substrates of mangrove environment in addition to carbon sequestration. Mangrove forests usually create thick, organically rich sediments as their substrata. Most of the substrata in the tropics except under deltaic environments consist of mangrove peat which mainly derives from mangrove roots. Carbon sequestration potential of the mangroves is 4 times greater than other tropical forests. This is because of high levels of below ground biomass and also considerable storage of organic carbon in mangrove sediment soils. *Rhizophora* forest has higher below-ground carbon sequestration than other types of mangrove forest and displays its greatest ability during the early stage of the forest. Carbon burial rate of *Rhizophora* forest was faster at 580 g C m^{-2} yr^{-1} than those of other types of mangrove forest.

The total above ground biomass for the world's mangrove forests may be over 3700 Trillion grams of carbon (Tg), and that sequestration of organic matter directly into mangrove sediments is in the range of 14 -17 Tg of carbon per year. Organic-rich soils ranged from 0.5 metre to more than 3 metres in depth and accounted for 49-98% of carbon storage in the mangrove systems. Thus mangroves are capable of accumulating and storing carbon in the soil in large quantities. The mangroves may have an important role to play in global carbon budgets and in the process of mitigating climate change.

Protecting the world from the adverse effects of climate change, the Sundarbans forests play a crucial ecological role by acting as a carbon sink. With 2118 square kilometres of total mangrove forest cover, the Indian Sundarbans absorb 41.5 million tonnes of carbon dioxide daily from the atmosphere, valued at around USD 79 billion in the international market.

This will ultimately result in controlling the rise of atmospheric temperature and the subsequent climate change.

Mangrove deforestation in the world generates emissions of 0.02–0.12 Pico grams of carbon per year—as much as around 10% of emissions from global deforestation of all forests. Thus failing to preserve mangrove forests could cause considerable carbon emissions and thus global warming. Therefore, mangrove restoration can be a novel counter-measure for global warming issue.

7.6.3. Mitigating the fury of cyclones

Mangrove forests protect the coast against cyclones and storms. The best example is the 'super-cyclone' that occurred on the 29th October 1999 with a wind speed of 310 kilometres per hour along the Orissa coast in India. This cyclone played havoc, largely in the areas devoid of mangroves. This event killed almost 10,000 people and caused a heavy loss of livestock and property. On the contrary, practically no damage occurred in the regions with dense mangrove forest. Had the mangrove forests been intact, more than 90 per cent of the deaths due to the 1999 cyclone would have been avoided. The protection value of one hectare of intact mangroves in Orissa was estimated about US$ 8,700. At the time, one hectare of cleared land was fetching only $5,000. In other words, in areas affected by storms and cyclones, the protection benefits of a hectare of land with mangroves can be nearly 2 times higher than the value of 'cleared' land. Thus, protecting mangroves as storm breakers generates more economic value to society. Mangroves also enhance fisheries and forestry production. These benefits are not expected with concrete coastal protection structures. The cyclone 'Nargis' hit the coast of Myanmar (Burma) on the 3rd May, 2008 caused devastating effect including death of over 30,000 people only in the areas devoid of mangroves. This is yet another strong reminder of the importance of mangroves in coastal protection against storms and cyclones.

7.6.4. Mitigating the fury of giant waves

Mangrove forests offer coastal protection against wave actions. This was proved by nature during the 26th December 2004 tsunami that occurred in the Indian Ocean area. Mangroves and *Casuarina* plantations mitigated the deleterious impact of the tsunami waves and protected the shoreline against damage (Fig. 7.12). The dense growth of mangroves in Sundarbans saved West Bengal in India and Bangladesh from the killer impact of the tsunami. Myanmar and Maldives suffered very less from the tsunami; one of the reasons for which was that there is still a lush growth of mangroves surrounding the coastline. In Thailand, mangroves that surround the Island of Surin off the west coast helped to break the lethal power of the tsunami.

Two villages in the lagoon of southern Sri Lanka showed the importance of mangroves in saving lives: the tsunami killed only two people in Kapuhenwala, surrounded by 200 hectares of dense mangroves and scrub forest, whereas it caused 6,000 death cases in Wanduruppa, surrounded by degraded mangroves.

Fig. 7.12: The 2004 tsunami broken a boat jetty in to pieces, while mangroves intact without damage in Parangipettai, south east India.

At higher magnitudes, the tsunami caused damage to mangroves. In Indonesia, Banda Aceh coastline received no protection from natural ecosystems, against the tsunami wave height of 30 metres. Another example is the tsunami of Krakatoa of Indonesia in 1883 with a wave height of 35 metres that penetrated up to 8 kilometres inland. The 2004 tsunami damaged four Nicobar Islands namely Camorta, Katchal, Nancowry and Trinkat. In Andamans, the tsunami caused considerable change in the mangrove stands of the islands; where, *Avicennia marina* and *Sonneratia alba* were not generally affected, while in some places, *Rhizophora* species got affected due to continuous submergence by seawater as a result of the tsunami waves. In South Andaman, 30-80% of mangrove stands got affected; however, in middle Andaman and North Andaman, mangroves were not affected due to elevation of land. The tidal waves damaged mangrove-associate species, *Acrostichum aureum* (Fig. 7.13) and *Fimbristylis littoralis*. Immediately after the tsunami, mortality of shellfishes such as prawns and crabs was observed in the vicinity of mangrove areas due to leaching of acid sulphate salts and water quality deterioration. The tsunami also destroyed seaweed habitats in South Andaman due to damage to mangrove vegetation. Regeneration of the habitat-specific seaweed populations depends on the restoration of mangrove plants.

Fig.7.13: *Acrostichum aureum*, the mangrove fern bioindicator of soil degradation

The mitigating effect of mangroves depends on two physical processes of tsunami - (i) wave attack, and (ii) towing flow. Response to wave attacks depends on vegetation characteristics, whereas the response to towing flow relies on 'drag force' caused by the mangroves, resulting in prevention of coastal erosion. Thus the protective role of mangroves depends on: (1) vegetation characteristics such as, density, height, species composition, density of forest, diameter of mangrove roots and trunks, and elevation of habitats, as well as status of ecological degradation of the forests; and, (2) tsunami wave characteristics such as, wave height, wave period and depth of water. Analytical model shows that 30 trees in10 metre square area in a 100-metre wide belt may reduce the maximum tsunami flow pressure by more than 90%, if the wave height is less than 4-5 metres. Therefore, conserving or restoring mangroves will save the coast from future events of natural disasters such as tsunami.

7.6.5. Controlling the flood

Mangrove systems offer coastal protection against the floods, often caused by tidal waves or due to heavy rainfall associated with storms. The ability of mangroves in flood control is due to the root system which has a larger spread out area and also ability to promote sedimentation.

Global warming leads to sea level rise and as a consequence of that, mangroves may shift more landward. However, in many parts of the world, mangroves are unlikely to migrate landward because of human settlement at the landward boundary and the width of the mangrove forests is likely to decrease due to sea level rise. The effects of sea level rise on any mangrove habitat will be influenced by local conditions such as the wetland type, geomorphic setting and human activities in the wetland. It is predicted that the sea level rise may range from 45 to 65 centimetres per 100 years. However, mangroves can survive the sea level rise of 8-9 centimetres per 100 year and hence the predicted faster rates may seriously threaten mangrove ecosystems. This view was challenged with historical records showing mangrove expansion under relative sea level changes nearly twice as great. It is also predicted that the sea level rise poses no serious threat to most mangrove forests.

Besides flood control, the mangroves prevent the entry of seawater inland and thus protecting the underground water systems, which are a source of drinking water supply to coastal population. The mangrove system reduces the salinity of the ground water and there is a very sharp difference in salt concentrations of groundwater at the interface between salt flats and mangroves.

7.6.6. Prevention of the coastal erosion

The mangrove systems minimize the wave action and prevent the coast from erosion. The reduction of waves increases with the density of vegetation and the depth of water. In the tall mangrove forests, the rate of wave reduction in a distance of 100 metre is as large as 20%, as demonstrated in Vietnam.

Mangrove forests are 'live seawall'-like natural formations and are very cost-effective as compared to the concrete sea wall and other structures for coastal protection. The mangrove forest with a width of 100 metres protected the sea dyke, lying behind the forest, for more than 50 years. In contrast, the rock fencing protected the sea dyke for only five years. This is because of the fact that the rock fencing is not long resistant to wave damage as compared to mangrove forest, as proved in the Red River Delta, Vietnam. The cost of planting the mangroves is just US$ 1.1 million but it has helped to reduce maintenance cost of the seadyke by US$ 7.3 million per year.

7.6.7. Trapping the sediments

One of the important functions of mangroves is trapping of sediments. The mangrove trees catch sediments by means of their complex aerial root systems and thus function as land expanders. The annual sedimentation rate ranges between 1 and 8 millimetres in the mangrove areas.

The mechanism of sediment trapping in mangrove habitats is shown in Fig. 7.14. The mangrove structures inhibit tidal flows, due to the friction force that is provided by the trees with complex aerial root structures. The soil particles are carried in suspension into mangrove forests from seawater by the incoming tide and the soil particles are left behind in the swamps and within the root system by the outgoing tides. Thus, the particles settle in the forests during the low tide, when the turbulence gets reduced. In contrast, the particles are held in suspension during the high tide, when the turbulence is high.

Several factors influence the sedimentation process. The density of mangrove species and their complexity of root systems are the most important factors for determining the sedimentation process. The sedimentary process varies in different types of mangrove forests: riverine, basin and fringe types. The process falls in decreasing order: Riverine> basin> fringe. The river-dominated system receives sediment supply from the upland areas. The tide-dominated fringe system contains abundant sediment from the sea, but the sedimentation gets disturbed by the tides. The interior basin mangroves are the sinks for sediments.

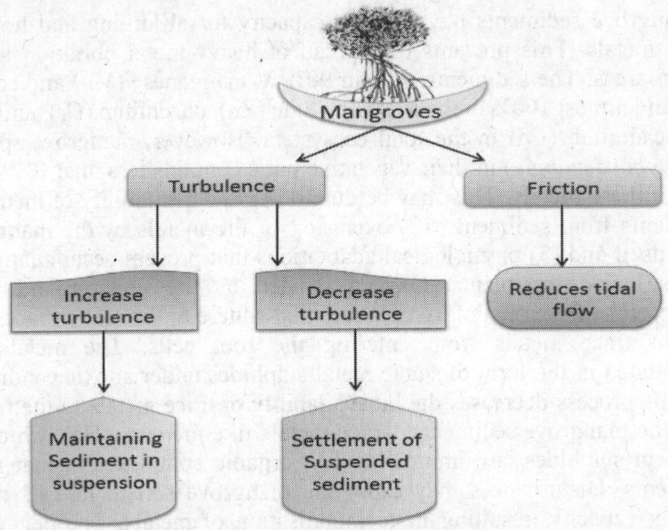

Fig. 7.14: Mechanism of sedimentation as induced by mangroves (From Kathiresan, 2003a)

7.6.8. Deepening the creeks

Water circulation through mangrove forests is important especially for the riverine forest, which often consists of tidal creeks and surrounding forest swamps. The water movement in the creeks is different from the mangrove swamps, as the former is deep, while the latter is shallow. The cause of water movement is the tide. The flow of water during the low tide is much greater than that of the high tide. For example, the riverine mangroves produce asymmetrical tidal currents which are stronger at the low tide than at the high tide. The fast low tide tends to flush out the material from the mangrove area and maintains the depth of creeks. When the area of forest swamp is reduced, the speed of the low tide is reduced, and the creeks to get clogged up. This is commonly observed in some Southeast Asian countries where deforestation of mangroves has reduced the navigability of the canals and river mouths.

7.6.9. Trapping and recycling of nutrients

Mangrove sediments have the ability to retain and recycle nutrients. Resident bacteria and benthic algae rapidly assimilate available ammonium and prevent its export. There is a significant loss of nitrogen and phosphorus from the soil due to reduced flow of water in severely damaged mangrove areas. Thus the mangroves act as a sink for nutrients. The mangrove systems also help in recycling of carbon, nitrogen and sulphur. It is perhaps the only biotic system that recycles sulphur efficiently in nature, and making it available in assimilable forms to other organisms.

Mangrove sediments have a high capacity for absorbing and holding heavy metals. This prevents the spread of heavy metal pollution in the coastal areas. The sediments contain 90% of manganese (Mn) and copper (Cu) and almost 100% of the iron (Fe), zinc (Zn), chromium (Cr), and lead (Pb), cadmium (Cd) in the total ecosystem. However, mangrove species (*Rhizophora mangle*) in their vegetative parts contains less than 1% of the total of these metals. This may be due to (1) low availability of metals to the plants from sediments, (2) exclusion of the metals by the mangrove plant itself and (3) physiological adaptations that prevent accumulation of metals inside the plants. Oxygen exuded by the underground roots promotes the formation of iron plaques that adhere to the root surfaces and prevent trace metals from entering the root cells. The metals are precipitated in the form of stable metal sulphides under anoxic conditions and this process decreases the bioavailability of trace metals to the plants from the mangrove sediments. Trace metals like mercury (Hg), which do not form sulphides, are immobilized in organic complexes in mangrove sediments. Disturbances may cause the mangrove soil to lose its metal binding capacity, resulting in to mobilisation of metals. The degrading mangroves then shift their site from 'sink' to 'source' of heavy metals.

Mangrove ecosystems produce large amounts of litter in the form of falling leaves, branches and other debris. The nutrient cycling begins when leaves fall from the mangroves and are subjected to leaching of nutrients and microbial degradation. Leaching alone removes a number of substances and can produce high levels of dissolved organic matter (DOM). Potassium is the most thoroughly leached element and up to 95% of the total potassium is removed in a very short time. Carbohydrates also leach quickly during early decomposition. Tannins, in contrast, leach very slowly and the high tannin contents may slow colonization of bacterial populations in the initial period of decomposition. As the tannins are eventually leached, the bacterial populations rapidly increase.

Bacteria and fungi contribute to decomposition of the mangrove material and to the transformation of cycling of nutrients. Fungi are the primary litter invaders, reaching their peak in the early phases of decomposition. This is because that fungus is capable of degrading tannins present in the mangrove litter. Bacterial colonies appear shortly after the litter is colonized by fungi. The bacteria grow quickly and can reach very high densities. The N_2-fixing azotobacters are one of the important groups in the decomposing litter and their activities increase the nitrogen content of the leaves 2-3 times in decomposing leaves. In fact, the decomposing mangrove litter attracts shrimp, crabs and fish (particularly juveniles) due to high content of nitrogen especially after one month of litter decomposition (Fig. 7.15).

A number of factors can affect the rate of litter decomposition and nutrient cycling. For example, litter decomposition rates vary among mangrove species. *Avicennia* leaves, because they are thinner and have fewer tannins, decompose faster than those of other species. *Avicennia* leaves also sink and begin to decompose immediately whereas the leaves of other species (*e.g.* *Sonneratia* and *Rhizophora*) may float for several days.

Decomposition is influenced by tidal height, rainfall and temperature. Decomposition of *Avicennia marina* is faster in lower intertidal zones with greater inundation. There is an exponential relationship between leaf decomposition rate and latitude with leaves decomposing most quickly at low latitudes. The seasonality can have important effects on organic cycling and nutrient export from mangrove systems. In general, mangrove debris decomposes faster in the rainy season than summer.

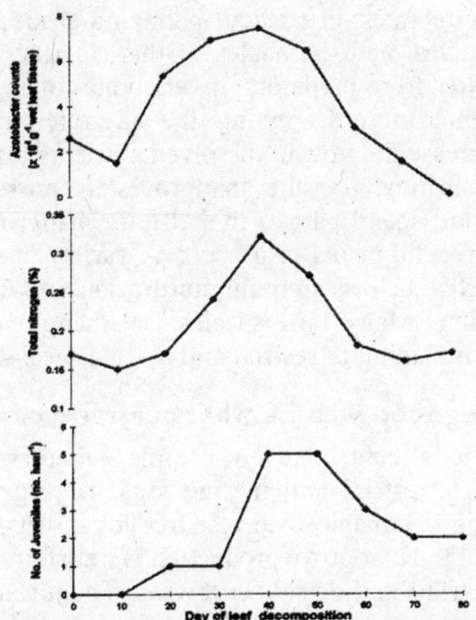

Fig. 7.15: Changes in the nitrogen – fixing azotobacter counts (1×10^4 g^{-1} leaf tissue), the total nitrogen content (% of leaf tissue) and the juvenile prawns (no. $haul^{-1}$) associated with decomposing senescent leaves of *Avicennia marina*, kept in nylon bags and submerged in mangrove waters

7.6.10. Supporting the fishes and wildlife populations

Mangroves are mainly detritus-based system. The detached parts of the mangrove plants when fall on the floor are called 'litter-fall'. These include leaves, stems, roots, flowers and fruits. Microorganisms found in the soil and water decomposes the fallen parts. During this process, nutrients are released which enrich the surrounding waters. The decomposing organic matter along with microbial biomass is known as 'detritus'. This is an important product produced in the mangrove ecosystems. It is rich in protein nitrogen and it serves as a nutritious food for a variety of organisms. The organisms feed on detritus or filter the detritus-particles from the water column. Such detritus–feeding fishes are preyed upon by larger carnivorous forms. Besides fish, the mangroves support a variety of wildlife such as the Bengal tiger, crocodiles, deer, sea turtles, monkeys, pigs, snakes, fishing cats, insects and birds. Mangrove leaves, flowers and fruits are also fed in fresh condition by insects, mollusks and crabs as well as larger mammalian browsers.

Mangrove plants are 'engineers', building and maintaining the physical structure of the habitat. Besides mangrove plants, algae are present as epiphytes on the leaf surface and trunks and on the surround benthic surfaces as well in the water bodies as plankton. Many smaller herbivores such as crustaceans and mollusks, feed primarily on the fine algal films that coat leaves and sediments. Roots and shoot branches are colonized by many benthic invertebrates, in particular grazing gastropod mollusks and filter-feeding bivalves and barnacles. Other marine species use the mangroves as shelter from predators. Invertebrates often dig burrows that allow oxygen to enter more deeply into the substrate reducing the anoxic conditions and increase the flow of dissolved nutrients from the soil. Many species such as shrimps use the mangroves as nursery, feeding and breeding ground and spend part of their life in mangroves. Ants protect many mangrove trees from insect grazers. A variety of species including bees, bats and birds such as hummingbirds, sunbirds and honey eaters facilitate pollination. Many bird species use mangroves as nesting or roosting grounds, including terrestrial and marine species.

7.6.11. Supporting food web in other coastal ecosystems

Mangrove habitats contribute to complex food webs and energy transfers between terrestrial and marine systems (Fig. 7.1). While the living vegetation is a valuable food resource for insects, crustaceans and vertebrates, most of the mangrove production is transferred to other trophic levels through litterfall and detrital pathways. The juveniles feed directly on mangrove detritus, on other small detritivorous invertebrates and on benthic microalgae growing in the mangrove ecosystem. The sub-adult fishes emigrate to offshore which is predominant with phytoplankton-based materials, while the juveniles are immigrated into the mangrove system which is predominant with detritus-based materials. Invertebrates also feed on a variety of cyanobacteria and microalgae that live on submerged portions of the mangroves and on leaf litter.

Mangroves are the key primary producers. Other primary producers are epiphytic vascular plants, algae, benthic algae and phytoplankton on mangrove habitats. In addition, mangrove forests receive considerable quantities of external organic material transported from upstream or offshore ecosystems. About 10% of the net productivity of mangroves is incorporated within local sediments, while 50% is consumed or decomposed, and 30% is exported leaving 10% unaccounted. The high level of nutrient export also supports productivity in adjacent waters including benthic and pelagic systems of other coastal ecosystems such as seagrasses and coral reefs.

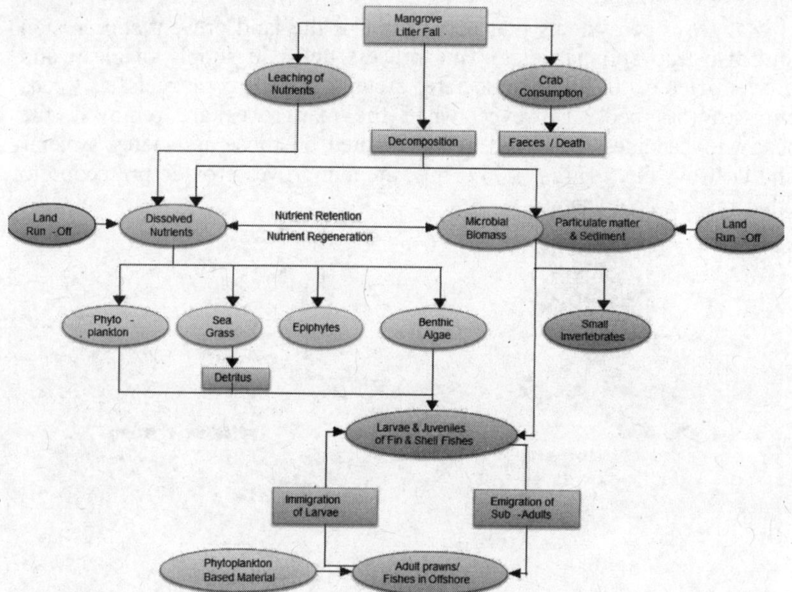

Fig. 7.16: A food web in a mangrove and coastal offshore ecosystems (Kathiresan and Bingham, 2001)

The microbes in the mangrove habitats are known to decompose mangrove litter. During litter decomposition the nutrients are released and these help in multiplication of primary producers such as plankton and microorganisms. The microorganisms multiply in the decomposing organic matter which built up a protein-rich detritus which is consumed by the detritivorous fishes especially their juveniles. During this process of food consumption, the flow of carbon and nitrogen from the primary producers to the consumers through microorganisms is taking place. This has been proved by using stable isotopes that the values of stable isotopes of $\delta^{13}C$

and $\delta^{15}N$ for most of the fishes examined are nearer to the range observed for microbes, followed by phytoplankton and microphytobenthos. Thus, mangroves do not directly contribute, but mangrove-derived organic material is exported to adjacent coastal ecosystems, favouring the growth and multiplication of producers such as bacteria, phytoplankton and microphytobenthos which form the basis of the food web. This has also been proved by using fatty acids and sterols as biomarkers that the abundance of the bacterial biomarkers decreased from mangroves to coral reefs through seagrass in a seascape reflecting a stronger contribution of the mangroves to microbial population in the offshore areas. Thus, mangroves contribute significantly to foodweb through the indirect pathway and thus the coastal ecosystems are interconnected.

7.6.12. Protecting other coastal ecosystems

Mangrove ecosystems that occur towards the land prevent soil erosion and also trap soil particles. This process helps in supply of clean and nutrient-rich water for the associated systems such as coral reefs, seaweeds and seagrass beds. However, when the mangroves are removed, the sediment becomes loose and gets deposited on those associated systems and destroys them (Fig. 7.17). Thus, the mangroves provide protection to other marine ecosystems.

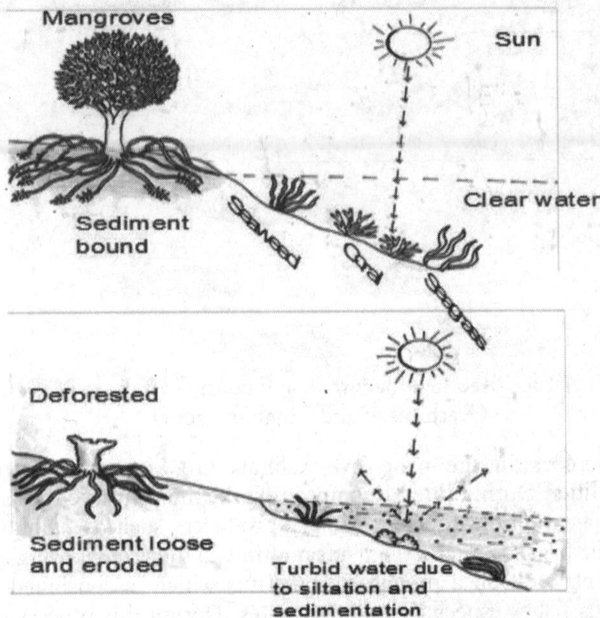

Fig. 7.17: Influence of mangroves and deforestation on seaweed, coral and seagrass ecosystems

- Mangroves protect the coast from solar UV-B radiation, 'green house' gases, cyclones, floods, sea level rise, wave action and coastal erosion.
- Mangrove swamps act as 'traps' for the sediments, and play vital role in recycling of the nutrients. The root systems of the plants keep the substrate firm and thus contribute to a stability of the coast.
- The mangrove forest increases the speed of the high tide and maintains the depth of water creeks against clogging.
- The ecosystem provides a source of food, breeding grounds and nurseries for many food fishes and shellfishes.
- They further help in protection of coral reefs, seaweeds and seagrasses.

Do you know?

- Bhitarkanika (Orissa, India) and Baimuru (Papua New Guinea) are the world's genetic paradises for mangrove diversity.
- Indonesia is the country having the largest extent of mangrove forest area in the world.
- Sundarbans is the largest single block of mangroves in the world, when the mangroves of India and Bangladesh put together. The Sundarbans is the only mangrove forest colonized by tigers.
- Vegetated coastal habitats – mangrove forests, salt-marshes and seagrass meadows –are called "Blue carbon sinks", which cover only 0.2% of seafloor, but contribute about 50% of the total burial of organic carbon in ocean sediments.
- With 2118 square kilometre of total mangrove forest cover, the Indian Sundarbns absorbs 4.15 crore tonnes of carbon dioxide, valued at USD 79 billion in the international market.
- Mangroves are 4 times more efficient in carbon sequestration than many tropical forests.
- One ton of mangrove firewood is equivalent to 5 tonnes of Indian coal.
- Worldwide, about 30% of all commercial fish species are mangrove-dependent.
- Annual loss of mangroves is 0.66% per year which is 5 times greater than the global forest loss.
- Clearing of mangrove forests contributes up to 10% of total carbon emissions from global deforestation of all forests.

7.7. USES OF MANGROVES

Mangrove forests are extremely important coastal resources, vital to our socio-economic development. They provide many goods and services to humans particularly fisheries, forest products, pollution abatement and coastal protection against natural calamities. Mangroves are important in: (i) protecting shores from wind, waves, water currents and natural calamities like flood, cyclone and tsunami; (ii) preventing soil erosion and siltation; (iii) protecting coral reefs, seagrass beds and shipping lanes; (iv) supplying wood and other forest products like honey, medicine *etc.*; (v) providing important nursery grounds, feeding and breeding sites for a variety of organisms including crustaceans, shellfish, finfish, birds, reptiles and mammals; (vi) supporting coastal fisheries and livelihood; and, (vii) serving as accumulation sites for sediment, pollutants, carbon and nutrients.

7.7.1. Timber and firewood

Mangrove wood has a high content of tannin and therefore it is used as timber because of its durability. The pneumatophores are used to make bottle stoppers and floats. Due to high calorific values, mangrove twigs are used for making charcoal and firewood. In Pakistan, about 100,000 people use annually 18,000 tonnes of mangrove firewood.

One ton of mangrove firewood is equivalent to 5 tonnes of Indian coal, and it burns fast producing high heat without generating smoke. Thus charcoal is the most important product largely derived from *Rhizophora* species in Thailand, Malaysia, Vietnam and Indonesia. In Thailand, about 90% of the felled timber is used for charcoal production. From Thailand and Indonesia, mangrove charcoal is exported to Singapore, Malaysia, Hong Kong and other Asian countries.

7.7.2. Logs for multiple uses

In India, mangrove wood is used for boat building for toys and match boxes. Poles are used as shuttering in the construction of houses, tourist cottages. The wood is also used for making traps for catching birds and fishes.

Nypa is the most valuable plant within the mangrove community (Fig. 7.18). The fronds of this plant are processed into shingles to be used as thatch for houses and as outer boundary walls. One hectare of *Nypa* plantation provides about 15,300 of palm leaves to be used each year. In addition, mats, baskets, hats and rain caps are also woven from the *Nypa* fronds. The shoots also serve as a substitute for cigarette and bidi wrapper. Young leaves are used as wrapper of food, while the ribs are used as fuel. The sap of young inflorescence is tapped for sugar production, alcohol distillation and vinegar production. The soft endosperm of fruits is edible and widely used in Thailand, Indonesia and the Philippines. The hard shells of the ripen fruits are used to make buttons.

Fig.7.18: *Nypa fruticans*, economically valuable and freshwater loving mangrove species

7.7.3. Wood-chip industry

Wood chips are a priced commodity of mangrove forests for export in Indonesia and East Malaysia. The stalk and fibers are processed into cellulose, paper and artificial silk (rayon). Major customers are Japan and Taiwan, whose cellulose industry is heavily dependent on the supplies of this product from Indonesia. Several Japanese companies have established paper mills and chipboard factories in Kalmantan and Sumatra. A chipboard production factory in Indonesia processed locally between 3,00,000 and 4,00,000 tonnes of chipboard till 1999.

7.7.4. Foliage as animal feed

Mangroves, especially *Avicennia* are cheap and nutritive source of feed for buffaloes, sheep, goats and camels (Fig. 7.19). These animals are allowed to graze in mangrove areas and camels are periodically taken to uninhabited islands with mangrove cover for grazing. This is very common in India, Pakistan, Persian Gulf region and Indonesia. To cite an example, about 16,000 camels are herded into the mangroves of Indus delta of Pakistan.

Camel herding is one of the activities practiced by the pastoral communities known as 'Maldharis' in Gujarat. The maldharis are in the habit of shifting along with their livestock to far away areas in search of fodder for their cattle. The degradation and non-access of mangroves has critically impacted their livelihoods.

Fig. 7.19: Goat grazing on the leaves of *Avicennia marina*

7.7.5. Decoction of bark for leather industry

Tannin is extracted from the bark of mangroves belonging to the family Rhizophoraceae and is used for tanning leather and fishing nets in India, Malaysia and Pakistan. The extracts of *Bruguiera parviflora, B. gymnorrhiza, Rhizophora apiculata, R. mucronata* and *Ceriops tagal* are used for tanning fishing nets and fish traps in northern Australia.

7.7.6. Natural beehives and extraction of honey

Mangroves attract honey bees and facilitate apiculture activities in some areas. The Indian Sundarbans provide employment to 2000 people engaged in extracting 111 tonnes of honey annually and this accounts for about 90% of natural honey production in India. In Bangladesh, an estimated 185 tonnes of honey and 44.4 tonnes of wax are harvested each year in the western part of the mangrove forest. The best quality honey is produced from *Aegialitis rotundifolia* and *Cynometra ramiflora.* The bulk of honey seems to come from *Ceriops* and *Excoecaria.*

During honey collection practices, local peoples of Sundarbans drive away the honey bees by smoking of leaves of *Phoenix paludosa*. This traditional knowledge is yet to be scientifically validated. Once the active substances are known, there is a promise for the development of insect repellents. Also, Royal Bengal tigers prefer to stay along the tree species and the reason of which is yet to be known.

7.7.7. Fisheries and livelihood

Mangroves provide both shelter and food to the juveniles of a wide variety of marine organisms; notable among these are shrimps/prawns.

Many coastal communities depend on mangroves for their livelihood as mangrove-fringed waterways are used for the exploitation of various species of fishes, crustaceans and other shellfish in traditional ways. To cite an example, 40,000 fisherfolk get an annual yield of about 540 million seeds of *Penaeus monodon* for aquaculture, from the Sundarban mangroves of West Bengal. More than 80% of marine catches are directly or indirectly dependent on mangroves and other coastal ecosystems worldwide. Shells of mangrove mollusks are also collected for the manufacture of lime. One hectare of mangroves can yield 767 kg of wild fish and crustaceans. Each hectare of a managed mangrove system gives revenue of about $ 11,300 a year which is at par with the most profitable intensive shrimp farming. Benefits from mangroves are 25-fold greater than that of paddy cultivation in India. It is widely believed that the mangroves are the 'roots of the sea' and, if there are no mangrove forests along the coast, then there will be no fish in the sea and the sea would behave like a tree without roots.

7.7.8. Pharmaceuticals

Mangrove extracts are used in indigenous medicine and this traditional knowledge is yet to be scientifically validated. *Bruguiera* species (leaves) are used for reducing blood pressures and *Excoecaria agallocha* for the treatment of leprosy and epilepsy. Roots and stems of *Derris trifoliata* are used for narcotising fishes, whereas *Acanthus ilicifolius* is used in the treatment of rheumatic disorders. Seeds of *Xylocarpus* species have antidiarrhoeal properties and *Avicennia* species have tonic effect, whereas *Ceriops* species produce hemostatic activity. Barks of *Rhizophora* species have astringent, antidiarrhoea and antemetic activities. Tender leaves of *Acrostichum* are used as a vegetable and a beverage is prepared from the fruits of *Sonneratia* spp.

7.7.9. Novel values of mangroves

Kathiresan and his team of students of the Annamalai University have shown that mangroves can serve a variety of purposes: - (i) as a source of black tea, (ii) as cholesterol-feed for prawns, (iii) as potential sources of mosquito repellents, (iv) as antiviral-drug- formulations especially against AIDS, cancer, diabetics and jaundice, (v) as a source of UV-screen compounds, (vi) biosynthesis of silver and gold nanoparticles and (vii) as a source of bacterial biofertilizers. Further studies on these aspects will prove a greater efficacy of the mangroves in clinical medicines and for other useful substances of possible economic utility. There is a greater potential for raising mangroves as "cash crops" in the future. Bioprospecting of mangrove ecosystems may lead to development of patents, which in the future can be an enormous source of revenue and employment opportunities.

Excoecaria agallocha is the toxic plant of mangrove forest. The plant produces milk like latex which causes blindness and skin blisters in humans. The blinding tree is worshipped by local people and it is the 'temple tree' of Lord Nataraja temple at Chidambaram, South India. This town is named after the local name of the blinding tree species ('Thillai'). The latex is traditionally used to cure skin diseases especially leprosy.

In Sundarbans, Royal Bengal tigers are known for killing people, by jumping at the backside of human. In order to escape from the tigers, local people wear a mask of human face on head. The tiger along with forest goddess and Muslim god is worshipped in the 'Banabibi' temples, located at the periphery of Sundarbans, with a belief that the tigers will not kill the people who worship in the temple. The Indian Sundarban areas without tigers are degrading, whereas the forest areas with tigers remain pristine.

Polyphenols such as tannins are commercially important plant products. As the mangroves are rich in polyphenolic compounds, the leaves have been attempted for making a beverage having similar qualities as black tea. In fact, the mangrove tea exhibits better property as a beverage, and it has no side effects or toxicity. The mangrove tea has also been proved to prevent oral cancer in animal model. The tannic acids and gallotannins present in the mangrove leaves in particular *Xylocarpus* species have been proved efficient in reducing silver/gold compounds and producing silver/gold nanoparticles. These nanoparticles have potential application in medicine due to their strong antimicrobial activity.

In the mangrove wetlands of Sundarbans and Orissa, two species of horseshoe crabs - *Carcinoscorpius rotundicauda* and *Tachypleus gigas* – are found during the premonsoon period of high salinity (Figs. 20a and 20b). These species exhibit excellent biomedical property, as they are potential sources of bioactive substances - the Limulus Amoebocyte Lysate (LAL) and Tachypleus Amoebocyte Lysate (TAL). These substances are highly sensitive and useful for the rapid and accurate assay of Gram-negative bacteria, even if these are present in extremely minute quantities up to the level of 10^{-10} g. They can also be used for detecting endotoxins in several pharmaceutical products.

(a) (b)

Fig. 7.20a: *Carcinoscorpius rotundicauda* **Fig. 7.20b:** *Tachypleus gigas*
(http://tidechaser.blogspot.in/2009/03/scorching.html)

- The mangrove based resources are forestry products (firewood, charcoal, timber, medicines, honey *etc.*) and fishery products (fish, prawn, crab, mollusc) and the mangroves serve as sites of eco-tourism.
- The mangroves are a rich source of chemicals of industrial and pharmaceutical values.

Questions for Review/Exercise*

(1) How are mangroves distributed in the world? Explain.

(2) What are the different types of mangroves? Explain with suitable diagrammes.

(3) What are the environmental conditions favourable for mangroves? Explain.

(4) Explain the modified root systems of mangroves?

(5) Write a short note on the anatomical adaptations of mangroves.

(6) "Mangroves are tolerant to salinity"- discuss.

(7) List out the ecological functions of mangroves.

(8) "Mangroves provide essential resources and services for the livelihood of coastal people"-justify.

(9) Discuss the pharmaceutical potential of mangroves.

(10) Write short notes: mangal, vivipary, pneumatophore.

* Field visits to mangrove forest areas, to observe morphological adaptations of mangroves and the organisms associated with the mangrove ecosystem and to analyse physical and chemical parameters in the soil and water samples.

Functional Marine Ecology 8

Marine functional ecology is the branch of ecology that deals with the functions of species in the marine environment. In this chapter, eight major functional aspects of marine organisms are discussed: (1) Primary production, (2) Carbon sequestration, (3) Detritivory and herbivory, (4) Predation, parasitism and pathogenesis, (5) Fouling and boring, (6) Competition and succession, (7) Dispersal and settlement and (8) Marine food chains (Fig. 8.1).

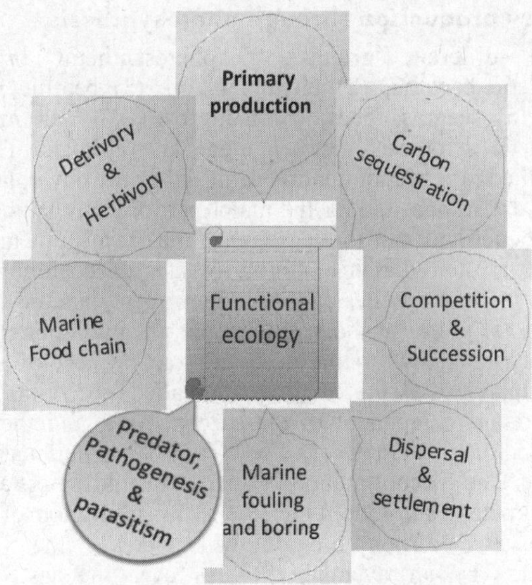

Fig. 8.1: Principal components of marine functional ecology

8.1. MARINE PRIMARY PRODUCTION

Primary production is the production of organic molecules from inorganic constituents. This is achieved through photosynthesis or chemosynthesis. The organisms involved in primary production are called 'primary producers' or 'autotrophs'. The photosynthesis occurs where light is available up to a depth of 200 metres. The chemosynthesis takes place in deep and dark parts of the oceans. Organic production is attributed mainly to the photosynthetic process, and very less extent to chemosynthesis.

In marine environment, algae are the predominant primary producers performing photosynthesis. During photosynthesis, the algae split up the molecules of carbon dioxide, liberate the oxygen, and combine the carbon with the oxygen and hydrogen of water to form simple carbohydrates. In short, the primary producers produce carbohydrate from carbon dioxide, using sunlight and water and releasing oxygen and water as follows:

Carbon dioxide + Water + Light energy from sun ⟶ Glucose + Oxygen + Water

Bacteria perform chemosynthesis. This produces carbohydrate through oxidation of inorganic molecules, such as hydrogen gas, hydrogen sulphide, or methane as source of energy, as follows:

Carbon dioxide + Water + Hydrogen sulphide ⟶ Glucose + Sulphuric acid

8.1.1. Primary production through photosynthesis

There are different groups of photosynthetic organisms : (1) photosynthetic bacteria, (2) phytoplankton, (3) benthic microalgae, (4) seaweeds, (5) seagrass beds, (6) mangroves, (7) salt marshes and (8) photosynthetic symbiosis between algae and animals. The primary production of the sea is predominantly achieved by the phytoplankton. The phytoplankton are responsible for the major part of the organic production in the sea and hence called as the 'pastures of the ocean'. The major groups of phytoplankton are diatoms, dinoflagellates, coccolithophores and prochlorophytes. Photosynthesis of marine phytoplankton contributes about 50% of total global primary production, playing a dominant role in the supply of organic matter to marine environment. Coccolithophorids are single celled algae, possessing calcite plates called 'coccoliths' (Fig. 8.2). They are important component of the carbon cycle, and they transport carbon to the sediments of the ocean floor. They bloom and make the water to appear white. The coccolithophorids flourish in cold waters and produce most of the dimethyl sulphide which helps in generation of cloud and rainfall. Increasing carbon dioxide will increase the biomass of coccolithophores, by an average of 40% over the last 220 years. Prochlorophytes are photosynthetic bacteria related to cyanobacteria and they are responsible for up to 50% of total oceanic primary productivity.

Near the shore, the primary production is carriedout mainly by the larger plants such as seagrasses, seaweeds and mangroves. But the larger plants occur only in less than 0.2% of the ocean. Some algal species called 'zooxanthellae' are endosymbionts to anemones and coral reefs. Symbiotic associations between animals and algae are widespread amongst sponges (Porifera), coral reefs and sea anemones (Cnidaria). The most important symbiotic relationship exists between dinoflagellates of the genus *Symbiodinium* and reef-building corals. The coral growth is fastest in regions where the rate of photosynthesis of the symbiotic algae is highest.

Fig. 8.2: Coccolithophores (http://www.bbc.co.uk/nature/life/Coccolithophore)

The production of organic matter by the process of photosynthesis is termed 'Gross Primary Production'. The gross primary production is defined as the total amount of organic matter synthesized by the photosynthetic organisms in a particular time period and in a specific volume of water or area of water surface. Some of the organic matter is utilized by the primary producers for respiration and excretion. The remaining of the organic matter becomes the plant tissue, and it is referred to as 'Net Primary Production'. The Net primary production is defined as the amount of organic material synthesized by the activities of photosynthetic organisms, in excess of that respired or excreted. There is another term called as "Standing Crop" and it is the biomass of an organism or group of organisms present per unit volume or per unit area at a given point of time. The average gross primary production of the world ocean is about 150 milligram carbon fixed in an area of one square metre for a day ($mgC/m^2/day$). This is equivalent to an annual total production of about $1.9x10^{10}$ tons. The average net primary production of western and eastern regions of the Indian Ocean is 240 and 190 $mgC/m^2/day$ respectively and these are equivalent to annual total productions of about $3.04 x10^{10}$ and $2.41 x 10^{10}$ tons respectively.

8.1.1.1. Role of picoplankton and nanoplankton in ocean

Picoplankton and nanoplankton dominate planktonic biomass and production in the open ocean. Picophytoplankton include heterotrophic bacteria, prochlorophytes and cyanobacteria, while nanoplankton include coccolithophorids. Their dominance indicates that the microbial food web is the most important trophic pathway in the ocean. The ocean is generally low in nutrients and the condition is termed 'oligotrophic'. This condition supports the primary producers to perform only low carbon fixation of 0.3-0.4 gram carbon fixed in an area of one metre square in a day. Oligotrophic ocean is a dynamic ecosystem covering 60% of total ocean surface. Despite low productivity, oligotrophic ocean may account for more than 30% of the total marine primary production. Bacteria may be the largest component of most planktonic assemblages, while autotrophic bacteria (cyanobacteria and prochlorophytes) and small flagellates constitute a significant fraction of the primary producers in many pelagic systems. In oligotrophic conditions, the bacteria account for 30% of the total particulate carbon and the autotrophic cells with size of less than 10 μm for 50-100% of the chlorophyll-a. In tropical waters, particulate matter of less than 10 μm comprises more than 50% of the total nutrient pool.

8.1.1.2. Primary production in relation to fisheries

Primary production is used as an effective tool for assessing fishery potential. There is a strong relationship between plankton production and commercial fish landing. The most important factor that limits primary production is the availability of nutrients in the ocean and the replenishment of the nutrient salts. Thus, data on the concentration of nutrients is important for forecasting the rate of primary production and thereby the fishery resources of a locality. The Peruvian coast is one of the important nutrient-rich upwelling areas of the world oceans and it supports the world's largest fishing industry, producing more than 10 million tons of anchovies per year that feed directly on the dense phytoplankton and associated zooplankton.

Primary productivity can be measured by different methods: (1) direct counting of phytoplankters, (2) chlorophyll estimation, (3) nutrient uptake measurement, (4) light and dark bottle method, (5) C^{14} method of determining photosynthetic carbon fixation and (6) remote sensing. Of these, remote sensing is the latest method for large scale assessment through satellite imaging and bio-optical methods of measuring photosynthesis, specifically from fluorometers. Net primary productivity (NPP) is then calculated from the absorbed photosynthetically active solar radiation (APAR) and 'e', is average light utilization efficiency, as follows:

$$\text{NPP} = \text{APAR} \times e$$

APAR can be derived from satellite measurements through the estimates of chlorophylls. 'e' cannot be accurately measured from space and must be determined with field measurements.

8.1.1.3. Factors influencing primary production

Photosynthesis is primarily dependent on natural light energy. The availability of light depends on wavelength, angle of incidence of sunlight and penetration of light in to seawater. Of the incident light that falls over the sea, only about 0.1% is utilized for the production of organic matter. The compensation depth is the depth at which oxygen released by photosynthesis equals to oxygen consumed by respiration of phytoplankton. Below the compensation depth, there is no phytoplankton growth and photosynthesis. The compensation depth varies with the availability of light and the transparency of water. The area above the compensation depth is the euphotic zone. This zone is less than 4% of the ocean, but about 95% of all marine primary production occurs in the euphotic zone. On an average, the euphotic zone covers a depth, varying from less than a metre to over 100 metres depending on the turbidity in the seawater. Compensation depth is only about 6 metres in the Gulf of Mannar and Palk Bay, on east coast of India, while it is at 50-60 metres on the west coast of India and at 90 metres in the blue water of the Laccadive Sea in India. There is a seasonal variation in photosynthetic primary production, which is more in summer than in winter. In other words, the compensation depth is higher in summer and lower in winter (Fig. 8.3).

The grazing by zooplankton is one of the major factors that influence the biomass of phytoplankton and the rate of primary production. The large difference found between the phytoplankton production and the standing crop is attributed to grazing. In the Antarctic areas the grazing activity is so intense that the biomass production of phytoplankton is only 0.5% of the primary production. When the density of phytoplankton increases, turbidity increases, penetration of light decreases and these events reduce the phytoplankton growth. In contrast, when the grazers reduce phytoplankton density, turbidity decreases, light penetration increases and these result in a large phytoplankton growth.

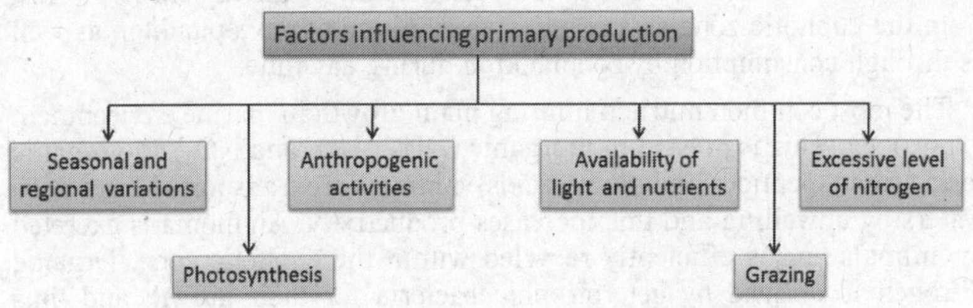

Fig. 8.3: Factors influencing primary production

Primary production varies with season and region. This is mainly due to variation in the levels of plant nutrients such as nitrogen and phosphate. In temperate latitudes, the phytoplankton productivity is high in spring, low in summer (Fig. 8.4) and this is correlated with availability of nutrients. Productivity is highest in northern latitudes and in southern waters due to nutrient-rich coastal upwelling. The warmer open oceans reduce the upward movement of nutrient-rich deep waters into the photic zone.

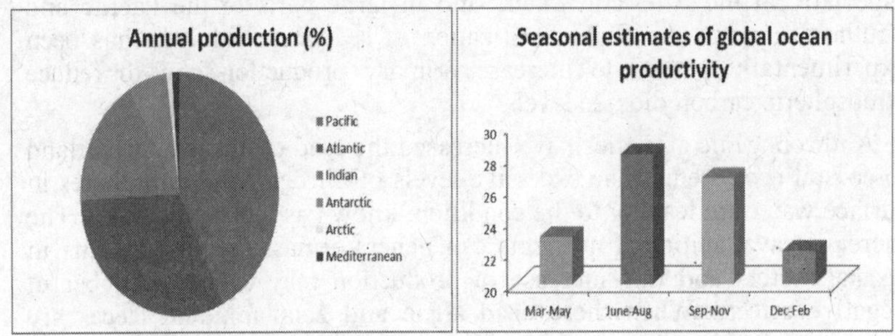

Fig.8.4: Estimates of Primary productivity of the oceans

There is great variation in the distribution of primary productivity in the world's oceans (Fig. 8.4). Annual production is the highest in the Pacific Ocean (19.7 x 10^{15} g carbon fixed; 42.8% of total) and lowest in the Arctic Ocean (0.4 x 10^{15} g carbon fixed; 0.9%). Annual production is the second largest in the Atlantic Ocean (14.5x10^{15}g carbon fixed; 31.5%) followed by the Indian Ocean (8 x 10^{15} g carbon fixed; 17.3%), the Antarctic oceans (2.9 x 10^{15} g carbon fixed; 6.3%) and Mediterranean sea (0.6 x 10^{15} g carbon fixed; 1.2%).

In general, the tropical waters are in short supply of nutrients and this limits the rate of primary production. During photosynthesis, the phytoplankton take in nutrients from seawater for the production of organic matter. This absorption decreases the concentration of the nutrients in the euphotic zone and this reduces organic production. A small amount of nutrients is regenerated by bacterial lysis of phytoplankton within the euphotic zone itself. But a considerable amount of phytoplankton is lost from the euphotic zone as a result of the sinking of phytoplankton as well as through consumption by zooplankton during day time.

The most common nutrient limiting plant growth in marine environment is nitrogen. This is present in inorganic (NH_4^+, NO_3 and NO_2) and organic forms (urea, peptides and amino acids). Nitrates are transported from deep waters by upwelling and this increases productivity. Ammonia is excreted by animals and is efficiently recycled within the euphotic zone. Organic nitrogen is utilized by heterotrophic bacteria for their growth and thus organic nitrogen is stored in the form of bacterial biomass in the euphotic zone. When these bacteria are attacked by viruses, the nitrogen is released in the ocean. In addition to nitrogen and phosphate, a number of minor nutrients, such as iron, manganese, copper, zinc, vanadium, cobalt and molybdenum are known to limit primary production. Silicon is used in the formation of siliceous shells of diatoms and silicoflagellates, and it does not normally act as a limiting factor, as it is rapidly regenerated in seawater. In contrast, the iron is rapidly oxidized in seawater and becomes insoluble. Therefore, iron deficiency limits phytoplankton growth especially in the Antarctic waters and in large parts of the Pacific and Southern Oceans. Artificial fertilization of seawater with iron has been experimentally proved to increase primary production and to reduce atmospheric carbon dioxide levels.

Anthropogenic activities have increased the load of nutrients from land to coastal ecosystems. The excessive levels of nitrogen and phosphates in surface water are leading to the condition, known as 'eutrophication'. The increased availability of nutrients can generate massive algal blooms in coastal waters and this increase of production may cause a number of negative effects. When these dead algae and zooplanktonic faeces are decomposed by bacteria through oxidation process, the oxygen concentration in the water close to seabed can become negligible. This zone of hypoxic condition is known as 'Dead zones' in which mass mortality of benthic invertebrates and fishes may occur. Other effects are: (1) loss of benthic vegetation from shading and harmful algal blooms and (2) inputs of freshwater with dissolved organic matter (DOM) from land, which may have a negative impact on phytoplankton productivity in shallow waters, through attenuation of light. Nutrients and DOM that enter coastal ecosystems from freshwater sources may cause ecological changes in the sheltered coastal ecosystems. This may not occur in most exposed ecosystems where nutrients and DOM became rapidly diluted.

Salinity change can result in osmotic stress on phytoplankton. Brackish and estuarine phytoplankton are saline tolerant due their ability of osmo-regulation in response to salinity changes. Alterations in salinity level frequently result in increased respiratory activity to maintain osmotic balance. The biological production exceeds respiratory consumption in the upper 100 m of the water column of oceans. However, the global warming has the potential to increase respiration at a faster rate than photosynthesis, so that in the future the oceans may capture lesser carbon dioxide than at present. However, some organisms like coccolithophorids are becoming much active by increasing their productivity under increased temperature and carbon dioxide. Over the last 220 years, there was a 40% increase in biomass of coccolithophorids.

8.1.2. Primary production through chemosynthesis

The chemosynthetic bacteria are present in the shallow marshes, lagoons and deepest oceans. They are free-living or more commonly as symbionts with animals. There are two types of bacteria: (1) the gamma proteobacteria commonly found as endosymbionts in clams, tube worms and snails, and (2) the epsilonproteobacteria found on the gills of deepwater shrimps. The chemosynthetic bacteria depend on sulphide or methane oxidation for their primary production. The microbes that use sulphides are known as 'thiotrophs' and those that use methane are called as 'methanotrophs'. These microbes occur in cold seeps and hydrothermal vents. In the cold seeps, dominant animals which harbour chemosynthetic autotrophs are bivalve clams and mussels in the genera *Bathymodiolus* and *Idas*. The chemosynthetic bacteria are also present in specialized cells at the surface of the gills, and receive their sulphides or methane from the surrounding waters. In the hydrothermal vents, specially adapted animals are polychaetes, mollusks and crustaceans and these can form symbiotic relationships with the chemosynthetic bacteria. Tube worms such as *Riftia pachyptila* have no mouth or gut and live close to the vents depending entirely on their symbiotic bacteria. Sessile worms and clams, mobile species also carry chemosynthetic symbionts. The branchial cavity of the shrimps is colonized by chemosynthetic bacteria.

- Primary production is achieved through photosynthesis or chemosynthesis.
- The photosynthesis occurs where light is available up to a depth of 200 metres in surface waters, shallow sub-tidal and intertidal areas.
- The chemosynthesis takes place from the shallow marshes and lagoons to the deepest oceans.
- Organic production in the marine environment is attributed mainly to the photosynthetic process, and very less extent to chemosynthesis.

- The primary production of the ocean is predominantly achieved by the photosynthetic marine phytoplankton especially picoplankton and nanoplankton.
- The photosynthetic organisms rely on carbon dioxide for their primary production, while the chemosynthetic bacteria rely entirely on sulphide or methane oxidation for their primary production.
- The bacteria perform chemosynthesis and these are free-living or more commonly as symbionts with animals.
- The most important factor limiting primary production is the availability of nutrients in the ocean and the replenishment of the nutrient salts. These factors are important for forecasting the rate of primary production and thereby the fishery resources of a locality.
- Primary productivity can be measured by different methods: (1) direct counting of phytoplankters, (2) chlorophyll estimation, (3) nutrient uptake measurement, (4) light and dark bottle method, (5) C^{14} method of determining photosynthetic carbon fixation and (6) remote sensing.
- Photosynthesis is primarily dependent on natural light energy. The availability of light depends on wavelength, angle of incidence of sunlight and the transparency of the water.
- The large difference noticed between phytoplankton production and biomass of phytoplankton is attributed to grazing by zooplankton.
- Primary production varies with the seasonal and regional variations in the levels of nitrogen and phosphate. In general, the tropical waters are in short supply of nutrients, and this limits the rate of primary production.
- Productivity is highest in northern latitudes and in southern waters in which there are coastal upwelling. The warmer open oceans limit the upward movement of nutrient-rich deep waters into the photic zone.
- The increased availability of nutrients can generate massive algal blooms in coastal waters. When these dead algae and zooplanktonic faeces are decomposed by bacteria through oxidation process, the oxygen concentration in the water close to seabed can become negligible and this hypoxic condition can lead to the death of aerobic organisms.
- Phytoplankton taxa vary greatly in their ability to osmoregulation in response to salinity changes.
- The biological production exceeds respiratory consumption in the upper 100 m of the water column of oceans. However, the global warming has the potential to increase respiration at a faster rate than photosynthesis, so that in the future the oceans may capture lesser carbon dioxide than at present.

8.2. CARBON SEQUESTRATION IN MARINE ENVIRONMENT

The process of removing carbon from the atmosphere and storing it in a reservoir is known as carbon sequestration. This process is remarkable in the ocean. About 2,000 trillion grams of carbon is annually transferred from the atmosphere to the oceans. About 50% of the carbon is stored within the top 400 metres of the ocean waters. However, only a portion of this carbon is actually stored permanently in the oceans and a large portion of carbon is released back within a few decades. The carbon is largely stored as dissolved inorganic carbon in the ocean water and the carbon does not penetrate deep sea to remain in the ocean for a long time (Fig. 8.5).

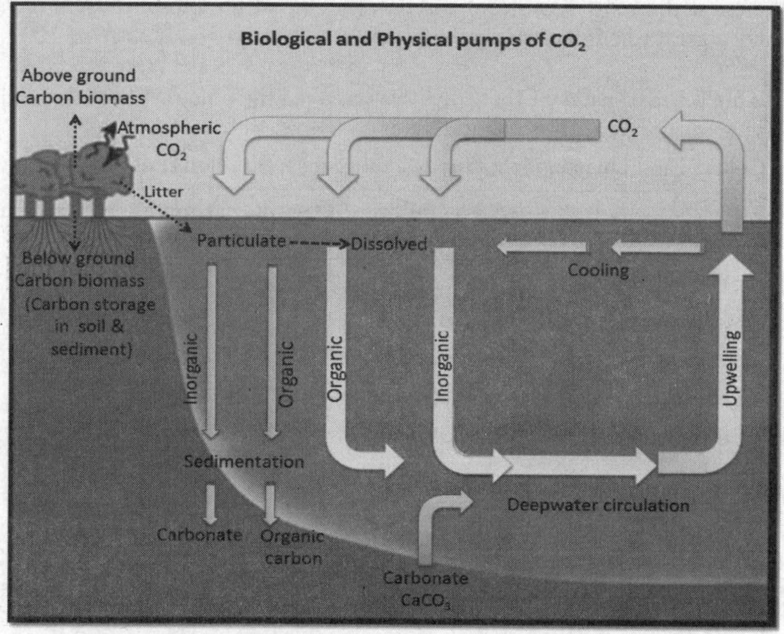

Fig. 8.5: Carbon sequestration in marine and coastal environments

The amount of carbon present in the deep ocean water is below the detection limit. Similarly, the amount of carbon buried over long periods of time in the deep sea sediment is only very little. The annual carbon burial in the deep sea sediments is about 6 trillion grams. This is 180 times lower than the rate of carbon burial by coastal vegetated habitats such as mangroves, saltmarshes and seagrasses. Therefore, coastal vegetated habitats are large carbon sink and they are considered as 'blue carbon sink'. The coastal vegetated habitats are strong in carbon storage in a long term basis. In contrast the capacity of the water column of the oceans to act as a sink for atmospheric carbon will become weakened in the future due to ocean acidification.

Vegetated coastal habitats are built by flowering plants and trees, but the coastal sea also contains vast areas covered by algal beds. Most macroalgal beds (including kelp forests) do not bury carbon, as they grow on hard rocky substrates. In addition to the flowering plants, terrestrial sources of carbon and planktonic source of carbon also accumulate in sediments of vegetated coastal habitats on the continental shelf and in estuaries. Thus the vegetated coastal habitats act as efficient carbon sinks, globally responsible for the annual burial of 329 trillion grams of carbon. The vegetated coastal habitats cover only less than 0.2% of the seafloor, but they contribute about 50% of the global burial of organic carbon in the marine sediments (Table 8.1). Therefore the vegetated coastal habitats are amongst the most significant carbon reservoir in the biosphere and they play a great role in oceanic carbon cycle.

Table 8.1: Estimates of the global area covered by 'blue carbon sinks' and the annual organic carbon burial rates
(Cebrián and Duarte, 1996; Duarte *et al.*, 2005a; Bouillon *et al.*, 2008).

Component of ocean	Area (million square km)	Organic carbon burial (trillion gram carbon per year)
Vegetated habitats:		
Mangroves	0.17	57.00
Salt Marsh	0.40	190.00
Seagrass	0.33	82.00
Total vegetated habitats	0.90	329.00
Depositional areas:		81.00
Estuaries	1.80	45.20
Shelf	26.60	126.20
Total depositional areas	28.40	455.20
Total coastal burial % vegetated habitats	330.00	0.72
Deep sea burial		6.00
Total oceanic burial		460.00
% vegetated habitats		0.71

There are three major proposals for enhancing the carbon sequestration potential of the ocean and mitigating the lethal effect of carbon dioxide in the atmosphere. They are: (1) ocean fertilization, (2) increasing ocean alkalinity, and (3) injection of carbon dioxide in to the deep sea.

OCEAN FERTILIZATION

Primary production in some areas of the ocean is limited by nutrients such as iron, silica, phosphorus or nitrogen. By increasing the availability of these nutrients, primary productivity can be increased resulting in an acceleration of natural rate of CO_2 uptake by the oceans to 2 gigatons of carbon annually.

INCREASING OCEAN ALKALINITY

Ocean alkalinity can increase the solubility of carbon dioxide in the ocean water. Ocean alkalinity can be increased by use of carbonate and thermally decomposed lime stones in the ocean water, thereby increasing the capacity of the water to absorb carbon dioxide.

INJECTION OF CO_2 IN TO OCEAN

In order to reduce the concentration of CO_2 in the atmosphere, CO_2 can be injected into the ocean for long term storage. CO_2 can be transported by ship or pipeline offshore and then injected into the water column at great depth (>1000 metres or deeper) where the CO_2 can dissolve and remain isolated from the atmosphere for centuries. CO_2 can be placed directly onto the sea floor at depths greater than 3000 metres, where the CO_2 will form long-lasting 'lakes' with low dissolution rates. Injection of CO_2 into the water column or on the sea bed may affect marine life and ocean chemistry (e.g. by increasing acidity). In view of the severe environmental impacts, the placement of carbon dioxide streams in the water column or on the sea bed is prohibited.

- The carbon is largely stored as dissolved inorganic carbon in the ocean water and the carbon does not penetrate deep sea to remain in the ocean for a long time. The annual carbon burial in the deep sea sediments 180 times lower than the rate of carbon burial by coastal vegetated habitats. Therefore, coastal vegetated habitats are large carbon sink and they are considered as 'blue carbon sink'.
- The vegetated coastal habitats cover only less than 0.2% of the seafloor, but they contribute about 50% of the global burial of organic carbon in marine sediments. Therefore the vegetated coastal habitats are amongst the most significant carbon reservoir in the biosphere and they play a great role in oceanic carbon cycle.
- There are three major proposals for enhancing the carbon sequestration potential of the ocean and mitigating the lethal effect of carbon dioxide in the atmosphere. They are (1) ocean fertilization, (2) increasing ocean alkalinity and (3) injection of carbon dioxide in to the deep sea.

Definitions

- Primary production is the production of organic compounds from atmospheric or aquatic carbon dioxide through the process of photosynthesis or chemosynthesis. The organisms involved in primary production are called 'primary producers' or 'autotrophs'.
- Gross primary production is the total amount of organic matter synthesized by the photosynthetic organisms in a particular time period and in a specific volume of water or area of water surface.
- Eutrophication is an increase in nutrients such as nitrogen and phosphorus that increase algal growth.
- Carbon sequestration refers to the capture and long-term storage of carbon in forests and soils or in the oceans, so that the build-up of carbon dioxide in the atmosphere will reduce.

8.3. MARINE DETRIVORY AND HERBIVORY

Primary consumption by detritivores and herbivores is one of the major components of marine food webs. Primary consumers feed on primary producers such as phytoplankton, macroalgae, and bacteria. The marine herbivory is performed by animals feeding mostly on single celled algae. Shallow-water surface sediment feeders consume both plant and detrital food sources. The deposit feeders dominate in the depths. Habitats where little or no current flow occurs will have very few suspension or filter feeders. Some organisms can change their feeding behaviours due to change in food supply. For example, when covered by the tide water a polychaete worm is a suspension feeder by using a pair of tentacles to collect suspended material in the water column; when the current velocity is low, the same worm becomes a deposit feeder.

8.3.1. Sediment and deposit feeders

Sediment determines the type of animals living and feeding within it. Deposit feeders such as worms, crabs, bivalves and sea urchins are present in fine sands and mud, while suspension feeders such as shrimps and polychaete annelids are present in coarse sediments. The deposit feeders are detritivores, which digest the organic matter and bacteria present in the marine sediments. The feeding rate increases with the organic content especially organic nitrogen of the sediment. This increasing feeding rate is also reflected in increased growth and fecundity of the animal. The deposit feeders such as sea cucumbers also help to remove waste products of oysters and mussels in intensive aquaculture.

8.3.2. Filter and suspension feeders

Filter feeders and suspension feeders can occur together in great numbers, and can reduce the abundance of phytoplankton and suspended solids, thereby make the water clear. This in turn, can promote light penetration and plant growth. Both filter feeders and suspension feeders collect dead particulate matter from water column by means of active or passive techniques. Active filter feeders have mechanisms which create water currents to bring food particles to them and these include sponges (Porifera), fan worms (polychaete, annelids), barnacles (crustacean), bivalve mollusks, sea mats (Bryozoa) and sea squirt (Urochordata). Passive filter feeders use a static device such as a tentacle held in water current to collect food matter and these include hydroids, corals and brittle stars. Turbidity is deleterious to marine suspension feeders, because it damages the soft feeding apparatus. In general, growth of filter feeders is directly related to the food available in the surrounding water.

Filter and suspension feeders are important to marine ecosystems. Majority of organisms on coral reef environment are filter feeders. They are sponges, fan worms, giant clams and hard and soft corals. This has a significant impact on primary producers present. For example, in the Great Barrier Reef of Australia, the reef filters and depletes 35% of the biomass and the net depletion of organic carbon is 15%.

8.3.3. Phytoplankton grazers

The phytoplankton are grazed by zooplankton, remarkably copepods. The copepods are more omnivores than herbivores, eating other zooplankton and organic particulate matter as well as phytoplankton. The feeding rate increases with phytoplankton population up to a level. Above this, ingestion remains constant. Krills are small mostly 1 to 2 centimetres long prawn-like decapod crustaceans which are important in food chains in both northern and southern hemispheres. They filter and feed on phytoplankton, and these krills are in turn, fed by the bigger marine filter feeders such as basking sharks and blue whales.

Marine sponges are major component of a trophic link between the benthos and the overlying water column through their active suspension feeding. Their high abundance on hard substrates, high filtration rates and retention efficiencies of particulate organic material make the marine sponges important for energy transfer and also for nutrient cycling within coastal ecosystems. Picoplankton are the primary diet of the sponges. The sponges have selective uptake of prokaryotic picoplankton (<2μm), as proved in a demosponge (*Callyspongia* sp.). Thus, marine sponges are key players in the transfer of carbon from the pelagic microbial food web in to the benthos.

8.3.3.1. Vertical migration of zooplankton

Zooplankters constitute a major link between primary production and high trophic levels in aquatic ecosystem. An interesting behaviour of zooplankton is their vertical movement within a water column known as 'diel vertical migration' (DVM). Zooplankton move to deeper and sometimes even anoxic waters during daytime and move upwards to surface waters during night time. A reverse movement also occurs with some zooplankton moving upwards during the day and moving downwards towards the night. Some zooplankton may not exhibit migratory behaviour at all. Predation plays a key role in DVM of zooplankton. The DVM helps zooplankton to avoid visual predator fishes. Normal migration behaviour is to move to the dark during day time and to move back at night when the predation by visual predators is reduced. In addition to predation, several factors influence DVM of zooplankton and they are light intensity, water transparency, temperature, chlorophyll levels, lunar periods and food availability.The zooplankton occupy depths with the highest food abundance, but also with the most optimal temperature conditions. In the absence of predation, zooplankton will distribute themselves in a water column according to the availability of food resources. Change in light intensity is important for DVM. At dawn and dusk, correlation exists between the measured descent and ascent velocities of zooplankton and the rate of the relative increase and decrease in light intensity respectively. Lunar periodicity may also act to modify the effect of change of light intensity on DVM. The behaviour of migrating upwards following full moon is lower than that on the nights following a new moon because of an increased risk of predation by visual hunters. Descend is related to water transparency. The zooplankton descend deeper on days with higher water transparency.

8.3.4. Surface algal grazing

Mobile marine herbivores are gastropod mollusks, echinoderms, sea urchins and fishes. Most of them eat biofilms of microalgae or young macroalgae. Fishes such as grey mullet, *Liza* spp., scrape the algae at high tide. In the Great Barrier Reef, the parrot fish grazes between 40 and 100% of a square meter area of reef crest within a month's time and the rabbit fish removes 25 cm^2 of algae in the same time and place. Fish herbivory shapes the coral reef algal communities. The grazers also eat 3 to 100% of seagrass net primary production. Dugongs are the biggest grazers of seagrass and the seagrass meadows are essential for dugong survival. The grazing effect of sea urchins is important and it has significant effect on macroalgae especially encrusting coralline algae. This type of grazing is important to maintain stability of coral reefs; otherwise these hard substrates will be totally colonized by seaweeds. Increasing herbivory will reduce algal dominance. High herbivore pressure is essential to enable coral reefs to recover from serious perturbations such as bleaching. A demerit of this grazing is bioerosion that is ingesting the coral material along with algae. An analysis of gut of sea urchins in Tanzanian coral reef has showed that approximately 80% of the contents are calcium carbonate derived directly from dead coral and also from eroded sediment trapped in the algal mat.

Dimethyl sulphide (DMS) is produced by marine algae and it is a major source of atmospheric sulphur. Marine organisms generate about half of the biogenic sulphur emitted to the atmosphere annually. The majority of this sulphur is produced as DMS, which is then rapidly oxidized to sulphur dioxides. These function as cloud condensation nuclei leading to raining and thus the production of DMS can potentially affect climate on a global scale. The precursor to DMS is dimethylsulphoniopropionate (DMSP), a cryoprotective and osmoregulatory compound present in many marine algae. DMS is produced as a result of cleavage of dimethylsulphoniopropionate by the enzyme DMSP lyase. Many factors are responsible for the release and transformation of DMSP into DMS and they are algal cell senescence, water column bacteria, phytoplankton-derived DMSP lyase and viruses. DMS is released when urchins feed on algae containing DMSP and that it functions as an herbivore deterrent.

- Primary consumption by detritivores and herbivores is one of the major components of marine food webs.
- Primary consumers feed on primary producers such as phytoplankton, macroalgae, and bacteria.
- The primary consumers are phytoplankton feeders, algal grazers, filter and suspension feeders, sediment and deposit feeders.
- There are two interesting aspects of herbivory: (1) vertical migration of zooplankton for getting food and to escape from predators and (2) release of Dimethylsulfide (DMS) from marine algae during algal grazing. This DMS is rapidly oxidized to sulphur dioxides and these function as cloud condensation nuclei leading to rain formation and the production of DMS can potentially affect climate on a global scale.

8.4. PREDATION, PARASITISM AND PATHOGENESIS

The secondary consumer is an organism that eats all or part of a primary consumer. They are basically carnivores of three types: predators, parasites and pathogens. Predators are a type of carnivores that kill their prey before eating it. Parasites are also carnivores, but they may not kill their host. Pathogens are a type of parasites. Scavengers are also a type of carnivores that eat dead animal.

8.4.1. Predation

The oceans are rich in predators ranging from the smallest zooplankton copepods to the biggest killer whales and giant squids. The predators may be sessile and sedentary (tube worms and sponges) or mobile (arrow worms and mantis shrimps). The predators may be benthic (flatfish and whelks) or pelagic (tuna fish). Conus shells and sea snakes are active hunters and sea anemones and octopus are slow predators.

Predatory behaviour changes between day and night. Lobster or eel fish will hide in a crevice during the day and catch food; and they come out of crevices in the night times to seek for prey. Coral species use their polyps at night to catch small living and dead organisms, whereas during the day, they rely on their symbiotic zooxanthellae to get the products of photosynthesis. Whelks are predatory gastropod mollusks drill holes with the help of radula in the shells of their prey, through which they are able to feed on the flesh of the prey such as barnacles or periwinkles.

There are heavy predations occurring with some species of nudibranch which eat sponges. The prey (sponge) and the predator (nudibranch) species are dispersed by planktonic larvae, however, the predator has a great potential for local increase when the prey is abundant, resulting in heavy predation pressure on prey.

Many predators in one place may result in competition for prey and decrease hunting efficiency. This will be overcome by several mechanisms, such as prey size, species selectivity, changing preference to prey, foraging for prey in different localities or depths or simply emigrating from a high density prey patch into other patch with fewer preys.

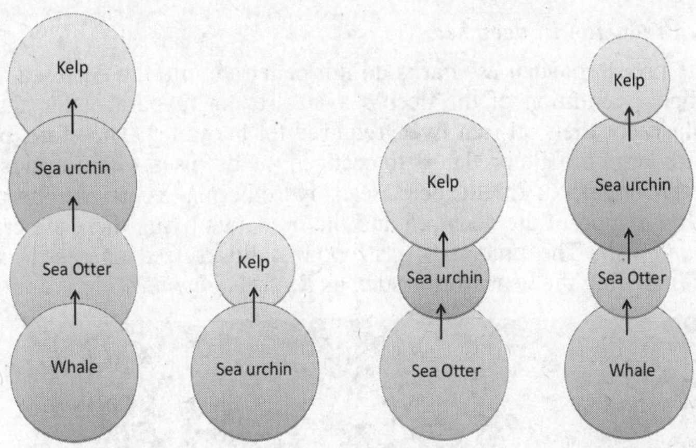

Fig. 8.6: The link between whales, sea otter, sea urchin and seaweed (arrow marks exhibit the feeding, population of individual group is indicated by relative size of circles)

There are links between killer whales, sea otters, sea urchins and seaweed (Kelp). The sea urchins reduce kelp by grazing. The sea otters eat urchins, helping more kelp to recruit and grow, but whales eat otters so that more urchins survive, and hence less kelp is produced. The net result of this process is if urchins are allowed to proliferate, seaweed will disappear. If grazing pressure of urchin is reduced by increased predation by sea otter, there is a recolonization by seaweed (Fig. 8.6). Other example is that the removal of large predators such as puffer fish and spiny lobsters increases attacks by the gastropod snail on gorgonian corals. One more example is the link between the reef fish and the crown-of-thorns starfish that eats live coral. Under normal condition, there is high species richness of predatory reef fish. These feed on the starfish that eat live corals and this in turn, allows live hard corals to proliferate. Due to the fishing pressure, the density of predatory fishes reduces, as a result, the density of starfish increases leading to loss of coral reefs. In the natural environment, there is a balance in population of the prey and predator. Another example is that the tiger shark preys on smaller sharks and rays which in turn, eat scallops. Removing an entire functional group such as the great sharks due to large scale exploitation around the world leads to increase of ray and small sharks, and decline of scallop-fishing industry.

8.4.1.1. Predation in deep sea

Large predators such as sharks do not penetrate into the deep sea. The oligotrophic condition of the deep sea is also not favourable for sharks particularly for their oil-rich liver, required for buoyancy. Therefore, bony fishes are large mobile predators to occur in the deep sea. Giant squids are also active predators in the deep sea. Hydrothermal vents are the only productive region of the deep sea and the predators living there are crabs, snails and fish. The predatory gastropod mollusks feed on sessile vent species including the vent worms such as *Riftia pachyptila* (Fig. 8.7).

Fig.8.7: Vent worms (*Riftia pachyptila*)

8.4.2. Parasitism

There are over 29,000 species of marine fish, but there are 100,000 fish parasite species. In other words, parasitism is an important issue as every one of fish is contracted with about four species of parasites. All taxa of animal parasites are present in the sea. Both ectoparasites (sealice and leeches) and endoparasites (flukes, nematodes and tape worms) are present in the sea. The world record for the size of a parasite is tapeworm with 30 metres in length recovered from the guts of adult sperm whales. In the Baltic Sea, the maximum infection by a parasite is as high as 50%. The dolphins are infested with endoparasites. Marine invertebrates have their own parasites, but also act as intermediate hosts for the life cycles of vertebrate parasites. Animals of the deep sea do exhibit parasitism. Crabs from hydrothermal vents in the Pacific Ocean are infested by nemertean worms situated on different parts of their bodies. Marine vertebrates have a nematode parasite namely *Anisakis simplex* in their guts. They are found in cetaceans (whales and dolphins), pinnipeds (seals and walruses) and fish eating birds as main hosts. Crustaceans such as shrimps, fish, squid and other macro invertebrates are the intermediate hosts. This will pass on to land through the consumption of raw fish by humans. This causes medical disorder known as 'anisakiasis' or 'anisakidosis', with the symptoms of stomach pains, diarrhoea, nausea, vomiting and allergic reactions like asthma. The endoparasites of all taxa can cause serious damage to their hosts. For an example, in the bull sharks digestive tract is affected with acute necrosis and gastroenteritis. Some copepods also infest the gills of fish such as gopies often with 100% prevalence, especially in tropical waters. These parasites reduce growth rate of host. The most important crustacean ectoparasites are sea lice mainly in the genera *Caligus* and *Lepeophtheirus* and they cause economic loss to the farming of salmonid fish such as salmon and trout.

Do marine organisms produce chemical weapons?

Biofilm protects bacteria from predation (grazing). The bacterial cells with biofilm produce anti-grazing chemicals such as alkaloid (violacein) which at nanomolar concentration inhibits the grazing of protozoa.

Octopuses eject out a cloud of purplish ink from its funnel. The ink is composed of melanin pigments derived from the amino acid tyrosine, released with a mucous binder. The ink makes the water dark and helps the octopuses to escape from predators.

Tetrodotoxin (TTX) is a toxin present in marine animals such as puffer fishes (Fig. 8.8), sunfishes, Xanthid crabs, sea stars and marine snails. Main function of TTX is in defense against predators and the toxin is also used to attack other animals.

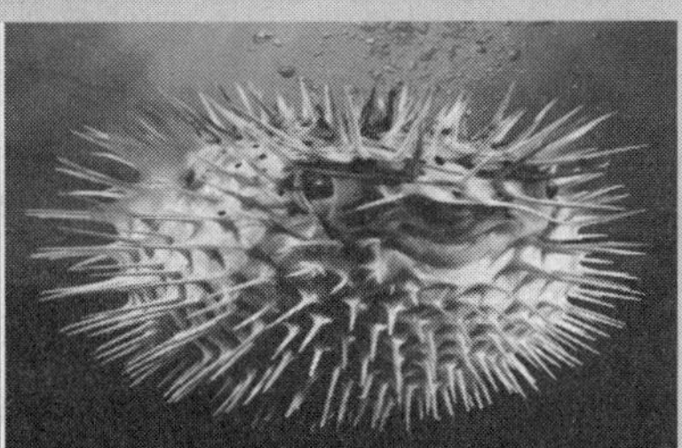

Fig.8.8: Puffer fish

(http://www.kenrockwell.com/trips/2010-07-maui/28b.htm)

The blue ringed octopus uses TTX to paralyse its fish prey. A tropical flat worm uses TTX to paralyse its prey of cowrie mollusks. This toxin is likely to be synthesized by symbiotic bacteria.

The animals that contain TTX are resistant to the toxin due to the presence of sodium channel protein that has single point mutations in its amino acid sequence.

8.4.3. Pathogens

Pathogens are biological agents such as viruses, fungi protozoa, bacteria that cause diseases. Seafans (gorgonids) are known to be attacked by fungal pathogens belonging to the genus *Aspergillus*. Coral reefs are infected by diseases. The white syndrome affects corals belonging to the genus *Acropora* in the Indo-Pacific region including the Great Barrier Reef and the coral tissue rapidly disappears at the rates of up to more than 100 cm^2 per day, leaving exposed coral skeleton. In the Caribbean, two coral diseases, white plague and black band diseases are of wide occurrence. Black band disease is prevalent when water temperatures are highest. The primary cause of the coral diseases is the bacterial infection by vibrios or alpha-proteobacteria or filamentous cyanobacteria, sulphur oxidizing bacteria or a consortium of microbes. However, there are genotypes of corals resistant to diseases.

- The secondary consumer is an organism that eats all or part of a primary consumer. They are basically carnivores of three types: predators, parasites and pathogens.
- Predators are a type of carnivores that kill their prey before eating it. Parasites are also carnivores, but they may not kill their host. Pathogens are a type of parasites. Scavengers are a type of carnivores that eat dead animal.

8.5. MARINE FOULING AND BORING

8.5.1. Marine fouling

The hard substrates exposed to seawater are subjected to the settlement of a large number of sedentary organisms called 'foulers'. The fouling organisms are algae and almost all the invertebrate groups such as protozoa, porifera, colenterata, polyzoa, annelida, crustacea, mollusks and the tunicates. Among crustaceans, the barnacles are the most dominant with about 100 species among the fouling community (Fig. 8.8). Next to barnacles, annelids (serpulids) are an important component among the fouling communities. In the fouling community, mollusks are represented by *Mytilus, Modiolus* and *Perna* and the mussels are the most important foulers.

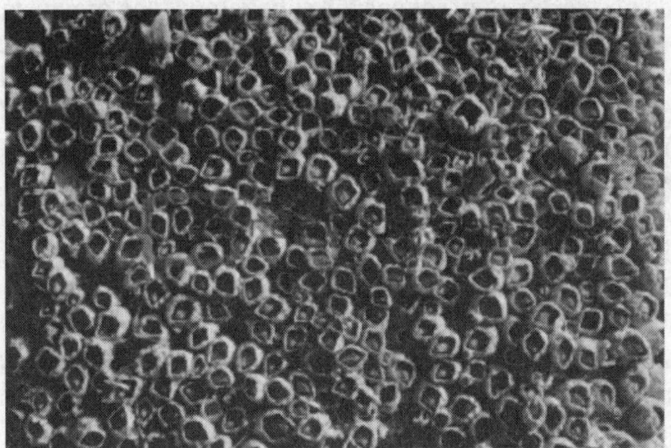

Fig. 8.8: Barnacles, a predominant fouler

The first and foremost process in biofouling is the formation of the primary film. This is essential for many foulers for settlement of their larvae. The primary film is composed mainly of diatoms, filamentous algae and bacteria, which serve as food for the larvae of foulers and substrate for firm attachment. The early colonizers of the primary film are mussels, bryozoa, hydroids, barnacles, tube worms or tunicates. The ultimate outcome is the establishment of a complex 'fouling community'.

The larvae settle under different tidal, temperature and light conditions. Some species of barnacles settle during day time on dark surfaces whereas cyprids prefer a low diffused light intensity. In tropical waters, fouling occurs throughout the year, while in temperate waters, it is more significant during the warmer months. The fouling also depends on reproductive cycles of the species. A decrease of foulers is usually leading to an increase in borer activity.

Fouling organisms are of great economic importance, since fouling of ships results in increased resistance to movement in water, reducing the navigation efficiency, lowering of speed and increasing fuel consumption. The underwater structures such as piling, pillars, buoys, cables and equipment are destroyed by the fouling organisms. The flow of seawater is interrupted by the growth of the foulers inside the pipes that are used to transport saltwater in ships, industrial plants, oil refineries, nuclear power plants and aquaria. The removal of the fouling organisms is a costly practice. The most extensively used antifouling toxic substance is copper. This is incorporated as the cuprous oxide in the antifouling paint. Chlorination is the best treatment for condenser system where painting inside narrow pipes is difficult and expensive.

8.5.2. Marine boring

Boring organisms are able to excavate hard substrates available in the sea. Boring is performed mainly by mechanical means or through the help of an acid substance which softens the substratum. The organisms involved in the boring activity are green, blue-green, red algae and fungi among the plants; and, porifera, annelids, arthropods, mollusks and echinoderms among the animals. Based on the substrate preference, the boring organisms are grouped into three categories: (1) rock borers, (2) coral reef borers, and (3) wood borers. Regarding the rocky borers, especially limestone rocks in the intertidal belt are riddled with holes and tubes by marine invertebrates such as sponges, polychaetes, sea urchins and mollusks. Regarding the coral borers, a variety of boring animals bore either on dead or live coral reefs and they are sponges, polychaete (Eunice), sipunculids, cirripedes and bivalve especially the giant clam, *Tridacna*.

Among the borers, wood borers are important in causing extensive damage to underwater woody structures. The chief wood borers are of two groups of marine animals: (1) bivalves and (2) crustaceans. There are only two families of wood-boring bivalves: (1) Teredinidae (shipworms, example: *Teredo, Lyrodus, Bankia*) and (2) Pholadidae (example: *Martesia* and *Lignopholas*). Among crustaceans, the most important ones are the isopods *Limnoria, Sphaeroma* and the amphiphod *Chelura*. In shipworm, infestation takes place when they are in the free swimming larval stage and the adult is confined to the burrow, but in the crustaceans, it is the adult that infests new wooden structures. Of the wood-borers, the shipworms are most important in causing extensive damage to wooden structures by excavating long, deep tunnels into the timber. The burrow is lined with a calcareous secretion. Boring by shipworm is made by the rasping movements of the shell valves, as operated by contractions of the adductor muscles. The foot serves to hold the animal in position in the burrow during boring. A constant hydrostatic pressure is maintained in the mantle fluid. The shipworms penetrate deep into the wood and making it porous and fragile, while the crustacean borers destroy the surface of the wood by producing minute holes. In general, the organisms bore wood for their protection and nourishment. Cellulose digestion in the shipworms is carried out by cellulolytic bacteria present in the gut of the shipworms. In tropical waters, the most destructive crustacean borer is the pill bug *Sphaeroma* and *Limnoria*. They make burrows with the help of their mandibles. They breed rapidly and spreading fast as a pest all along the coasts. Besides the wood boring animals, wood infesting bacteria and fungi especially of the Ascomycetes and the Deuteromycetes actively participate in the wood destruction. The fungi digest the wood and facilitate the attack by crustaceans and mollusks. The wood infested with fungal spores provides nutrients for fast reproduction of the wood boring organisms. On the positive side, marine wood borers do play a useful role in cleaning the coastal habitats by decomposing the trash wood materials.

- The hard substrates exposed to seawater are subjected to the settlement of a large number of sedentary organisms called 'foulers'. The most important fouling organisms are algae and almost all the invertebrate groups.
- The primary process in biofouling is the formation of the primary film, essential for many foulers for settlement of their larvae. The primary film is composed mainly of diatoms, filamentous algae and bacteria, which serve as food for the larvae of foulers and substrate for firm attachment.
- Fouling organisms are of great economic importance, since fouling of ships results in increased resistance to movement in water, reducing the navigation efficiency, lowering of speed and increasing fuel consumption.
- Boring organisms are able to excavate hard substrates mainly by mechanical means or through the help of an acid substance which softens the substratum.
- The organisms involved in the boring activity are green, blue-green, red algae and fungi among the plants; Porifera, annelids, arthropods, mollusks and echinoderms among the animals.
- Based on the substrate preference, the boring organisms are grouped into three categories: (1) rock borers, (2) coral reef borers and (3) wood borers.

8.6. COMMENSALISM AND SYMBIOSIS

8.6.1. Commensalism

In this association, at least one partner is benefited without severely harming the other. The commensalism also includes other associations such as epizoism, endoecism and inquilinism. In commensalism, food is obtained with the help of the hosts and in this aspect, commensalism differs from epizoism, endoecism and inquilinism. The sucker fish attach with the dorsal surface of the head of sharks and other large fishes through sucking disc, a modified form of dorsal fin of the sucker fish. When the host obtains food, the sucker fish releases its hold and swims forward to snatch the food. Spider crabs and sea-anemones are living in close association. The spider crab gets protection and shelters under the tentacles of the anemones and the anemone gets food from the claws of the crab. There is a permanent association between the damsel fish *Amphiprion* and large sea anemones (Fig. 8.9). The fish are comfortable to live among the stinging tentacles of anemone for protection and the anemone by using the tentacles get food material for the fish. A large number of polychaetes are in association with echinoderms and they exhibit food sharing relationship.

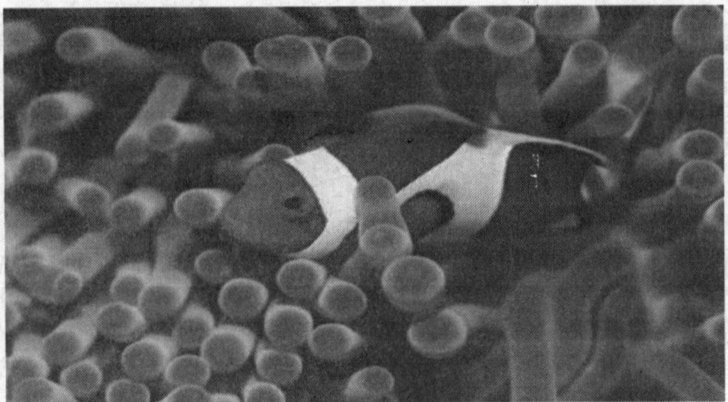

Fig.8.9: Clown fish (*Amphiprion*) and sea anemone

8.6.1.1. Epizoism

Many sedentary marine animals are epizoic attached on hard shells. Ciliates, hydroids, bryozoans, polychaetes and cirripeds attach themselves on gastropods, crustaceans or on the calcareous tubes of other sessile forms. There are also unattached forms such as copepods which roam about on the surface of a variety of invertebrates.

8.6.1.2. ENDOECISM

It is a partnership in which one animal takes shelter within the tube or burrow of another animal for protection and food. An example of commensal is nereis worms which live in the shells inhabited by hermit crabs. In this association, the worm occupies the upper coils of the shell. When the crab is feeding, the worm moves forward and extends its head outside the shell to get food from the crabs's mandibles.

8.6.1.3. Inquilinism

In this type of commensalism, one lives within the other for protection. This association is stronger than endoecism and epizoism. Some teleost fishes live inside sedentary animals. The pearl fish dwells in the intestine of a sea cucumber. Young fishes enter through head, but adults through cloaca pore of sea cucumber. Some cardinal fishes live in association with large conch shells of the genus *Strombus*. They find shelter in the mantle cavity and emerge out of the host during hours of darkness.

8.6.2. Symbiosis

Symbiosis is a partnership between two different types of organisms for mutual benefit. There are two types of symbiosis: (1) 'physiological symbiosis' a strong association and (2) 'cleaning symbiosis' a loose association.

Physiological symbiosis is the true symbiosis. The association between the two organisms is so strong that one or both of the partners may not survive alone. The best examples of such an association in the marine environment are that between certain algae and a number of marine invertebrates such as protozoan, sponges, coelenterates, tuberllarians and mollusks. The association is beneficial to both the partners. The algae benefit by deriving protection and procuring carbon dioxide and nitrogenous products from the host for performing photosynthesis, while the host obtains additional oxygen supply through algal photosynthesis and removes the wastes with the help of algae. The photosynthetic activity of algae also helps in calcium deposition in reef-building corals.

Animal changes behaviour for algal partner

Jelly fish (*Cassiopea xamachana*; Fig.8.10) usually lies up side down on the seafloor in shallow water exposing algal symbiont (Zooxanthellae) to light.

The giant clam, *Tridacna* contains millions of zooxanthellae located within their siphon tissue and the siphon is projected upward to sunlight for photosynthesis by the algal symbiont.

Fig.8.10: Jelly fish (*Cassiopea xamachana*)
(http://www.visualphotos.com/image/1x9092970/upside-down_jellyfish_cassiopea_xamachana_with)

Cleaning symbiosis is a relationship in which the cleaners remove epizoites, parasites and decaying tissues from the hosts in order to get food from the cleaned material. Thus this association is a mutually beneficial; the hosts are cleaned while the cleaners get food. Many fishes, turtles and sea-urchins are cleaned by fishes, shrimps and crabs. The cleaner shrimps are abundant in tropical oceans, adapted for cleaning to have a long snout. Most of the cleaners are brightly coloured to attract the hosts. The goby fish *Elacatinus ocenopsis* is a common cleaner found in reefs. This small, striped yellowish fish has a sucking disc formed by fused pelvic fins. The goby enters freely in to the mouth and gill cavity of many host fishes such as *Epinephelus striatus, Pomacanthus paru, Pomacentrus* sp., and *Acanthurus* sp.

Microbial associations are interesting and they occur in almost all the marine organisms. To cite a few: (1) intracellular bacteria in *Noctiluca scintillans*, causing red tides, (2) luminous vibrios in light organ of squids, light producing fishes and mammals, (3) nitrogen-fixing bacteria in boring mollusks, (4) endosymbiotic bacteria with marine nematodes. The remarkable event in the history of marine microbiology occurred in the year 1977 on discovery of microbes in hydrothermal vents at the sea bottom. The microbes living there are reported to be chemoautotrophic bacteria utilizing hydrogen sulphide and endosymbiotically associated with a giant worm *Riftia*.

- Commensalism is an association in which at least one partner is benefited without severely harming the other. The commensalism also includes other associations such as epizoism, endoecism and inquilinism.
- Symbiosis is a partnership between two different types of organisms for mutual benefit. There are two types of symbiosis: (1) 'physiological symbiosis' a strong association and (2) 'cleaning symbiosis' a weak association.

8.7. COMPETITION AND SUCCESSION

8.7.1. Competition

Competition occurs within a trophic level, whereas predation, parasitism and grazing take place between trophic levels. Competition is an interaction that takes place between individuals for the purposes of space, food, refuges, mates and territories. As long as these requirements are sufficient, there is no competition and the species coexist in the same habitat. Only when there is a short supply of space and food, then the competition occurs. This competition may occur with members of their own species ('intraspecific competition') or with members of different species ('interspecific competition'). An example for interspecific competion is that the barnacles compete for space and food with another filter feeder such as a sponge or a fan worm. This competition is prevalent as organisms enter a new habitat, and sequentially replace each other and this is termed as 'ecological succession' and this leads to the climax finally.

8.7.1.1. Intraspecific competition

This occurs between individuals of a single species for a limited resource. Intraspecific competition regularly occurs in the long-terms, in the absence of other species. For example, macroalgal settlers often colonize rocks in large numbers, grow more slowly and experience high mortality. This mortality rate is higher at high nutrient levels, because the young algae grow faster and compete more for space and light. For sessile organisms, competition for space becomes more intense with age and the size. The intraspecific competition is reduced by changing the feeding habit of the competitors. The common periwinkle *Littorina littorea* is an algal grazer which can change its feeding as scavenger when dead animal material is available.

8.7.1.2. Interspecific competition

Interspecific competition is competition between different species. In many cases, interspecific competition occurs between multiple species, especially if they are closely related, such as coral reef fish. The exotic Pacific oyster is able to overgrow the native rock oyster at low and mid intertidal levels because of its 60% faster growth rate and at the high intertidal level the Pacific oyster suffers up to 80% mortality. Therefore, the native species is likely to be replaced by the exotic species in the lower intertidal, but the native oyster *Saccostrea* can hold its own in the higher intertidal region. The coral reefs are systems with low nutrients and therefore, food limitation may result in weak interspecific competition but strong intraspecific competition. Bryozoa are often competing with corals. Bryozoa is able to overgrow corals when coral is weak by bleaching. Competitors at different trophic levels interact with each other to share resources. This competition is avoided by (a) fish such as salmon migrating away, in response to declining food supplies; (b) fish such as sand lance change food prey from zooplankton to epibenthic species; (c) fish such as herring feed at different depths at different times of day.

8.7.2. Succession

Succession is the process by which an area is colonized by a sequence of species. This leads to stable climax community. In the case of plankton, the succession in temperate and Mediterranean regions shows annual cycles. During spring and summer, species of phytoplankton replace one another. In the case of soft sediments, the species richness increases with succession stages, the abundance of single species is most pronounced earlier in succession, and peaks at the intermediate level of succession. Succession in sand and mud can be inhibited by certain groups of organisms such as lugworms through a process called 'bioturbation'. The lugworm, *Arenicola marina* is sediment-feeding polychaete annelids that live in U-shaped tunnels in intertidal beaches. They are bioengineers inhibiting succession.

8.7.2.1. Succession in deep sea

Succession in the deep-sea occurs in hydrothermal vents and cold seeps. Hydrothermal vents are colonized by the large giant tube worms such as *Riftia* and *Ridgia* species, or vent mussel, *Bathymodiolus* species and these are pioneer species that provide community structure and habitat diversity. To tube worms, 24 invertebrate species are associated, and they are gastropods, polychaetes, few crustacean and bivalves. The aggregations of species associated with *Riftia* communities vary with productivity. Artificial polyvinyl chloride (PVC) tube clusters which mimicked young tube worm communities showed only low productive communities one year after the experimental set up. Thus, the succession is not just on physical complexity, but also nutrient load. Whale-falls are sources of organic enrichment on the deep-ocean floor. Once this large whale reaches on the seafloor, other species accumulate on the carcass over a period of months and years.

- Competition occurs within a trophic level; in contrast predation, parasitism and grazing take place between trophic levels.
- Competition is an interaction that takes place between individuals for the purposes of space, food, refuges, mates and territories. As long as these requirements are sufficient, there is no competition and the species coexist in the same habitat. Only when there is a short supply of space and food, then the competition occurs.
- This competition may occur with members of their own species ('intraspecific competition') or with members of different species ('interspecific competition').
- This competition is prevalent as organisms enter a new habitat, and sequentially replace each other and this is termed as 'ecological succession' and this finally leads to the climax.

8.8. DISPERSAL AND SETTLEMENT

The success of any species or population is based on egg production, juvenile dispersal, settlement and subsequent recruitment. This is therefore, an important aspect of functional ecology. A large number of marine organisms produce larvae in their life cycle. Larvae are morphologically different from the adult and the larvae undergo a profound change in morphology when they become adult. There are three patterns of development in marine organisms: (1) planktotrophic larvae, (2) lecithotrophic larvae, and (3) direct development forming juveniles or non-pelagic larvae (Table 8.2). Many marine benthic communities are established by larval settling from the water column. The larval settlement depends on suitable substrates. Many larvae settle in the living place of adults and this is attributed to the chemical attractant namely 'pheromone' secreted by the adult. Larvae respond to physical factors, such as light, pressure and salinity. Many free-floating larvae are positively phototactic in the early stages of their larval life. This condition ensures that they will reside in the upper layer of waters where dispersal is greatest. At the time of settlement, the larvae become negatively phototactic and migrate towards the bottom. To ensure settlement into the proper environment, larvae respond to various environmental stimuli, of which some chemicals are most influential.

Table 8.2: Different patterns of development in marine organisms

Aspects	Planktotrophic larvae	Lecithotrophic larvae	Direct development of juveniles or non-pelagic larvae
No. and size of eggs produced	Great in number, Very small in size	Few eggs with large size	Few eggs with large size
Yolk content of egg	little	high	Large
Nutrition	Feeding on plankton	Not feeding on plankton	Not feeding on plankton
Duration of plankton life	Long time	Less time	no free swimming larval stage
Advantage	Large number of young ones are produced; high dispersal rate	Less predation	Planktonic mortality is absent
Disadvantage	High predation, low survival and recruitment	Poor dispersal due to a short time in plankton	Dispersal is absent
Dominant Occurrence	Tropics	Polar waters	Polar waters
Most common occurrence	Shallow waters	Deep sea	Deep sea

Sex reversal in fish

Change in sex in marine fish is interesting. Many fish start life as males and turn into females as they grow (e.g. the giant clams *Tridacna gigas*, clown fish *Amphiprion*), but fish such as groupers (*Epinephelus*; Fig. 8.11) begin life as females and turn into males as they grow.

Fig.8.11: *Epinephelus* sp.
(http://blogs.crikey.com.au/firstblog/2008/10/26/potato-cod-epinephelus-tukula/)

8.8.1. Dispersal

A species or population disperses from one place to another for many reasons: genetic mixing and out-breeding, reduction in competition, colonization in new habitats, avoidance of predators and parasitic infections. Many marine species have pelagic larvae which spend different times in the form of plankton, while others have no free-living larval stage. Female octopus lays eggs that hatch into small octopus, and there is no larval stage. Many sharks give birth to fully formed pups. The organisms without larval stages have the risk of dying from predation. There are three stages of movement by dispersal: (1) departure stage (emigration), (2) travelling stage, and (3) settling stage (immigration).

8.8.1.1. Departure stage

Brooders and broadcasters are two categories, depending on how they care for their young ones before departure from the parent. The eggs and the larvae hatching from the corals are brooded inside the parent coral. Broadcast spawners may release hatched larvae or fertilized eggs which hatch in the water or they may even shed sperms and unfertilized eggs separately into the water where fertilization takes place, away from the adult. The resulting larvae from brooders or broadcasters may spend a considerable time in the water and travel long distance before settling. In the pelagic environment, huge losses of eggs are expected due to predation. Therefore, large numbers of eggs are produced to ensure that at least a few survive in the larval stage to become juveniles. An adult barnacle produces between 20,000 and 30,000 eggs. Deep water broadcasters produce much larger eggs than similar species occurring near the surface. Broadcasters release the eggs into the water (pelagic) or lay close to the substrates (demersal). The pelagic eggs have long range of dispersal, but demersal eggs have limited dispersal. Most benthic marine invertebrates are sedentary or sessile as adults, so their pelagic larval phase is the primary means of dispersal.

There are two types of marine larvae: the planktotrophs have to feed on plankton while the lecithotrophs need not feed and rely on the energy stores. The lecithotrophic larvae do not have food requirement, hence, they are able to survive in nutrient-poor or very cold environment such as Polar Regions. Planktotrophic larvae require more protein, and hence actively feeding, while lecithotrophic larvae require more lipid in particular triglyceride. This requirement of lipid is more in the larvae of plankton species rather than benthic species.

8.8.1.2. Spawning and hatching

Mobile species such as fish, squids, sea stars and spider crabs are forming large aggregations and simultaneously release eggs and sperm for external fertilization into the water. The timing of this aggregation occurs at the phase of the moon, tidal height and water temperatures. One advantage of releasing larvae at night is their protection from visually hunting predators, while a high tide would disperse the larvae away from their site of release. Many coral species undergo external fertilization, and so have to broadcast their unfertilized eggs and sperm simultaneously as a mass spawning event at one particular time of year especially in late spring and early summer at last quarter of moon. The bryozoans release larvae once in a year as water temperature rises.

8.8.1.3. Travelling

Once larvae are released from the parent or hatched from the egg, the larvae have to travel to a sufficient distance to find a suitable place for settlement and recruitment as a juvenile. This depends on larval duration and horizontal and vertical movements in the water and also on retention of larvae where they are kept closer to their place of release.

Planktonic larval duration (PLD) is extremely variable. Sponges have larvae which exist for 2 or 3 days, while the flat fish has PLD of 40 days. In the damselfish on the Great Barrier Reef, PLD varied from 0 to 28 days. PLD depends on environmental factors such as food availability and water temperature. Warm water and food abundance accelerate larval development, and consequently PLD will be lower. Coldwater increases PLD. The PLD influences population genetics. A long PLD provides the potential for long distance dispersal, and so genotypes are able to mix freely.

The dispersal of eggs or larvae is determined by current speed and direction. Diel vertical migration is very common amongst marine planktonic organisms. Zooplankton are able to swim up and down in a water column according to environment conditions such as light, temperature, phytoplankton abundance and the presence of predators. So the dispersal of propagules in the sea is a combination of active swimming and passive drifting of vertical and horizontal movements.

8.8.1.4. Retention

Many marine fish and corals remain in the vicinity of their site of origin. This is a process known as 'retention'. The planula larvae of the octocoral are brooded inside the parent, and only spend a few hours after release before settling back on the substrate.

8.8.1.5. Connectivity

Connectivity is the potential for populations at isolated sites to be linked with each other. This connectivity has significance for species and populations. Their nursery and adult grounds are spatially separated, but interconnected. Marine populations with dispersive stages are considered as meta-population with interconnected sub-populations. In healthy ecosystems, many habitat patches are connected by dispersal of the larvae of both sessile and mobile adults of a species easily moving between patches. Coral reef system is an example. Connectivity reduces between individual patches of corals, when the patches disappear due to pollution, disease and storm damage. Species with limited dispersal are vulnerable to recruitment failure.

8.8.2. Settlement

Young organisms settle in the habitats where they can recruit into the adult population, and begin the life cycle again. This settlement takes place in the most suitable habitats at the most appropriate time and in the optimal environmental conditions for better survival and establishment of marine organisms. Settlement timings of the marine larvae at the end of travelling is long or short, and it depends on biological (presence of predators or competitors), physical (wind, tides or currents) or chemical (water or substrate chemistry), or a combination of all the three.

Juvenile crabs settle on complex habitats such as mussel beds or algae or sand. This recruitment reduces the risk of predation by providing the hiding place for young settlers. The coralline algae produce and support biofilms of bacteria which attract settling animals. Some urchin prefers to settle in the seaweeds *Sargassum* sp., and *Dictyota dichotoma*. The biofilms are produced on the surface of the seaweeds, and these are used as food by the urchins.

Barnacle cyprid larvae prefer rocky settlement sites especially the rougher and irregular rocky surface. Barnacle cuticles produce a settlement–inducing protein complex which is a glycoprotein. This acts as a pheromone to induce settlement on newly arriving cyprids. The coral larvae of *Acropora* species prefer living horizontal substrates rather than dead vertical ones. Reef fish larvae can detect and recognize the sounds of their home reefs and swim toward them. Thus the settlement of fish larvae can be influenced by sound.

- The success of any species or population is based on egg production, juveniles dispersal, settlement and subsequent recruitment.
- A species or population disperses from one place to another for many reasons: genetic mixing and out-breeding, reduction in competition, colonization in new habitats, avoidance of predators and parasitic infections.
- There are three stages of movement by dispersal: (1) departure stage (emigration); (2) travelling stage; and (3) settling stage (immigration).
- Young organisms settle in habitats where they can recruit into the adult population, and begin the cycle again. This settlement takes place at the most appropriate time, in the optimal environmental conditions to maximize survival, and in the most suitable habitats.
- Settlement timings of the marine larvae at the end of traveling is long or short, and it depends on ecological (presence of predators or competitors), physical (wind, tides or currents) or chemical (water or substrate chemistry), or a combination of all the three.

8.9. MARINE FOOD CHAINS

The primary source of energy is sunlight, utilized by autotrophs during photosynthesis for production of organic matter. This is then passed through the food chain. The shape of food chain is pyramid-like, since at each level, there is a loss of energy and materials. There are three pyramids: (1) pyramid of numbers, (2) pyramid of biomass and (3) pyramid of energy flow. The pyramid of numbers exhibits the reduction in the number of organisms through successive trophic levels. In marine environment, thousands of diatoms provide the food of a copepod, and hundreds of such copepods in turn, offer the food for a fish. Thus there is a reduction in the number of organisms. A similar trend is displayed by the biomass of organisms and the energy flow. In the case of the pyramid of biomass, 1000 kilograms of phytoplankton are required to produce 100 kilograms of copepods, and these in turn produce 10 kilograms of smaller fish which in turn produce 1 kilogram of larger fish. In the case of the pyramid of energy flow, the energy utilized by the organism can be measured from the oxygen consumed during respiration and this energy is reduced during its transfer through different trophic level. The efficiency of energy transfer is low when steps involved the food chain are more complex. This loss of efficiency with increasing steps in the food chain is largely due to a small transfer of energy between the steps and a major amount of energy is used on the maintenance of the organism at each step.

One organism may belong to more than one trophic level, at many sites. Hence, ecologists prefer the term 'foodweb' to indicate the complex trophic interrelationships among organisms of an ecosystem (Fig. 8.12).

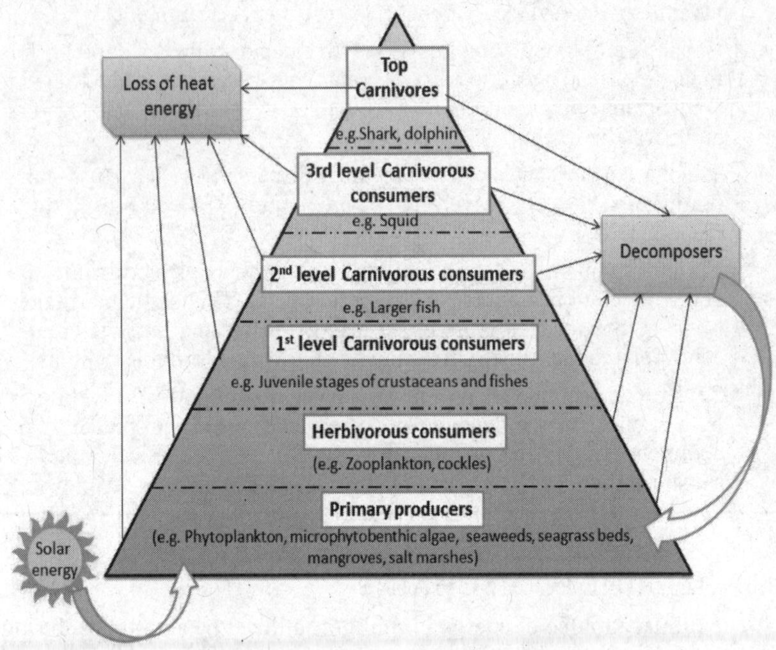

Fig. 8.12: A general marine food chains

The trophic structure of an ecosystem consists of four important components: (1) producers, (2) consumers, (3) decomposers, and (4) transformers. The primary producers are the autotrophs which produce organic compounds from inorganic constituents using the solar energy. The photosynthetic and chemosynthetic bacteria are also producers; the former use sunlight for energy, while the latter obtains energy by the chemical oxidation of organic compounds. The consumers are directly or indirectly dependent on the organic matter synthesized by the producers. There are different types of consumers: (1) primary consumers, (2) secondary consumers, (3) tertiary consumers and (4) top level consumers. The primary consumers are the small animals which feed on the autotrophs. The secondary consumers are carnivorous animals that eat the primary consumers. The tertiary consumers are the large carnivores that feed on other carnivores. The top level consumers are the largest animals, in particular marine mammals eating mostly on tertiary consumers. Besides these, there are other consumers such as parasites, saprophytes and scavengers. The decomposers are bacteria and fungi, playing an important role in the marine ecosystem. Bacteria are mainly involved in the decomposition of plant and animal bodies, while fungi are important in the breakdown of plant material. These heterotrophic organisms breakdown dead organic material and release inorganic nutrients. These nutrients are, in turn, utilized by bacteria and phytoplankton for their multiplication and this process of regeneration is known as transformation. In a balanced ecosystem, all the four processes namely production, consumption, decomposition and transformation of organic matter, occur in a cyclic pattern known as the 'organic cycle' or the 'food cycle' of the sea.

The trophic relationships that exist in the marine environment can be grouped under two components: (1) pelagic food chains, and (2) benthic food chains.

8.9.1. Pelagic food chains

The phytoplankters are the primary producers, representing the first trophic level in coastal waters, while picoplankton and nanoplankton constitute the first trophic level in the open ocean. They are consumed by herbivorous zooplankters and this is termed as 'secondary production' or 'zooplankton production'. The copepods and the euphausids are chief consumers of phytoplankters and they form the second trophic level in the food chain of the pelagic zone. The pelagic coelenterates, chaetognath, polychaete and pteropods are feeding mostly on copepods. They together with many fish larvae form the third trophic level in the pelagic food chain. The cephalopods and fishes are the predominant high level carnivores of the pelagic zone. Teleosts such as snappers and barracudas feed mostly on other carnivorous fishes. In addition, many sharks and toothed whales such as dolphins and porpoises feed mostly on fishes and squids, and thus they are at the top level of the pelagic food chain. In the pelagic zone, the organisms are densely populated in the upper epipelagic zone where primary production is high. The organisms in the deeper zones are carnivores or scavengers. There is a remarkable reduction in the number of organisms with increase in depth and this is due to non-availability of sufficient food to support them. There is a continuous horizontal transfer of nutrients, organic matter and organisms between habitats of the pelagic environment through water currents. This occurs between the freshwater and ocean through estuaries, the offshore and near-shore coastal waters, and the seagrasses and rocky reefs.

8.9.2. Benthic food chains

In the coastal waters, the bottom usually contains high organic matter. In contrast, the bottom of Deep Ocean has only small quantities of organic matter. The organic matter is largely dependent upon the depth and it decreases with depth.

On the rocky shore environment, larger algae act as primary producers. They are grazed by many mollusks and sea urchins. Many suspension feeders such as bivalves and sea squirts inhabiting the rocky shore trap the organic detritus by means of ciliary mechanisms. Carnivores of the rocky shore include sea anemones, Nereid, nudibranchs, whelks, star fishes, cephalopods and fishes which feed on a wide variety of invertebrates.

On soft substrates of sandy and muddy shores, the primary producers are diatoms and also chemosynthetic bacteria. The organic matter produced by these bacteria is consumed mainly by nematodes, which in turn are consumed by other organisms. Thus in the benthic zone, the nematodes take up the position of zooplankton of the pelagic zone. The consumers that occur in the soft substrates are microfauna, meiofauna and macrofauna. The micro and meiofauna consume on bacteria, diatoms and protozoans. The macrofauna such as polychaetes, crabs, bivalve mollusks and echinoderms feed on organic detritus either by filtering suspended organic particles or eating the sediments along with detritus. The carnivores are not many in sandy and muddy substrates. Some fishes, isopods and polychaetes are predators.

Coral reefs have the most complex food chain of the benthic zone. The corals are carnivorous feeding on zooplankton. The coral polyps are consumed by the butterfly fish, parrot fish, surgeon fish, puffer fish and also the crown-of-thorns star fish. The coral reef ecosystem is represented with the herbivores such as worms, crustaceans, mollusks, echinoderms, and some fishes and also with the suspension feeders such as crustaceans and bivalves. Most of the fishes inhabiting the coral reefs are carnivores. The coral reefs are also associated with carnivorous invertebrates such as sea snails, nudibranchs and star fishes. The predators are eels, barracudas, grey sharks, and octopuses which feed mostly on other fishes.

- The primary source of energy is sunlight, used by autotrophs during photosynthesis, to produce organic matter which is passed through the food chain. At each level, the reduction in biomass is 90% from that of the previous level. The shape of food chain pattern is pyramid-like, since at each level there is a loss of energy and materials. There are three pyramids: (1) pyramid of numbers, (2) pyramid of biomass and (3) pyramid of energy flow.
- The trophic structure of an ecosystem consists of four important components: (1) producers, (2) consumers, and (3) decomposers, and (4) transformers.
- The photosynthetic and chemosynthetic bacteria are producers; the former use sunlight for energy, while the latter obtains energy by the chemical oxidation of organic compounds.
- The consumers are directly or indirectly dependent on the organic matter synthesized by the producers. There are different types of consumers: (1) primary consumers, (2) secondary consumers, and (3) tertiary consumers.

- The trophic relationships that exist in the marine environment can be grouped under (1) pelagic food chains, and (2) benthic food chains.

Questions for Review/Exercise*

(1) Describe: Functional ecology, measurement of primary productivity, Blue carbon sink, Vertical migration of zooplankton, Inquilinism, Competition, Succession

(2) Explain the process of primary production through photosynthesis? Add a note on its importance in fisheries productivity.

(3) Discuss the significance of chemosynthesis in deep sea environment.

(4) List out the factors influencing primary production?

(5) Explain the mechanism of carbon sequestration in marine and coastal environments.

(6) Explain the proposals for increasing carbon sequestration in the ocean.

(7) What is the major source of atmospheric sulphur and how is it produced?

(8) Explain the link among whales, sea otter, sea urchin and seaweeds with suitable diagramme.

(9) Parasitism is an important aspect of functional ecology- Justify?

(10) How are marine fouling and boring processes important?

(11) What is symbiosis? Explain its types with suitable example.

(12) Discuss the dispersal and settlement processes of marine organisms.

(13) Compare the different patterns of development in marine organisms.

(14) Write a short note on marine food chains.

* Field visits to marine environment, estimate the primary productivity, observe on interactions among marine organisms and make a note of fouling, boring, larval settlement on different substrates.

Threats to Marine Environment 9

Marine environment is essential for existence and survival of human life. However, all the world's oceans are affected by human activities. Shallow coastal and estuarine habitats are much vulnerable to threats. In 2004-2005, three billion people have lived within 200 kilometres of the coastline that is almost half of the global population. In addition, 14 of the world's 17 largest mega-cities are located along the coasts. About 40% of these cities accommodate populations of 1 million to 10 million people, imposing pressure on the coastal resources for water, energy, space, sanitation and infrastructure. This results in depletion of marine natural resources. The marine environment is often subjected to natural or man-made threats. There are two types of threats: (1) sudden threats such as dredging, chemical explosives, storms, oil spill, solid waste disposal and tsunami and, (2) continuous threats such as over-exploitation of fishes, industrial and sewage pollutions, land reclamation, invasive alien species and climate change (Fig. 9.1). The impact of natural threats is severe, if temperature, salinity, sea level rise and solar-UV radiation act together.

Marine biodiversity loss is an important result of the threats. The global rate of biodiversity loss among the birds is more than doubled in 10 years between 1990 and 2000. The declines were severe in East Asia, South Asia and Pacific regions. Albatrosses are the most threatened sea bird. Out of 21, 19 species of the sea bird are now globally threatened and the remaining are near threatened. Many petrel bird species are also globally threatened. Both albatross and petrel species face threats at their breeding and feeding sites of the marine environment. The depletion of prey fishes affects the population of birds in the area. Apart from sea birds, the oyster reefs are one of the most affected marine habitats on earth. It has declined more than 90% from historic records. The oyster reefs are functionally extinct by more than 99% in 37% of estuaries and 28% of marine eco-regions. Ocean acidification will pose a significant threat to shellfish. There are many threats to marine environment and they are summarized (Fig. 9.5).

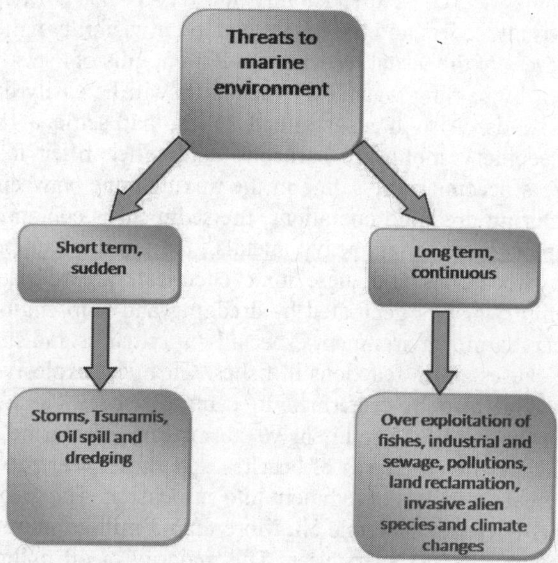

Fig. 9.1: Major types of threats to marine environment

- The marine environment is often subjected to natural or man-made threats.
- The impact of natural threats is severe, if temperature, salinity, sea level rise and solar-UV radiation act together.
- All the world's oceans are affected by human activities. The man continues to disturb the ocean through fishing, hunting, pollution, shipping, tourism and invasion of new species.
- Shallow coastal and estuarine habitats are much vulnerable to the threats.

- Marine biodiversity loss is an important result of the threats. Apart from sea birds, the oyster reefs are one of the most affected marine habitats on earth.

9.1. SUDDEN THREATS

The short term and sudden threats are caused by dredging, chemical explosives, storms, oil spills and solid waste disposal. Dredging can affect marine benthic communities and their habitats leading to the loss of feeding and breeding grounds. Fish and other organisms can be directly sucked into the dredger and killed. Dredging produces increased suspended sediment. This will reduce the visibility in the water body. Many estuarine fishes are visual feeders and hence, a reduction in visibility will reduce the hunting success of the visual feeders. In addition, gills of fish which obtain their oxygen by passing water over their gills will be easily damaged by suspended solids. Also the suspended solids that settle at bottom are affecting sedentary mollusks, barnacles and other filter feeders. The dredged areas become more saline in the mixing zone between fresh and seawater. During dredging operations, the sediment is contaminated with toxic compounds such as heavy metals, antifouling compounds and aromatic hydrocarbons and these toxic chemicals are released into the water column. Noise is generated by dredging and ship engines and the noise affects marine environment especially in estuaries and shallow seas. This noise causes alarm reactions in fishes. Chemical explosives are used in military and also by fishermen to capture fishes. These explosives damage the marine life. The intensive storms can kill marine organisms, but also change the structures of beaches and shores, destroy coral reefs and bring great quantities of sediment into suspension. The greatest release of pollutants is caused by crude oil. More than 3 million metric tonnes of oil contaminate the sea every year. The majority of oil pollution in the oceans comes from land. Runoff and waste from cities, industry and rivers carry oil into the ocean. Ships cause about one third of the oil pollution in the oceans when they wash out their tanks or dump their waste water. Spilled oil can harm the environment by causing physical damages to habitats and poison effect on marine life. Solid waste disposal is yet another threat that results in sedimentation and heavy metal pollution. Dredged material, fish processing waste, inert materials of natural origin, low level of radioactive waste and vessels or aircraft are now permitted for disposal into the sea.

9.2. LONG TERM AND CONTINUOUS THREATS

9.2.1. Over-exploitation of fishes

Technological advances in fishing gear and ships have resulted in the significant reduction in many fish stocks. The major fisheries are concentrated in the waters over-lying the continental shelves. The commercially important marine animals are drawn mainly from four groups: (1) bony and cartilaginous fishes, (2) marine mammals, (3) mollusks and (4) crustaceans. In the year 1997 the total marine fisheries landings were 85.6 million metric tonnes, but this figure was lower than the highest catch obtained in 1989. Most of the fish stocks on the continental shelf areas of the world are fully exploited or over-exploited. Overall, 80% of the world fish stocks are fully exploited or over-exploited. Only 20 % of stocks are moderately exploited or under-exploited. Most of the stocks of the top ten exploited species worldwide, together account for about 30% of the world marine capture fisheries production and these are fully exploited or over-exploited. The areas showing the highest exploited stocks are the Northeast Atlantic, the Western Indian Ocean and the Northwest Pacific. Therefore, effective management is highly important in saving the fish stocks of the world's ocean.

The maximum sustainable yield (MSY) is the largest number of fishes that can be harvested in the next year without reducing stocks. The estimated MSY from the world's oceans varies between 60 million tonnes and 290 million tonnes. Productivity has declined in all the major fishing areas of the planet due to over-exploitation and natural changes in the environment. The decline of the Peruvian anchovy's fishery is due to over-fishing combined with environmental changes known as 'El Nino'. The decline of the whales is due to over-fishing for the purpose of food, oil, cosmetics, candles and fertilizers. Fishing affects the marine ecological communities directly by removal of species. This reduces the species diversity, and alters the size, species structure, reproduction and life history characteristics of the fished populations and ultimately imbalances the marine ecology. Also the fishing has significant effects on the non-target species. Not only capture by the fishing gear, but also exposure on deck during commercial catch and sorting operations the individual fish may suffer due to thermal shocks and exposure to air. This can lead to high mortality rates among the discarded species. Thus by-catch is a source of destruction to marine communities largely to crustaceans. Fishing gear itself has a deleterious effect on the habitat. Capturing fishes for the aquarium trade has resulted in significant losses, particularly to coral reefs. Unlawful fishing and dynamite fishing can cause continued depletion of organisms.

Over-exploitation affects the socio-economic status of fishermen who are dependent on the marine environment. Small-scale fishers all along the mainland coast are displaced in many places due to industrial development and pollution. In many places, fishermen are unable to catch fishes especially in dead zones, which are formed due to dumping of sewage and industrial wastes near major cities. The small scale fishers use more nets and stay for more days at sea. Another problem is that more crew work on each vessel, so profit from each trip is reduced. In the mechanized trawl sector, depletion of local fish stocks has led to migration of crew to other places where they face risks such as low wages and conflict with local fishers. In many places, illegal fish catches are very high due to involvement of foreign trawlers (Fig.9.2).

- Advancement of fishing gear and ships has resulted in the significant reduction in many fish stocks. Fishing affects the marine ecological communities directly by removal of species. This reduces the species diversity, alters the size, species structure and reproduction of the fished populations, thereby affecting the marine ecology. The fishing has drastic effects on the non-target species largely to crustaceans. Fishing gear itself has a deleterious effect on the habitat. Unlawful fishing and dynamite fishing cause continued depletion of organisms.
- Over-exploitation of fishes affects the socio-economic status of fishermen. Small-scale fishers all along the mainland coast are displaced in many places due to industrial development, pollution and formation of new dead zones with poor catch of fish.

Fig.9.2: Over-exploitation of fishery resources

9.2.2. Industrial pollution

The industries such as power stations, chemical works, refineries and desalination plants draw large volumes of seawater and discharge it back into the sea. In thermal and nuclear power stations, the water is used for cooling purpose and the discharged water is about 10°C warmer than the normal temperature. This increase in temperature reduces solubility of gases particularly of oxygen. This may seriously affect the marine life for survival, growth, metabolism, swimming activity, behaviour, reproduction and rates of gonad development, egg development, hatching success and also may change the distribution of fishes. The fishes are also attracted towards the thermal discharges, for getting dead animals as food. The macro-invertebrates of sandy beach community change the growth when temperature exceeds 2°C above normal and a significant loss in diversity occurs when temperature exceeds 8°C above normal. Tidal power generators have negative impacts on eggs, larvae and juveniles of organisms that are passing closer to rotating turbines and other machinery. The tidal power generators can also have indirect effects on organisms by changing salinity, sedimentation and intertidal habitats. The power generating systems are vulnerable to biofouling by a variety of organisms. In order to control this, chlorine is injected into the system as a biocide. This usage causes reduction of phytoplankton and planktonic crustaceans. Among planktonic crustaceans, the copepod communities closer to the power stations are reduced in species diversity and body size, thereby affecting the marine food chains in the locality.

In the case of desalination plants, the effluent is warmer than the normal temperature and also is higher in salinity. This high salinity increases the density of the water and will sink to the seafloor. The high saline water reduces its oxygen solubility and affects the bottom-living fauna. Thus salinity and temperature act together to affect the marine life especially plankton. The hermit crab can not tolerate high range of temperatures.

Mass death of marine mammals

Mass mortalities of marine mammals occurred during 1980-1990. This was attributed to environmental pollutants such as mercury, polychlorinated biphenyls (PCBs) dioxins and organochlorine pesticides (DDT, Dieldrin) as well as tributyl tin (TBT).

The pollutant in particular TBT is an immune suppressor, impairing the functions of T and B cells in the immune systems. This increases susceptibility of marine mammals to infectious diseases.

The chemical pollutants such as metal contaminations are the most serious threats that influence structure of marine communities. Copper and tributyltin are the two compounds of human activity that pollute the sea. When copper in water exceeds concentration of more than 50 μg per litre, coral settlement and regeneration will not occur. Tributyltin (TBT) is used as wood preservative in marine antifouling paints. It is toxic to fish and other marine life such as gastropod and mollusks. The major effect of TBT is imposed in mollusks, where females develop male reproductive organs and become sterile. This chemical is mostly prohibited, but it is still high in levels around the coastal ocean in particular Mediterranean Sea.

- Power stations, chemical works, refineries and desalination plants all take large volumes of seawater and discharge it back into the sea. This severely affects eggs, larvae and juveniles of organisms.
- The power systems are vulnerable to biofouling. In order to control this, chlorine is injected into the system as a biocide. This usage causes reduction of phytoplankton and planktonic crustaceans.
- The chemical pollutants are the most serious threats to marine life.

9.2.3. Sewage pollution

Sewage is an ever increasing ecological problem. It is defined as waste matter released from domestic or industrial establishments. This problem is increasing throughout the world with incidence of water-borne diseases. Suspended solids may prevent respiration of benthic flora and fauna. The microbial decomposition of these sewage wastes releases ammonia that is toxic to fishes. Sewage contaminated seawater may cause gastrointestinal diseases in tourists due to the bacteria and viruses present in the sewage. Shellfish grown in sewage-contaminated waters may cause food poisoning, as mussels and oysters can accumulate human pathogens through their filter feeding apparatus. Sewage pollution also results in economic loss due to loss of tourism and cost involved in removal of the debris.

Will dead zones induce climate change?

The number of hypoxic dead zones in shallow coastal waters has doubled since 1960s. The input of excess nutrients, in particular nitrate and phosphorus from fertilizer, stimulates algal blooms. Bacterial decomposition of dead algal matter results in depletion of O_2 producing hypoxia and affecting the population of fish and invertebrates. The expansion of dead zones will increase heterotrophic denitrification leading to release of nitrous oxide, the green house gas. This may lead to ozone depletion, global warming and climate change.

Dead zones are formed by high organic sewage pollutants from urban and agricultural sources. The dead zones are hypoxic or oxygen-deficient areas. The numbers of such zones are increasing from 149 in the year 2003 to 200 in 2006 in the world. Dead zones result when oxygen in ocean waters is reduced to support the marine life. The largest one in the Western Hemisphere is located during every summer off the Louisiana and Taxas coasts. The mobile organisms in the dead zones escape, but the sedentary and sessile species such as clam, starfishes and worms are killed. The dead zones have large amount of dead phytoplankton, zooplankton and fecal pellets along with oxygen-consuming bacteria. By reducing the amount of fertilizers that leach from agricultural lands the problem can be partially solved.

- Sewage is a major ecological problem. Suspended solids may prevent respiration of benthic flora and fauna. The microbial decomposition of these sewage wastes releases ammonia that is toxic to fishes.
- Sewage pollution also results in economic loss due to loss of tourism and cost involved in removal of the debris.
- Dead zones are formed by high organic sewage pollutants from urban and agricultural sources.

9.2.4. Land reclamation

Coastal wetlands have been converted into agricultural land, aquaculture farms and industrial complexes for airports, power generation, oil refining and other chemical process. These habitats are important nurseries for many marine animals. In many parts of the world, more than 50% of the salt marsh habitats have been lost. In the Philippines, 80% of the mangroves were lost for development of aquaculture.

9.2.5. Invasive alien species

An 'introduced species' is a plant or animal that has been introduced by humans into an area where it is not normally found. When the introduced species becomes established in the new environment, it is an 'invasive species'. This species is able to adapt to new environmental conditions and multiply, it can become a threat to native species and disrupt the ecosystem. Marine invasive species are one of the threats to global biodiversity. The number and severity of outbreaks and infestations of invasive alien species are increasing with negative effects on biodiversity and productivity of the sea. The human introduction of non-native species has made profound changes in the coastal waters. The species are introduced for commercial gain, aquarium trade or accidental introduction due the constant movement of ships and goods.

9.2.5.1. Means of introduction

There are two major ways by which marine species are introduced globally. Most species are introduced by ballast water transfer and hull-fouling. Ballast water is the seawater that is used to stabilise and balance cargo ships. When the ship has no cargo, it must be filled with water to stabilise it. When it is full of cargo, the ballast tanks are emptied.

Therefore, the water is loaded at one port, taking lots of marine species with it and unloaded at another port, carrying these species and dumping them in new marine environments. Some of these species are able to tolerate the new environment, establish themselves and so become invasive. It is estimated that 7,000 species are carried around the world in ballast water everyday and 10 billion tonnes of ballast water is transferred globally each year. Hull-fouling is the fouling of underside of a ship vessel, which provides the surface for settlement and attachment of foulers. When the vessels move, these species are transported to other areas of the world. This applies to all vessels, including cargo ships, sailing ships and fishing vessels. Thus in the marine environment, ships are considered as the major vectors for the transfer of invasive species through their attachment and ballast water. In island nations such as Hawaii and Seychelles, hull-fouling introduces as many organisms as ballast water. The success of the invasive species is attributed to their characteristics such as fast growth, prolific vegetative reproduction, flexible habitat requirements, tolerance to environmental stresses, free from predators and parasites.

Fig.9.3: *Caulerpa taxifolia,* a green seaweed growing as an invasive species in Mediterranean regions

(http://ucanr.org/repository/cao/landingpage.cfm?article=ca.v057n04p102&fulltext=yes)

Marine species can be introduced by means of aquaculture, live seafood, aquarium introductions, use of plants for fish and bait nets. Some species are introduced in captivity for food purposes and they escape into the wild and get established. An example is *Oreochromis mossambicus*, a tilapia fish that was introduced in the 1950s as a food fish in the Seychelles and is now officially an introduced species in the country. Some species have been accidentally introduced when they have been brought into a country for aquarium use. An example is green seaweed, *Caulerpa taxifolia* (Fig. 9.3) that was introduced into the Mediterranean by mistake from an aquarium and now overgrows 13,000 hectares of seabed and competes with native plant species for food and light. Plants can be used for fish bait and fishing nets. If these nets and bait are used in an area where the plants packed on them are not normally found, the plants may become introduced into the new area.

9.2.5.2. *Ecological impacts*

The invasive species compete with native species and they threaten diversity and abundance of native species. These may result in lower biodiversity and an unhealthy ecosystem. The 1998 bleaching event killed 80-90% of the coral reefs in the Seychelles, leaving them more vulnerable to invasions. A healthy ecosystem will be more resistant to establishment of introduced species. Marine invasive species affect marine protected areas, as these areas are often located in close proximity to introduction points such as ports. The climate change disrupts ecosystems processes favouring introduced species to get established through competition with native species.

Prior to 1992, the nutrient-rich Hudson River estuary on the Atlantic coast supported abundant phytoplankton populations. The introduction and population explosion of the zebra mussel (*Dreissena polymorpha*) has depleted the biomass of phytoplankton in the estuary. Excessive removal of the phytoplankton by zebra mussel reduces the phytoplankton-feeding zooplankton species and ultimately results in depletion of fish resources. The zebra mussel also out-competes native mussel species for food and space, leading to a decline in native mussel populations. Mussels can filter large amounts of water and reduce the available food in the water column. Their filtering activity increases water clarity and hence, the light penetration is high allowing seaweeds to proliferate. This can change the benthic community structure. Zebra mussel causes significant biofouling problem. To control this, higher levels of biocide chemicals are required and this may cause pollution in the environment.

Fig.9.4: *Prosopis* species, an invasive species in mangrove habitats

There is a growing problem of invasive species which disrupt ecological balance and dynamics of mangrove ecosystem. To cite few examples from India, (i) aggressive colonization of *Prosopis* species (Fig. 9.4) in south east coast of India, (ii) strangulating of the mangroves by *Derris trifoliata* in Sundarbans and (iii) profuse growth of the aquatic weeds such as *Eichhornia crassipes* and *Salvinia* in mangrove waters under low saline conditions. The aquatic weeds cause hindrance in navigation, choke irrigation systems and reduce aesthetic value of water bodies. Mangrove species itself sometimes pose a problem, when it is introduced as an exotic species. For example, the palm, *Nypa fruticans* was introduced from Singapore to Nigeria in 1906 to control coastal erosion. However, the palm spreaded extensive areas and replaced the native mangrove species such as *Rhizophora* in Nigeria. Hence, the Federal Ministry of Environment of Nigeria has developed the 'Nipa Palm Control programme' to control the invasive species. This reveals that the native mangrove species have to be tried for plantation, instead of introducing a new species in to an area.

9.2.5.3. Economic impacts

If marine invasive species damage the abundance of native species and the ecosystem, the aesthetic quality of the environment will be affected. This may impact the tourism industry, which accounts for a large income. They can also affect native fisheries by reducing the number of fishes being caught, resulting in less income for fishermen. The marine invasive species cost millions of dollars for damages and eradication efforts. In addition to affecting people's livelihoods and jobs, marine invasive species can cause serious health problems. For example the bacteria that cause cholera can be carried in ships thorough ballast water, which killed more than 10,000 people in South America. Introduced species such as the zebra mussel in the U.S., have clogged water pipes that are used for nuclear power and sewage treatment plants and public water supplies, causing huge expenditures.

- When introduced species get established in their new environment, they are invasive species. These species can become a threat to native species and disrupt the ecosystem. The number and severity of outbreaks and infestations of invasive alien species are increasing with negative effects on biodiversity and productivity of the sea.
- Introduction of species may be for commercial gain, aquarium trade or accidental introduction due the constant movement of ships and goods. Most species are introduced by ballast water transfer and hull-fouling.

9.2.6. Climate change

Global warming is mainly due to increased atmospheric level of carbon dioxide. Increased carbon dioxide emissions are due to man-made activities such as use of fossil fuels. The global warming increases the melting of the polar ice caps. This will result in major changes in ocean currents and large changes in ocean productivity. The melting of the ice caps will also cause an increase in seawater level which will have a direct impact on coastal wetlands. The global estimate of sea level rise is around 2 millimetres per year and this will inundate large areas of low-lying coastal habitats. One potential impact is called 'coast squeeze', in which the littoral zone is reduced because man builds coastal structures to stop the inundation of occupied land and thus prevents the formation of new intertidal habitat.

The global warming may have long term effects. The coldest habitats will disappear and extinction is likely to start from the Polar Regions first. The warmest habitats will migrate poleward. For example, tropical marine ecosystems such as coral reefs and mangroves may migrate to colder latitudes. Tropical ecosystems may suffer from high temperatures. If global warming continues in the same manner, neither human beings nor the other aerobic organisms would manage to survive in the 23rd century and the earth will be the home of thermophilic, chemotrophic and anaerobic organisms. This hypothesis is supported by scientific data.

The northern Indian Ocean is one of the hotspots of climate change. This region warms faster than 90% of global oceans. Sea surface temperature has increased by 0.2 to 0.3°C along the Indian coast in the last 45 years, and is likely to increase by 2.0 to 3.5°C by 2099. The projected sea level rise is 9-88 cm by the year 2100. Coastal wetlands are at high risk from the predicted effects of global climate. Therefore, salinisation occurs in brackish and many freshwater coastal aquatic ecosystems such as low lying lakes, wetlands and lagoons. The increased wind speed and coastal upwelling may result in higher concentration of chlorophyll-*a*. These changes are likely to influence the structure and function of marine ecosystems. The phytoplankters grow faster at elevated temperature, but the decay sets in earlier. Occurrence of harmful algal blooms may be more frequent which cause considerable mortality of fish. Coral bleaching is likely to be a regular event in the future and the reefs will become remnant between 2050 and 2060 in the Indian seas. Mangroves in tropical regions are extremely sensitive to global warming and sea level rise, and therefore, the extent and composition of mangroves may undergo major changes. The marine fisheries sector has to face the threat from climate change. If small-sized fish species of low commercial value are able to tolerate climate change, they may replace large-sized fish species of high commercial value. The commercial fisheries are already showing declining trends due to over-fishing and other non-climatic factors. Distributional changes of fishes will result in considerable changes in marine biodiversity, ecosystem structure and function. Small-scale fisheries will be the most vulnerable to climate change as the adaptive capacity of the fisheries is low.

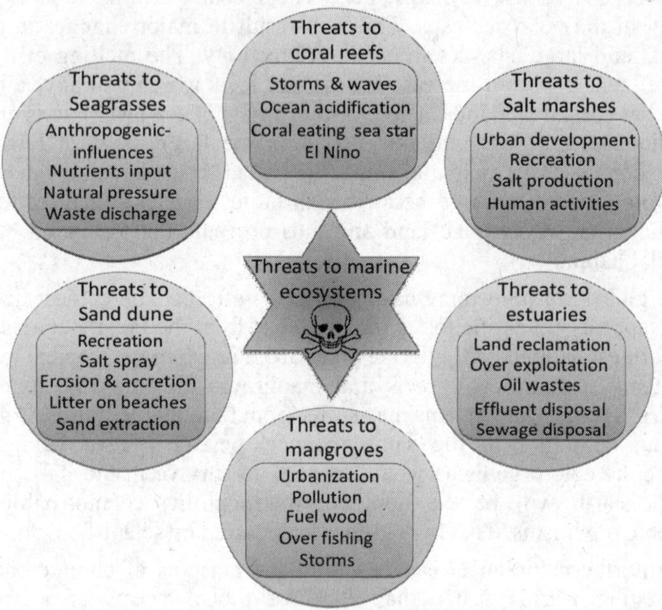

Fig. 9.5: Major threats to different coastal and marine ecosystems

In order to overcome climate impacts, most effective actions are required on different aspects: (1) control of over-fishing, (2) adoption of code of conduct for responsible fisheries, and (3) management of integrated ecosystem-based fisheries. Fishing mortality is an important impact of climate change, and hence reducing the fishing mortality is a means of reducing the impacts of climate change. Projections on fish distribution and their catches have to be developed for planning better management adaptations. It is also necessary to reduce the CO_2 emission with the increasing number and size of fishing boats.

There are seven measures that can contribute to the management adaptation with climate change: (1) evaluating the adaptive capacity of important fish groups, (2) identifying adaptive fishing and post-harvest practices to sustain fish production and quality, (3) supporting energy efficient fishing craft and gear, (4) cultivating halophytes which are tolerant to climate change, for production of food, medicines and biodiesel, (5) increasing awareness of 'climate literacy' among the stakeholders, (6) continuous monitoring of weather, and (7) developing strong coastal protection structures.

- Marine fisheries have to face the threat from climate change. Occurrence of harmful algal blooms may be more frequent which cause fish mortality.
- Coral bleaching will be a regular event in the future. Mangroves in tropical regions are extremely sensitive to global warming and sea level rise, and therefore, the extent and composition of mangroves may undergo major changes.
- The climate change may replace small-sized fish species of low commercial value for large-sized fish species of high commercial value.
- Small-scale fisheries will be the most vulnerable to climate change as the adaptive capacity of the fisheries is low.
- In order to tackle climate impacts, most effective actions are required on (1) control of over-fishing, (2) adoption of Code of conduct for responsible fisheries and (3) management of integrated ecosystem-based fisheries.

Do you know?

- The total length of the world's coastlines is about 315,000 miles, enough to circle the Equator 12 times.
- Less than one half a per cent of marine habitats are protected, compared with 11.5% of global land area.
- Death and disease caused by polluted coastal waters costs the global economy US$12.8 billion a year.
- Plastic waste kills up to 1 million sea birds, 100,000 sea mammals and countless fish every year (Fig.9.6).
- Over 80% of all pollution in seas and oceans comes from land-based activities.

Coastal erosion in India

Rivers are the major source for the sediments at river mouth regions. In India, there are 14 major, 44 medium and 162 minor rivers with a total length of 40,000 km releasing a large quantities of nutrients and alluvial soil to the coastal sea. The rivers discharge sediments to sea along the Indian coast is about 1.2×10^{12} kg. The construction of inland dams and irrigation barrages has considerably reduced the sediment load brought to the sea. Due to the fall in the influx of sediments and concentration of wave energy, many coastal areas experience erosion. Sediment transport along the east coast of India is towards the north, and that along the west coast of the country is towards the south. The construction of ports and coastal structures has resulted in the erosion towards the north in many areas along the east coast. The presence of human settlements and industries along the estuaries and the discharge of effluents into many estuaries have resulted in their degradation.

Fig.9.6: Plastic pollution in Ocean
(http://www.shft.com/reading/oceanic-plastic-pollution-enters-the-food-chain)

9.3. THREATS TO CORAL REEFS

Coral reefs are the vulnerable ecosystem in the world. The coral reefs throughout the world are suffering from both natural and human disturbances. The time taken for recovery of reefs ranges from 7 to 100 years. About 20% of coral reefs world wide have been destroyed, while 24% are in danger and further 26% are under long-term danger of collapse. The loss of coral reefs seriously affects over 500 million people who are dependent on coral reefs for food, coastal protection, building material, livelihood and income from tourism. Disturbances to the coral reefs are bleaching, fishing, pollution, waste disposal, coastal development, sedimentation, SCUBA (Self Contained Underwater Breathing Apparatus) diving, anchor damage, predator outbreaks, invasive species and diseases such as white-band disease and black-band disease.

There are three natural causes for mass mortality of reefs: (1) the storms and waves, (2) the outbreaks of the coral-eating sea star *Acanthaster planci* and (3) the meteorological and oceanographic phenomenon known as 'El Nino'.

9.3.1. Storms and waves

Most coral reefs are vulnerable to storms and waves. Heavy winds cause extreme wave action that kills many flora and fauna, also breaks coral reefs into rubbles. Also freshwater runoff kills many flora and fauna in semi-enclosed bays and lagoons by lowering salinity, and depositing large amounts of sediments. The 2004 tsunami waves caused the most serious damage to the coral reefs in Andaman and Nicobar Islands.

9.3.2. Out-breaks of the coral-eating sea star (crown-of-thorns star fish)

The mass mortality of coral reefs is caused by the coral-eating sea star *Acanthaster planci*. The predators that feed on the sea star are the puffer fish and the polychaete worm. These predators eat away the excessive population of sea star and keep the population under control. When these predators reduce in population, the sea star gets exploded in population. Another reason is that juvenile recruitment of sea star is enhanced by a combination of low salinities, high nutrients and high temperatures. In the Great Barrier Reef of Australia, the sea stars are controlled by a special method. In this method, the sea stars are injected with sodium bisulphate (dry acid) solution by trained divers and this effort kills the sea stars within a few days. This chemical is non-toxic to other marine life. But, this control method is extremely costly.

9.3.3. El nino

El Nino is another cause of mass mortality of coral reefs. This El Nino is a massive influx of nutrient-poor warm water into the nutrient-rich cold water off the coasts of Ecuador and Peru. Mass bleaching of coral reefs have been recorded with higher frequency since the 1980s and are linked to El Nino. Coral bleaching is now a worldwide phenomenon and it is a great threat to coral reefs. During this, coral reefs expel the zooxanthellae and become white. This bleaching effect reduces growth and reproduction and also causes death of coral reefs. The dead corals are further degraded by macroalgal overgrowth and bioerosion. The coral bleaching occurs in response to increased sea surface temperatures and solar UV-radiation, as a result of increasing solar radiation in the atmosphere. Even temperature increase of 1 or 2°C above average over a period of a month can cause mass bleaching of coral reefs. This bleaching problem becomes serious due to predicted increases in sea surface temperatures. The solar UV-radiation produces harmful oxygen radicals, which ejects the zooxanthellae and causes mortality of coral reefs.

9.3.4. Ocean acidification

The increase in the concentration of atmospheric carbon dioxide (CO_2) poses a threat to coral reefs. The average pH of seawater is around 8.2 and 8.3. Atmospheric carbon dioxide, at high levels, increases the concentrations of carbonic acid. This is a weak acid dissociates into bicarbonate generating hydrogen ions (H^+), which leads to a decrease in pH, and it is called 'ocean acidification'. However, the levels of carbonate ions which are essential for formation of coral reef are low. A 40% reduction in calcium carbonate (aragonite) in the world oceans is predicted by 2100. This leads to reduction in calcification of coral reefs. This situation weakens the strength of coral skeletons and makes it more susceptible to breakage and bioerosion. Even a small decrease in pH may have a serious effect on marine organisms that use calcium carbonate for building their skeletal structures. Under doubling of atmospheric carbon dioxide concentration, calcification rates of corals may decrease by 10-30%. The calcification rate of *Porites* in Australia has declined by about 14% since the year 1900, due to the reduction in the ability of coral reefs to deposit calcium in seawater of low pH. Thus the ocean acidification will reduce the availability of carbonate minerals in seawater, important building blocks of marine plants and animals. It is expected that by 2100, 70% of coral reefs will be exposed to corrosive waters. Arctic Ocean and Antarctic Ocean will become unsaturated to essential carbonate minerals by the year 2032 and 2050 respectively, with disruptions to large marine food webs.

Experimentally it has been proved that ocean acidification affects the interaction between the two most abundant bottom-dwellers in coral reefs: (1) reef-building corals and (2) seaweeds. These two groups compete on reefs for both space and light. Seaweeds have the upper hand, when there is low herbivory. When seaweeds are kept in control by herbivores and the reefs are healthy, then corals and seaweeds are in a competitive balance. The ocean acidification is likely to enhance the ability of seaweeds to out-compete coral reefs, leading to coral death. However, active biological processes and small-scale temporal and spatial variability in ocean pH may render marine biota more resistant to ocean acidification than hitherto believed.

9.3.5. Human activities

Human activities can directly cause mass mortality of reefs through dredging, pollution and over-fishing. The reefs in Hawaiian Bay were destroyed through sewage pollution. In certain areas of India, coral reefs are actively mined for use in building and also damaged by destructive fishing practices such as trawl operation for commercial fishes and dynamite fishing for ornamental fishes). In the Philippines, dynamiting for fishes is a major cause of destruction. Oil pollution adversely affects coral reefs especially the branching species such as *Acropora*. In Jamaica, the reefs have become seaweed-dominated due to removal of seaweed-eating sea urchin and fishes by over-fishing. Coral cover declines at frequently dived sites compared with non-dived sites. The coral reefs are disturbed by human activities such as tourism, bait-digging, walking on intertidal areas and SCUBA diving.

- Coral reefs are vulnerable ecosystems in the world. There are three known natural causes for mass mortality of reefs: (1) the storms and waves, (2) the outbreaks of the coral-eating sea star *Acanthaster planci* and (3) the meteorological and oceanographic phenomenon known as 'El Nino' as well as ocean acidification.

- Human activities can directly cause mass mortality of reefs through tourism, SCUBA diving, dredging, pollution and over-fishing.

9.4. THREATS TO MANGROVES

The mangrove forests continue to disappear all over the world. Global mangrove cover is estimated at 15.2 million hectares with the largest areas in Asia and Africa followed by North and Central America. About 90% of the global mangroves are growing in developing countries and they are nearing extinction in 26 countries. The long term survival of mangroves is at great risk and the services offered by the mangroves are likely be totally lost within 100 years. About 20% or 3.6 million hectares have been lost since 1980. The annual loss in the years 2000-2005 was about 102,000 hectares. The rate of loss of mangrove forest was 0.66% per year during 2000-2004 periods and this is still 3-5 times greater than overall rates of global forest loss. Although, some 12,000 marine protected areas safeguard 25% of all mangrove areas and the remaining 75% are still unprotected. Thus the mangroves are disappearing fast.

Are mangrove biodiversity globally threatened?

Globally mangroves are declining rapidly as they are cleared for coastal development (Fig. 9.7). The major causes of mangrove decline are conversion to aquaculture, agriculture, urban, industrial and tourism developments. Little is known about the effects of mangrove area loss on individual mangrove species, local and regional populations. Mangrove species are relatively low in number (70). In the world, eleven of the 70 mangrove species (16%) are at elevated threat of extinction. The Atlantic and Pacific coasts of Central America are highly vulnerable and about 40% of the mangrove species present in the areas are threatened with extinction. Across the globe, mangrove species are found primarily in the high intertidal and upstream estuarine zones. These areas are the most threatened areas because they are often the first cleared for development of aquaculture and agriculture. The loss of mangrove species will have devastating economic and environmental consequences for coastal communities. In addition to mangroves, about 40% of animal species restricted to mangrove habitats are at elevated threat of extinction. Several species in the mangrove ecosystems are at high risk of extinction and may disappear in the near-future if existing protective measures are not enforced.

Fig. 9.7: Mangrove habitats at man-made threat by cutting and road constructions

Mangrove plant species can indicate certain environmental changes and habitat conditions. They can be used as 'indicator species' to monitor the changes in the mangrove habitat (Table 9.1). For an example, the presence of the mangrove fern *Acrostichum* sp., indicates that the mangrove habitat is degrading.

Table 9.1: Indicator species of mangrove habitat

No.	Indicator species	Environmental conditions
1	Salt marsh species (*Suaeda* sp., *Salicornia* sp.)	High salinity
2	*Avicennia marina*	Aridity, high salinity, high organic load
3	*Acanthus* sp.	Low salinity
4	*Nypa fruticans, Heritiera fomes, Sonneratia caseolaris*	Freshwater inflow
5	*Acrostichum* sp.	Acid sulphate soil, degradation of mangrove habitat

9.4.1. Causes of mangrove degradation

Mangroves are degraded by nature-induced changes. Tropical storms are common in the Bay of Bengal and the Caribbean region. To cite an example, a cyclone has destroyed about 8.5 million trees in Bangladesh, which is equivalent to 66.3 million cubic metres of timber in the year 1988. The damaged forests take a very long time to recover.

Propagules are abundantly produced in most of the mangrove forests, but their dispersal, survival and establishment are serious problems. Therefore, it is necessary to assess difficulties for natural regeneration of mangroves and also to implement correction measures for facilitating the dispersal and establishment of mangrove propagules.

Diseases cause devastating damages to mangroves. For example, in the Sundarbans, "top dying" disease has damaged about 45 million *Heritiera fomes*, which is locally called as 'Sundari' meaning 'beautiful tree'. This is about 20% of the entire forests in Bangladesh. The top dying disease is believed to be caused by many factors such as increased soil salinity due to reduced water flow, reduction in periodic inundation, excessive flooding, sedimentation, nutrient imbalances and cyclone-induced stress.

Biological pests and parasites affect the mangroves. Significant damage is caused by grazing of buffaloes, sheep, goats and camels in dry coastal areas of Asia and the Middle East. In a number of areas, mangrove fern (*Acrostichum aureum*) colonize rapidly in the gaps of mangrove cover and can prevent re-establishment of large trees. The spread of the invasive species such as *Prosopis juliflora* is a threat to the mangroves in southeast coast of India. This is mainly due to poor flushing of the intertidal areas and dry conditions of the soil. Young plants of mangroves are damaged by barnacles and leaf eating crabs of the sesarmid family. Caterpillars infest the seedlings of *Rhizophora* and inhibit sprouting.

In general, seedling and juvenile mangrove plants are susceptible to barnacle fouling which inhibits the gaseous exchange and food production, which can subsequently result in reduced growth and survival. However, adult plants are tolerant to barnacle fouling. The stomata and lenticels of the mangrove plants are covered by barnacles. The fouled plants can obtain less CO_2 and H_2O from photosynthesis, resulting in reduced production of chloroplasts and biomass. As a result, fouled mangrove plants often have lower survival and growth rates than non-fouled plants. Barnacle fouling is considered an important issue in the successful management of mangrove plantations for conservation purpose.

9.4.2. Causes of mangrove destruction

Globally mangroves are destroyed by man-made activities, which pose significant threats and they are detailed briefly as follows:

9.4.2.1. Urbanization

Coastal mangroves have been cleared for human colonisation in many areas such as Singapore, Jakarta, Bangkok, Yangon (Rangoon), Kolkata (Calcutta), Mumbai (Bombay) and Lagos. Roads are often built across tidal flats in arid countries, causing the drying out and hyper-salinisation on the landward margin. In India, Mumbai is one of the best examples for the mangrove destruction due to urbanization (Fig. 9.8). All the seven islands of Mumbai were reclaimed and linked to a continuous landmass after destroying mangroves in the process.

Fig. 9.8: Mangroves (*Avicennia marina*) growing along the highly sewage polluted waters with backdrop of urbanization

9.4.2.2. Agriculture

Sundarbans is the best example for the conversion of mangroves for agriculture. This reclamation process was initiated in 1770, and it continued till recent past. In the largest delta region of the world, existing between India and Bangladesh, 1, 50,000 hectares of mangroves were destroyed during the past 100 years, and these were mainly reclaimed for agriculture.

The mangrove areas are deforested and reclaimed with rainwater to drain the salt content of the soil and these areas are protected from seawater intrusion by constructing embankments. Once the salt is leached to a sufficient level, the land is cultivated with paddy or coconut.

9.4.2.3. Aquaculture practices

In recent decades, one of the major drivers of mangrove conversion has been the growth of aquaculture, particularly shrimp farming. The aquaculture practices are typically made in intertidal areas and are responsible for the most widespread losses of mangroves. A large scale destruction of mangroves was made for aquaculture in countries such as Indonesia, the Philippines and Bangladesh. To cite an example, in the Philippines, between 1968 and 1983, about 2, 37,000 hectares of mangroves were lost for shrimp pond construction. This is almost half of the total national mangrove area of the country. One major issue associated with the farms located in mangrove habitats is acidification of pond waters that kills aquatic organisms. In Ecuador, the decline in mangrove areas was largely due to the construction of 21,587 hectares of shrimp ponds, compared to only 1,157 hectares for urban expansion in 1969-84. In

Thailand, around 50% of the total denuded mangrove area of 1,71,472 hectares were converted into aquaculture ponds during the period 1961-87. Intensive aquaculture systems are often placed above the high tide. Although they may not impact directly upon mangroves, the out-flowing water of high organic and inorganic pollutants from these ponds pollutes the ecosystems and fishery resources. In many cases, the aquaculture ponds have been abandoned due to high pollution and shrimp disease issues. The abandoned aquaculture ponds have the problem of acidity and to remove the acidity it takes 5-10 years to flush out the pond by using aerated seawater. Then only, mangrove restoration is possible in that area.

Coastal aquaculture is an important source of livelihood options and employment opportunities in the coastal areas. It should be developed sustainably and ecofriendly manner, using the traditional practices. In this situation, an integrated aquafarming with mangrove silviculture that uses no chemicals and artificial feed may be of great help in ensuring livelihood security of poor fishers without damaging the environment. Such integrated aquafarming models have to be developed and demonstrated.

9.4.2.4. *Over-fishing*

Over-fishing is a problem in many areas, both locally within mangroves, but also in some offshore fisheries that depend on mangroves (Fig. 9.9). Capture of wild-caught penaeid shrimps from all mangrove areas showed a peak in capture during the early 1980s, with gradual decline since then. In 2006 the catch was 30% lower than that of 1980. Over-fishing can affect the forest itself. The heavy trampling by fishers during the fishing operations can damage roots and prevent regeneration. Unregulated use of fishing nets of small mesh size in the mangrove waters for the collection of tiger prawn (*Penaeus monodon*) seeds are also resulting in tremendous loss of diversity of faunal components, as in the case of the Sundarbans. It has been observed that for catching 8,106 tiger prawn seeds, collectors destroy juveniles of 13, 04,937 other prawns, 60,581 fish, 2,46,097 crabs, 8,391 mollusks and 66,792 unidentified meroplankton. In other words, the people use only 0.25% of the catch and discard the rest as unwanted by-catch mostly dead, greatly reducing available stock for other fisheries.

Fig.9.9: Prawn seed collection in Sundarbans, West Bengal

9.4.2.5. Cutting for timber, fuel and charcoal

Mangroves have been cleared for timber, charcoal and firewood. Because of higher calorific value, the mangrove twigs are used as firewood. The mangrove wood is rich in phenols and hence is highly resistant to deterioration and it is widely used as timber for construction purpose.

9.4.2.6. Prevention of freshwater flow and tidal flow

Mangroves are best developed in areas that receive freshwater run-off and tidal water flushing. Embankment construction or siltation at the river mouths restricts the inflow of tidal water in mangrove swamps. Dam and barricade constructions in upstream areas for diverting water for irrigation purposes have resulted in poor flow of fresh water into mangrove swamps. The poor flows of tidal water and fresh water result in high salinity of mangrove areas and reduce the growth of mangroves. Dam building in the Indus Delta in Pakistan has reduced water flows at the mouth by 90% and this has led, in turn to losses of mangrove areas as well as significant increases in erosion at the coast due to lack of new sediment inputs. The Sundarbans has experienced a great threat of species loss due to the construction of Farraka Barrage in the upper reaches as well as building dykes and bunds arresting normal flow of water. In Colombia, large parts of mangrove forests along the lagoon of Cienaga Grande de Santa Marra, have died within 3-decades. In Pichavaram, south India, the mangroves are dying in many sites, due to hyper-salinity and other associated factors such as increasing of temperature, poor precipitation, and poor flushing of mangrove soil by tidal waters, as illustrated diagrammatically in Fig. 9.10.

Reduction in freshwater flow is a serious threat to mangroves especially in India. Reduction in freshwater flow affects areas of the mangrove wetlands, their floristic diversity, as well as biomass. It also affects fishery resources of the mangrove wetlands. Reduction in freshwater inflow has changed the species composition of mangroves in Sundarbans: reduction of freshwater loving species such as *Nypa fruticans* and *Heritiera fomes*. Wildlife species such as Javan Rhino, water buffalo and swamp deer disappeared in the last centuries due to lack of freshwater. Dense and tall trees of mangroves (*Avicennia officinalis, Excoecaria agallocha* and *Lumnitzera racemosa*) constituted nearly 90% of the population of the Godavari mangroves in India in the 1950s, but now they constitute only 37% of the population and are replaced by saltmarsh bushes (*Suaeda maritima* and *S. nudiflora*). Considering this seriousness, an interdisciplinary research studies are initiated to find out how much of freshwater, a mangrove ecosystem requires to sustain itself in the present state.

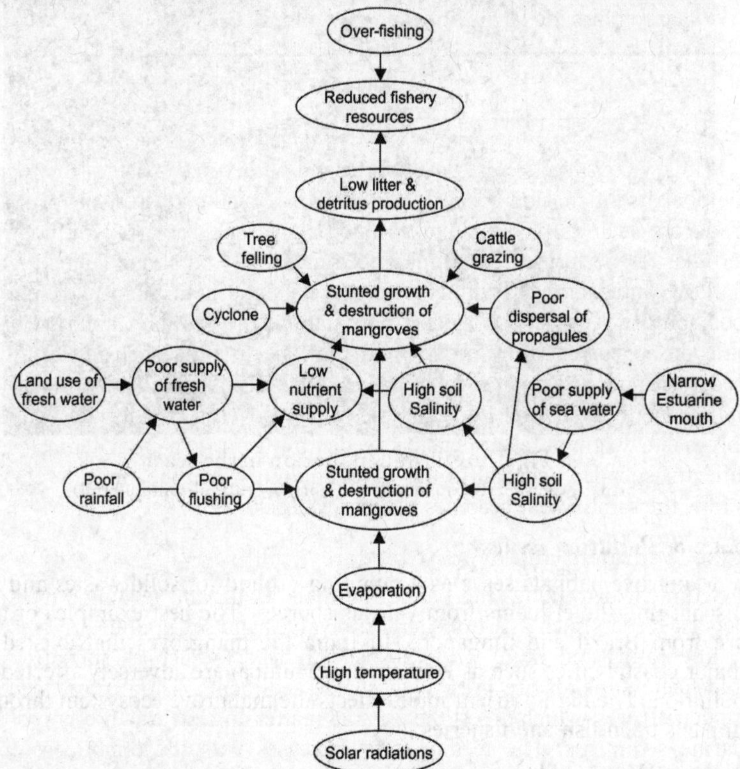

Fig. 9.10: Possible factors responsible for degradation of mangroves and depletion of fishery resources at Pichavaram situated in southeast coast of India. This specific area has already lost 75% its green cover within the last century and about 90% of the forest area is degrading (Kathiresan, 2000)

9.4.2.7. *Oil pollution*

Oil or gas exploration, petroleum production and accidents by large oil tankers cause significant damage to mangrove ecosystems. This is caused by toxic effects or the physical smothering of roots, or combination of the two. To cite an example, Nigeria's richest oil wells are situated close to inshore where rich mangroves once existed. Similarly oil tanker accidents in the Gulf of Mexico and in the Caribbean areas resulted in oil spill that severely damaged the coastal systems (Fig.9.11). As a result, the entire mangrove ecosystem got affected, causing defoliation of trees, mortality of all sessile and benthic organisms as well as water fowls. Once the mangrove forest is affected by oil pollution, it will take a long time of at least 10 years for recovery of the forest.

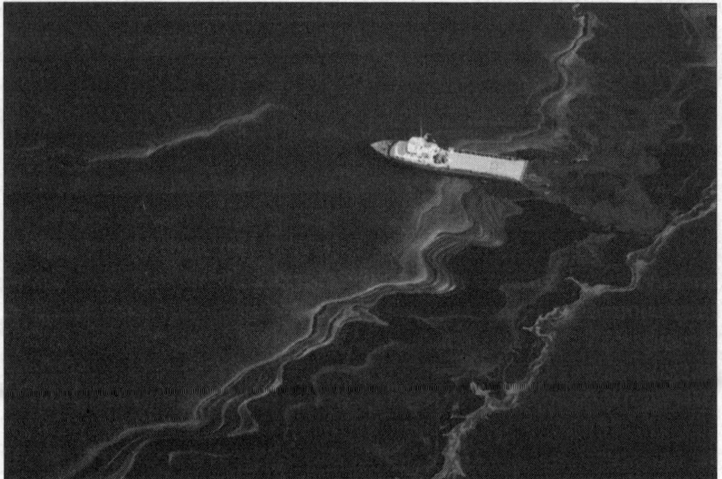

Fig.9.11: Oil spill/pollution in the sea
(http://giveusliberty1776.blogspot.in/2010/05/gulf-oil-spill-101.html)

9.4.2.8. *Pollution issues*

Mangrove habitats serve as a dumping ground for solid wastes and for discharging the effluents from various sources. The best examples of this are from Brazil and Singapore. In India the mangroves that existed in major coastal cities such as Kolkata and Mumbai are adversely affected by pollution. The high nutrient input affects the mangrove ecosystem through impacts upon fish and fisheries.

9.4.2.9. *War problems*

The best example for this kind of mangrove destruction occurred in the Vietnam War. During the war period (1962-1971), about 71 million liters of defoliant chemicals were sprayed over the coast that resulted in the destruction of 1,04,000 hectares of mangrove forest.

9.4.2.10. *Mining operations*

This was a serious problem in countries such as Thailand. Until the late 1980s, about 4, 27,000 hectares along the Thailand coasts in the vicinity of mangroves were mined for tin.

9.4.2.11. *Plantation practices*

In plantation programme, only few species of mangroves which are fast growing are used and this practice will reduce biodiversity in the course of time. Forest officials are to be trained properly on the eco-biology of mangroves, plantation techniques, use of the remote sensing tools for identification of potential areas for mangrove plantation and assessment of changes that take place in the mangrove forests.

- In the world, 16% of mangrove species and 40% of animal species restricted to mangrove habitats are at elevated threat of extinction. Across the globe, mangrove species found in estuarine areas are most threatened.
- Mangroves are degraded by nature-induced changes: tropical storms; poor dispersal, survival and establishment of seedlings; pests, parasites and diseases; grazing by farm animals; invasive species; infestation by barnacles, leaf eating crabs, caterpillars and filamentous algae.
- Mangroves are destroyed by man-made activities: urbanization, agriculture, aquaculture, industrial and tourism development, over-fishing, cutting for timber, fuel and charcoal, prevention of freshwater flow and tidal flow, oil pollution, pollution issues, war problems, mining operations and unscientific plantation practices.

9.5. THREATS TO ESTUARIES

Estuaries are important centers of human activities. Of the 32 largest cities in the world, 22 are located on estuaries (Boston, Baltimore, Charleston, Norfolk, San Francisco, Seattle, etc.). Human activities have led to a decline in the health of estuaries, making them one of the most threatened ecosystems on the Earth. Globally there are 1200 major estuaries covering 5,00,000 km^2. Estuaries are very fragile ecosystems where conditions are extreme with variable temperatures, salinities and sediment dynamics. The primary cause of estuarine damage is human exploitation, which has caused 95% species loss, in addition to habitat destruction. Most mammals, birds and reptiles in estuaries got depleted by 1950s. Oysters were the first invertebrate resource to degrade from estuaries. Man-made changes have destroyed 65% of seagrass and wetland habitat and deteriorated water quality.

Estuaries are increasingly under threat from global challenges of habitat loss, climate change and man-made inputs. Many estuaries exhibit declining water quality and eutrophication resulting from increasing man-made activities. Agriculture and aquaculture activities, expansion of urbanization and industrialization are the major causes of decline in the quality of estuarine habitats. Tidal currents and water residence time in the estuarine ecosystems control their hydrological properties and play a critical role in nutrient availability. Another anthropogenic influence is associated with urban sewage pollution, and high nutrient loads. The increased eutrophication, algal blooms and macroalgal mats have developed in many estuaries and such changes may lead to hypoxia that may affect the estuarine life.

9.5.1. Land reclamation and urbanization

Globally the shallow areas have been reclaimed by filling them for cultivation of rice and other agricultural crops, for the expansion of ports and harbours, for the construction of new buildings, roads and for the development of parks and recreation facilities. Industrial complexes are often established around rivers and estuaries and this resulted in urbanization and growth of human population leading to increased demand for living space, freshwater supply and disposal of wastes.

9.5.2. Sewage disposal

The main component of sewage is organic matter. Normally, the sewage is passed through treatment plants, before it is released into the estuary. These plants have limited capacity. The result is that the extra sewage is discharged into the estuaries either partially-treated or untreated. It leads to high bacterial load, high biological oxygen demand and deoxygenation of water which affect the estuarine life. In dry months, when freshwater discharge decreases, the oxygen level diminishes leading to high pollution load in the estuary.

9.5.3. Industrial effluent disposal

Industrial wastes find their way into the estuaries. In many cases, wastes are discharged directly with or without treatment. Thus, the estuarine organisms get exposed to a variety of toxic substances with varying concentrations. Lethal concentrations can destroy most of the estuarine life, while their sub-lethal concentrations may affect growth, reproduction and life histories of different organisms. Continuous discharge of non-degradable wastes such as heavy metals (Cd, Co, Hg, Pb, Zn) can create pollution problems because these are retained by the organisms (fishes, crustaceans and mollusks), and can influence higher trophic levels of the food chain. Similarly, pesticides and insecticides used in pest control and agriculture become persistent in the estuaries and can also accumulate in seaweeds and shellfish and thus affecting the predators (fish, birds and even man) with adverse consequences.

9.5.4. Oil waste

Two types of oil wastes reach the estuaries (1) spillage from ships while unloading the oil cargo, and (2) the effluents from refineries or petrochemical complexes. Some toxic materials present in the hydrocarbons may affect estuarine life directly or when these get deposited at the bottom leading to adverse effect on commercial fisheries.

9.5.5. Over-exploitation

Many estuaries have luxuriant growth of mangrove forests and swamps around them. Mangrove vegetation has been cleared in many cases and the swamps have been converted into brackishwater aquaculture farms. These are largely used for intensive or semi-intensive shrimp farming. Large scale shrimp seed collection practices in the coastal areas and estuaries lead to destruction of larval forms and juveniles of different organisms resulting in depletion of faunal resources. It is estimated that the seed collectors are destroying 540 million seeds of brackish water fish along with much higher number of other invertebrate larvae after retaining only the seeds of tiger shrimp in Sundarbans. The aquaculture activities around estuaries resulted in accumulation of organic and inorganic wastesto estuaries causing eutrophication. The over-exploitation of fishery resources in the estuary may lead to depletion of resources.

- Estuaries are increasingly at threat from global challenges of habitat loss, climate change and anthropogenic inputs. Many estuaries face declining water quality and eutrophication resulting from increasing man-made activities. Agriculture and aquaculture activities, expansion of urbanization and industrialization are the important causes of decline in the quality of estuarine habitats.
- Continuous discharge of non-degradable wastes such as heavy metals (Cd, Co, Hg, Pb, Zn) can create pollution problems because these are retained by the organisms (fishes, crustaceans and mollusks), and can influence higher trophic levels of the food chain. Similarly, pesticides and insecticides used in agriculture or pest control become persistent in the estuaries and can also accumulate in seaweeds and shellfish and thus affecting the predators (fish, birds and even man) with adverse consequences.

9.6. THREATS TO SEAGRASS ECOSYSTEM

Seagrasses are a vulnerable resource and subjected to destructive disturbances all over the world. Seagrasses cover about 0.1-0.2% of the global ocean. They are disappearing globally at a rate of 110 km^2 per hectare per year since 1980. The rate of loss is 0.9% per hectare per year before 1940 and is 7% per hectare per year since 1990. It is due to both natural pressure and anthropogenic influences such as inputs of nutrients, discharge of industrial and other wastes.

Natural causes of seagrass loss are geological and meteorological events and biological interactions. Earthquakes cause rise in shoreline and subsequent exposure of the seagrass vegetation. The ash and debris rains that accompany volcanic eruptions may destroy seagrass beds. Above all, hurricanes and cyclones cause damage to the seagrass meadows. These wind storms coincide with huge waves and strong currents over the seafloor. This may uproot seagrasses and erode the sediment surfaces. Seagrass loss is also due to grazing by waterfowl. Mortality of the vegetation is due to low winter temperatures and prolonged exposures to sunlight. Sea urchins grazing on seagrass beds and small limpet which consumes *Zostera marina* are also responsible for destruction of seagrass. Besides these, the "wasting disease" caused by the marine slime mould *Labyrinthula zosterae* is also responsible for large scale seagrass mortality.

The main cause of seagrass destruction is human disturbance, most notably eutrophication, mechanical destruction of habitat and over-fishing. Excessive input of nutrients (nitrogen and phosphorus) is directly toxic to seagrasses. The nutrients stimulate the growth of epiphytic and free-floating macro algae and micro-algae. This reduces light penetration and lowers the seagrass photosynthesis. Decaying seagrass stimulates algal blooms and this can also cause a complete change from seagrass to algal dominance. More over, over-fishing of large predatory fishes can indirectly increase algal growth by reducing grazing control performed by crustaceans and gastropods. Boating operations can make disrupting noises and sometimes the blade propeller can damage the seagrass.

- Seagrasses are vulnerable to natural pressure and anthropogenic influences. The main cause of seagrass destruction is human disturbance, most notably eutrophication, mechanical destruction of habitat and over-fishing.

9.7. THREATS TO SALT MARSH ECOSYSTEM

Salt marsh habitats are threatened by various threats. In the past, salt marshes were considered as coastal 'wastelands,' resulting in considerable loss through land reclamation for agriculture, urban development, salt production and recreation. The indirect effects of human activities such as nitrogen loading also play a major role in the salt marsh areas.

Wave action can damage and erode the marsh. Recreational use can damage salt marshes by trampling. For centuries, livestock such as sheep and cattle grazed on the highly fertile salt marsh land. Intensive grazing is a serious problem in some areas and it reduces the height of the vegetation, the diversity of plant and invertebrate species. Land reclamation for agriculture has resulted in many changes such as shifts in vegetation structure, sedimentation, salinity, water flow, biodiversity loss and high nutrient inputs.

Oil pollution can damage salt marsh vegetation while it usually recovers. Water pollution from sewage and fertilizers can lead to eutrophication. There is an excessive growth of green algae, which may cause local problems of smothering on salt marshes.

Major cities such as Boston, San Francisco, Amsterdam, Venice and Tokyo have expanded out into areas of salt marshes. The remaining marshes surrounding these urban areas are under heavy pressure from the human population. Human impacts such as sewage, urban run-off, agricultural and industrial wastes are running into the marshes from nearby sources. Salt marshes are nitrogen-limited. Any increase in level of nutrients will affect the vegetation structure and function of salt marshes.

- Salt marsh habitats are threatened by land reclamation for agriculture, urban development, salt production and recreation. The indirect effects of human activities such as nitrogen loading also play a major role in the salt marsh area.
- Water pollution from sewage and fertilizers can lead to eutrophication. There is an excessive growth of green algae, which may cause local problems of smothering on salt marshes.

9.8. THREATS TO COASTAL SAND DUNES

Coastal sand dune system is at constant threat due to human pressure, recreation activities, litter on beaches, sand extraction and construction activities. Some of the natural disturbances that affect the coastal dunes are on-going abrasion, salt spray, erosion-accretion and wave attack. Some countries, particularly the United States, Australia, Canada, New Zealand, the United Kingdom, and Netherlands, have developed significant programmes for dune protection through the use of sand dune stabilization.

- Coastal sand dune system is at constant threat due to pressure of human population, recreation activities, litter on beaches, sand extraction and construction activities.

Questions for Review/Exercise*

(1) Define:

  (a) Ballast water
  (b) Tributyltin (TBT)
  (c) Coast squeeze
  (d) Coral bleaching
  (e) Indicator species
  (f) SCUBA diving
  (g) Dead zones
  (h) Crown-of-Thorns Starfish

(2) What are the major threats to marine environment? Explain.

(3) Write a short note on ecological and economic impacts of invasive species in marine environment.

(4) What is ocean acidification? What is its impact on reef-building corals and seaweeds?

(5) List out the major causes of mangrove destruction and degradation.

(6) "Estuaries are increasingly under threat from global challenges of habitat loss, climate change and anthropogenic inputs"-justify

(7) Why is the seagrass ecosystem declining? Explain.

(8) What are the major threats to salt marsh and sand dunes? Explain.

* Regular field visits to marine environment to observe the destruction and degradation of ecosystems, and to analyse the factors responsible for the issues.

Conservation and Management of Marine Environment 10

- 10.1. Regulatory approach for marine management
  - 10.1.1. Marine Protected Areas
  - 10.1.2. Marine Fisheries Management policies
  - 10.1.3. Coastal Management Policies
  - 10.1.4. International cooperation
- 10.2. Participatory approach for marine management
  - 10.2.1. Types of Stakeholders
  - 10.2.2. Types of Participation
  - 10.2.3. Strategies for Participation
  - 10.2.4. Civil society awareness
  - 10.2.5. Training and Capacity building
  - 10.2.6. Coastal rehabilitation
  - 10.2.7. Coastal livelihood
  - 10.2.8. Income generation and financial sustainability
  - 10.2.9. Participatory monitoring
  - 10.2.10. Research and development in marine management
- 10.3. Management of Coral reefs
  - 10.3.1. Protection of coral reefs
  - 10.3.2. Monitoring of coral reefs
  - 10.3.3. Coral research studies
  - 10.3.4. Capacity building and awareness creation
  - 10.3.5. Restoration and rehabilitation of coral reefs
- 10.4. Management of mangrove ecosystems
  - 10.4.1. Silviculture management
  - 10.4.2. Legal protection
  - 10.4.3. Marine Protected areas
  - 10.4.4. International protection
  - 10.4.5. Restoration and afforestation
- 10.5. Management of Coral reef resilience and resistance to climate change
  - 10.5.1. Resistance factors
  - 10.5.2. Resilience factors
  - 10.5.3. Tools for enhancing coral resilience
  - 10.5.4. Strategies for enhancing coral resilience
- 10.6. Management of mangroves for resilience to climate change (sea level rise)
  - 10.6.1. Impacts of climate change on mangroves
  - 10.6.2. Strategies to mitigate climate change effects

Oceans are a great storehouse of resources such as food fishes, minerals (metals, oil, natural gas, chemicals) and energy (waves, water currents and tides). The oceans are used for transporting goods, recreation purposes and a dumping place for all waste materials. Man-made pressures on marine resources continue to increase due to explosion of human population. In order to meet food requirements, the production of fish has to increase by 50% from the current level to the year 2050. Globally there are serious declines in marine living resources, losses of coastal habitats, elevated pollution levels and poor water quality in many areas. This deterioration of the marine environment is accelerated by the impacts of climate change such as sea level rise, sea temperature rise and ocean acidification. Local economies are adversely affected by over-exploitation of the ocean resources. Therefore, management of the ocean resources is of utmost importance.

A model of ocean management is given in the figure 10.1. The main theme of ocean management is to reduce the pressure on marine resources. There are two principal approaches of ocean management: (1) Regulatory approach and (2) Participatory approach. The regulatory approach of management is based on policies and legal measures of individual countries, whereas the participatory approach of management is based on involvement of all stakeholders of the marine resources.

These two approaches can be evaluated by continuous monitoring. This will lead to improvement of management strategies through remedial measures and research intervention. Most research interventions in the field of marine conservation and management have been undertaken for coral reefs and mangroves. This can also be applicable to other marine habitats: Arctic, Antarctic, tropical or temperate, shallow or deep.

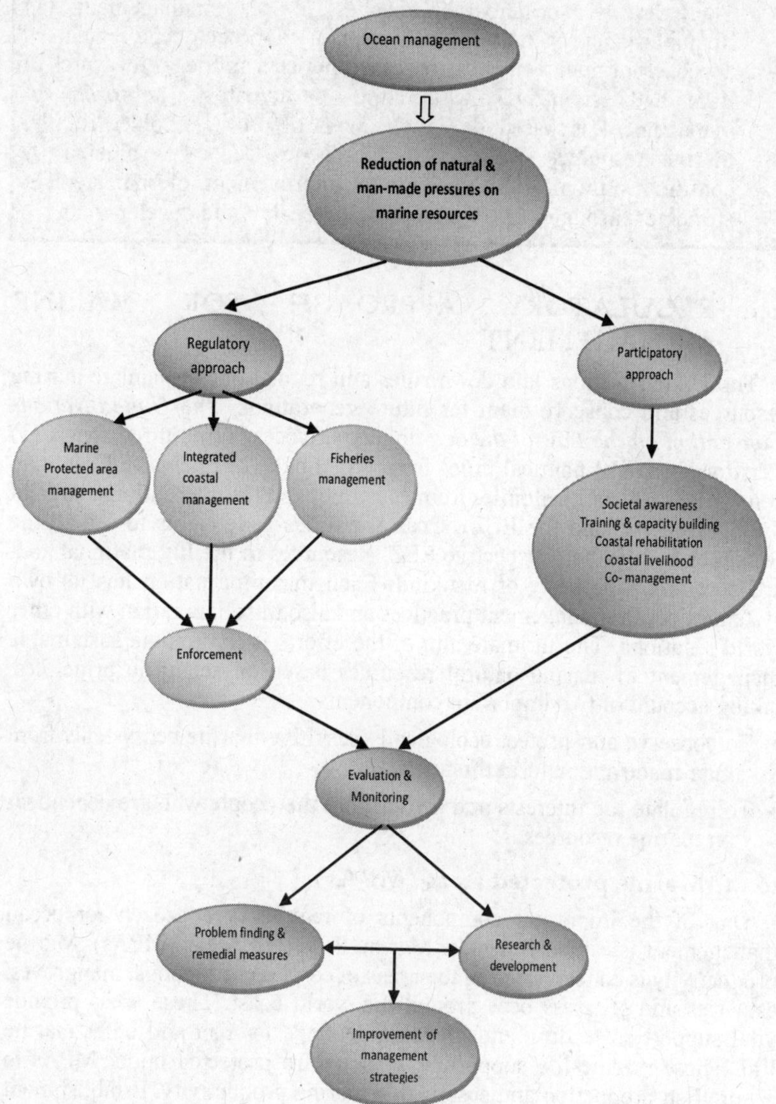

Fig. 10.1: An outline of ocean management based on regulatory and participatory approaches

- The main objective of ocean management is to reduce the pressure on marine resources.

- There are two principal approaches of ocean management: (1) Regulatory approach and (2) Participatory approach. The regulatory approach of management is based on policies and legal measures of individual countries, while the participatory approach of management is based on involvement of all stakeholders of the marine resources. These two approaches can be evaluated by continuous monitoring, leading to improvement of management strategies through remedial measures, research and development.

10.1. REGULATORY APPROACH FOR MARINE MANAGEMENT

The United Nations laid down rules and regulations to manage marine resources and conserve them for future generations. The *United Nations Convention on the Law of the Sea* divided the oceans into three zones: (1) Territorial sea (12 nautical miles from coastline), (2) Exclusive Economic zone (EEZ; 200 nautical miles from coastline) and (3) International Area of the Sea bed (beyond the EEZ). Coastal nations have rights to use all the resources within their respective EEZ. Resources in the International area are a common property of mankind. Each maritime nation has its own system of ocean management practices and also in collaboration with other world's nations. The ultimate aim of the efforts is to promote sustainable management of marine natural resources based on scientific principles, taking account of two important components:

- To conserve and protect ecologically sensitive marine ecosystems from man-made and natural threats; and,
- To regulate the interests and activities of the people who are dependent on marine resources.

10.1.1. Marine protected areas (MPAs)

One of the important components of regulatory approach for ocean management is establishment of Marine Protected Areas (MPAs). Marine productivity is concentrated in the areas of coral reefs, lagoons, mangroves, estuaries and seagrass beds around the world coast. These areas provide vital support as feeding and breeding grounds for fish and other marine life. These marine-life supporting systems are protected under MPAs to ensure fish production and sustainable marine productivity. Establishment of MPAs is one among the best conservation strategies. Its aim is to protect and conserve the natural marine ecosystems in their pristine condition. The International Union for the Conservation of Nature (IUCN) has defined a Marine Protected Area (MPA) as "*any area of inter-tidal or sub-tidal terrain, together with its overlying water and associated flora, fauna, historical and cultural features, which has been reserved by law or other effective means to protect part or all of the enclosed environment*".

The Marine Protected Areas are primarily classified into following three categories:

Category-I: This covers National Parks and Sanctuaries and having entire areas in intertidal/sub-tidal or mangroves, coral reefs, creeks, seagrass beds, algal beds, estuaries and lagoons.

Category-II: This includes Islands, which have major parts in marine ecosystem and some part in terrestrial ecosystem.

Category-IIIA: This includes sandy beaches beyond intertidal line but occasionally interacting with the seawater.

Category-IIIB: This includes ever green or semi ever green forests of Islands.

In order to reduce the human pressures on marine habitats, many areas have been identified all over the world and designated as "Marine Protected Areas". Over 4000 MPAs exist in the world today. Only 1% of ocean surface is protected and managed legally as MPAs, in contrast 15% of land surface is managed under "Protected Areas". Various components of marine conservation and restoration are combined into management packages for the marine protected areas.

10.1.1.1. Impacts of MPAs

MPAs help in enhancing the populations of marine organisms, saving rare species, maintaining food web structures, ensuring ecosystem stability, providing income for people and facilitating educational and research for promotion of marine management. The fishes are larger in size, higher in fecundity and rich in standing biomass within MPAs as compared to outside. This ensures a healthy recruitment of fishes for future generation. All these benefits are dependent on the duration of protection and extent of areas under MPAs. Older MPAs contain more fish than new ones. Larger protected areas have more abundance and biodiversity than smaller protected areas. Thus the MPAs increase fish catches of large size and biodiversity by maintaining part of a region, free from fishing pressure.

Some important MPAs in the world are briefed below:

In the case of **Darwin mounds, UK** (Fig. 10.2), there was no legislation in place to protect the mounds since they are located in deep offshore waters. As a result, deep-sea trawling caused a serious damage to fish resources of the mounds. In 2004, permanent prohibition of fishing on the mounds came into force. As a result the Darwin Mounds is the first offshore area managed for conservation.

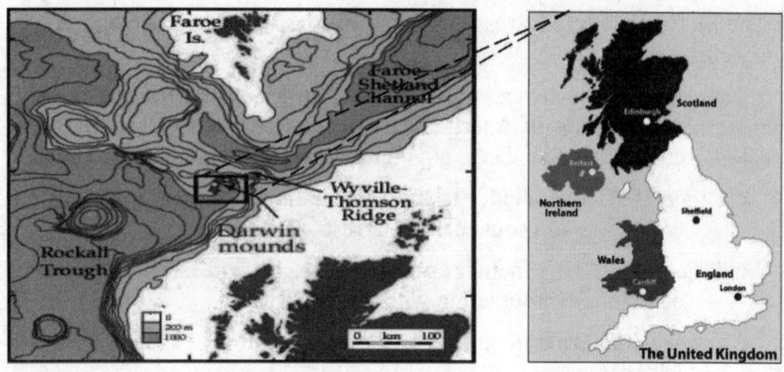

Fig. 10.2: Darwin mounds, United Kingdom
(http://www.noc.soton.ac.uk/gg/research/seafloor_env/darwin_mounds.php)

Mafia Island, Tanzania is a marine protected area. This area provides livelihoods of over 15,000 people. Therefore, fishing is allowed within the area but with a variety of prohibitions that limit the activities of fishermen especially for harvesting turtle and dugong, as well as dumping of wastes. Permit systems are set up to restrict the exploitation of octopus and lobsters and the collection of sea cucumbers using SCUBA is banned. Boat surveillance is carried out by the local community on a regular basis. Confiscation of fishing gear is levied for illegal fishing within the marine park area.

Florida Keys, USA contains different habitats including seagrasses, sand blanks, mangroves and coral reefs. There are five types of zones in the sanctuary: (1) Sanctuary Preservation Areas, (2) Ecological Reserve Areas, (3) Special-Use (research only) Areas (4) Wildlife Management Areas, and (5) Management Areas. These areas have different rules and regulations for restricting consumptive activities and allowing other activities which do not damage resources. Special protection is given for increasing the fish stocks and densities of the commercially important species.

Great Barrier Reef, Australia (Fig. 10.3) is the largest MPA in the world. Prior to 2004, the extent of "no take" zones was around 4.6%, and the remaining very large part of the reef areas were allowed for fishing and other uses. In 2004, the plan was modified which increased the 'no-take' zone to a 33.4% of the MPA in order to increase the level of protection.

Gulf of Mannar, India (Fig. 10.3) is the first Marine Biosphere Reserve in the South and Southeast Asia. This is unique to have a rich marine biodiversity with as much as 3,600 species in 19 islands with three inter-linked ecosystems: coral reefs, mangroves and seagrasses. Considerable efforts have been taken in awareness creation among the fisher folk, research and monitoring to collect baseline data on biological resources and environmental aspects of the area.

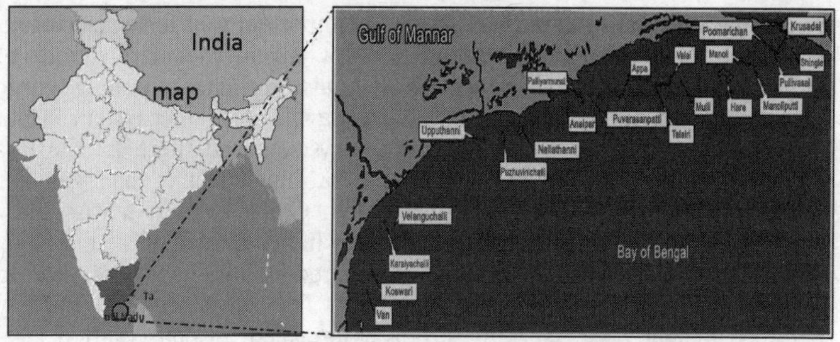

Fig. 10.3: Gulf of Mannar, India

(http://www.casmbenvis.nic.in/Gulf-of-Mannar-coral.html)

10.1.1.2. Surveys and monitoring for MPA status

One of the most important aspects of MPA is habitat heterogeneity. The structurally complex sites have a large number of fish species. The areas are mapped and surveyed and then the scientific and conservation merits of areas are prioritized. To select the locations of new MPAs, criteria such as economic, social and scientific aspects are considered.

10.1.1.3. Zoning of MPAs

The zoning MPAs into different areas with different pressures and purposes is a main aspect of marine conservation and management. Zoning defines spatial objectives and accompanying restrictions in a format understandable to all the stakeholders. To cite an example, three zones (A, B, C) are demarked in the case of Marine Reserve on St Barts of North eastern Caribbean area. All the three zones prohibit fishing with traps, spear and net fishing, collecting corals or plants, lobster fishing and littering. Zones B and C do not allow SCUBA diving. Zone C is restricted to only scientific observations.

There are also vertical zoning rather than horizontal zoning. To cite an example, the Huon Commonwealth Reserve in South east Australia experienced severe damage of rare and endemic benthic species by deep water fishing. In order to prevent this and to protect the benthic habitats, two zones have been created following the categories laid down by the International Union for the Conservation of Nature (IUCN). The first zone is managed as a resource protected area in which fishing is still allowed within limits, whereas the second zone is a strict nature reserve managed mainly for science. Some fishing is allowed in both zones, but only up to a depth of 500 metres in the sanctuary zone since the protected regions are located deeper than this.

10.1.1.4. Constraints of MPAs

Marine Protected Areas (MPAs) are an important tool for conservation of marine biodiversity. This alone can not ensure the maintenance of biodiversity and ecosystem processes. It is also difficult to manage large extent of MPAs effectively. The functioning of marine ecosystem is not better understood than that of terrestrial systems and there is also less technical capacity and expertise available for their effective management. Cost of managing MPAs is also very high. Only very few MPAs show adequate success. The reasons for failure are linked to lack of interest, lack of funding, lack of ecological baselines required for a long term monitoring and management for changing pressures. The rules are not properly enforced in many MPAs. Only strong enforcement of conservation laws and regulations can significantly enhance fish stocks.

- In order to reduce the pressures of human activities on habitats in the sea, many areas have been identified all over the world, designated as "Marine Protected Areas" (MPAs) and managed on scientific principles. Only 1% of ocean surface is protected and managed legally as MPAs.
- MPAs help in enhancing the populations of marine organisms, saving rare species, maintaining food web structures, ensuring ecosystem stability, providing income for people and facilitating education and research for promotion of ocean management. All these benefits are dependent on the duration of protection and extent of areas under MPAs.
- A few well known MAPs are Darwin mounds, UK (the first offshore area managed for conservation), Mafia island, Tanzania , The Florida Keys, USA, the Great Barrier Reef, Australia (the largest MPA in the world) and the Gulf of Mannar, India.
- Failures of MPAs are linked to lack of interest, lack of funding, lack of ecological baselines required for a long term monitoring and management for changing pressures. More importantly, improving MPA governance cannot be achieved without generating sufficient capacity, political will and resources for the enforcement of conservation laws and regulations.

10.1.2. Marine fisheries management policies

Each country has its own Marine Fisheries Policies for ensuring sustainability of fish population and securing livelihood of the people dependent on it. The policies are dealing with the coastal fishing in the territorial waters (which extend up to a distance of 12 nautical miles from the shore) and also dealing with the deep sea fishing (beyond the distance from the shore to the exclusive economic zone that extends up to a distance of 200 nautical miles). The Governments have attempted to conserve fish population through three methods: (1) dictating periods and time of fishing, (2) imposing bans for fishing at a specific season of fish breeding, and (3) specifying size of nets and type of gear to be used for fishing.

Among these, imposing a fishing ban is of recent importance. To cite an example, the Government of India through State Governments implements a fishing ban ('fishing holiday') every year. The annual closed fishing period in Indian fisheries varies and this ban lasts for 45-60 days with each state government using a different time period based on breeding season of fishes. Because the closed fishing season is not uniform for the Indian coast, trawler owners from one state do fishing in other states. These problems call for a uniform closed season by the state Governments, for the East Coast and the West Coast of India. The fishermen are given compensation in terms of monetary benefits by the government during the ban period. The artisan fishers are most benefited rather than trawler-fishers. There is dearth of information on biological attributes of fisheries in each of the region. The fisheries departments of each state Governments are the primary enforcement authority for fisheries management in territorial waters (0-12 nautical miles). However, majority of the states do not have patrol vessels to enforce a fishing ban. The coast guard is providing supplementary support in some states during the ban period, but it is impossible for the coast guard to monitor the vast numbers of trawlers operating along the Indian coastline. Illegal fishing by domestic mechanized trawlers in the inshore artisan zone has been reported in all coastal states of India. Indian fishermen are of the opinion that "fishing ban" every year is not working as fishermen put more effort of catching the fishes and shrimps before and after the ban period. Many fishermen are also of the opinion that both number of fishing vessels and hours at sea should be restricted for fishing trawlers during the breeding season to prevent capture of juvenile fish and shrimps.

- Each country has its own Marine Fisheries Policies for ensuring sustainability of fish population and fishers' livelihood security. The policies are dealing with the coastal fishing in the territorial waters, which extend up to a distance of 12 nautical miles from the shore and also dealing with the deep sea fishing beyond a distance of 12 nautical miles from the shore to the exclusive economic zone that extends up to a distance of 200 nautical miles.
- The Governments have attempted to conserve fish population through three methods: (1) dictating periods and time of fishing, (2) imposing bans for fishing at a specific season of fish breeding and (3) specifying size of nets and type of gear to be used for fishing.

10.1.3. Coastal management policies

Coastal areas are highly vulnerable to human pressure and developmental activities. Most of the nations are attempting to protect the coastal areas from developmental activities. Government of India has proposed the Coastal Regulation Zone Notification-2011. Accordingly there are four coastal management zones, as briefed below.

1. **Coastal Regulation Zone-I (CRZ-I)** consists of the following ecologically sensitive areas and the geo-morphological features that play a primary role in maintaining the integrity of the coast:
   - Mangroves (in case area is more than 1000 square metres, a buffer area of 50 metres shall be provided);
   - Corals and coral reefs and associated biodiversity;
   - Sand dunes;
   - Mudflats which are biologically active;
   - National parks, marine parks, sanctuaries, reserve forests, wildlife habitats and other protected areas including Biosphere Reserves encompassing salt marshes; turtle nesting grounds; horse shoe crabs habitats; seagrass beds; and nesting grounds of birds.

   No new construction shall be permitted in CRZ-I in the ecologically sensitive areas except the projects related to the Department of Atomic Energy, pipelines, transmission lines, installation of weather radar for monitoring of cyclones movement and prediction by the Indian meteorological Department, construction of trans-harbour sea link and roads without affecting the tidal flow of water, between low tide line and high tide line.
2. **Coastal Regulation Zone-II (CRZ-II)** consists of the areas which are developed up to or close to the shoreline and falling within government administrative limits.
3. **Coastal Regulation Zone-III (CRZ-III)** consists of all other open areas including the coastal seas but excluding those areas classified as CMZ-I, CMZ-II and CMZ -IV.
4. **Coastal Regulation Zone-IV (CRZ-IV)** consists of Islands of the Andaman and Nicobar and Lakshadweep.

- Coastal areas are highly vulnerable to human pressure and developmental activities. Hence, world nations are attempting to protect the coastal areas from developmental activities.

- Government of India has proposed the Coastal Regulation Zone Notification-2011. Accordingly there are four coastal management zones, of which Coastal Management Zone-I (CRZ-I) consists of the ecologically sensitive areas and the geomorphological features that play a primary role in maintaining the integrity of the coast.

10.1.3.1. Coastal aquaculture policies

One of the key drivers of mangrove conversion in recent decades has been the growth of aquaculture, particularly shrimp aquaculture. One major issue associated with the farms located in mangrove habitats is acidification of pond waters that kills aquatic organisms. Although they may not have impact directly upon mangroves, the out-flowing water with high organic and inorganic pollutants from the ponds pollutes the ecosystems and fishery resources. However, coastal aquaculture is an important source of livelihood options and employment opportunities in the coastal areas. Realizing this, many countries have their regulations for coastal aquaculture practices. To cite an example, India has provided guidelines for the coastal aquaculture practices. It has specifically instructed that mangroves, agricultural lands, salt-pan lands, ecologically sensitive areas such as sanctuaries and marine parks should not be used for shrimp farming. A minimum distance of 50-100 metres shall be maintained between the shrimp farm and adjoining land. All shrimp farms should maintain 100 metres distance from the nearest drinking water sources. The shrimp farms should not be located across natural drainage canals/flood drain. Shrimp farms should be located at least 100 metres away from any human settlement in a village of less than 500 populations and beyond 300 metres from any village of over 500 populations. For major towns and heritage areas it should be around 2 kilometres. Water spread area of a farm shall not exceed 60 % of the total area of the land. The rest 40 % can be used for other purposes. Plantation should be done wherever possible.

10.1.4. International cooperation

10.1.4.1. CBD (Convention on Biological Diversity)

"Biological Diversity" means the variability among living organisms from all sources including, terrestrial, marine and other aquatic ecosystems. This includes diversity within species, between species and of ecosystems. Realising the global importance of biodiversity, the International Convention on Biological Diversity (CBD) was signed by the Heads of over 150 governments at Rio de Janeiro in June 1992. The major objectives of the convention are: (1) the conservation of biological diversity, (2) the sustainable use of its components and (3) the fair and equitable sharing of the benefits arising out of their use. Different nations have developed 'Biodiversity Act' for saving and sharing the genetic wealth of our planet including marine biological resources. India has also enacted the Biological Diversity Act - 2002.

10.1.4.2. IUCN (*International Union for the Conservation of Nature*)

IUCN is an international organization dedicated to natural resource conservation. It was founded in 1948 as the world's first global environmental organization. IUCN is the world's oldest and largest global environmental network. This is a leading authority on the environment and sustainable development with more than 1,000 government and NGO member organizations, and almost 11,000 volunteer scientists in more than 160 countries.

IUCN helps the world to find pragmatic solutions for our environment and development problems. It supports scientific research and manages field projects all over the world. It also brings governments, non-government organizations, United Nations agencies, companies and local communities together to develop and implement policy, laws and best practices. Its headquarters is located in the Lake Geneva area in Gland, Switzerland. It is funded by governments, bilateral and multilateral agencies, foundations, member organizations and corporations. IUCN has an official observer status at the United Nations General Assembly.

Global Marine and Polar Programme (GMPP) of IUCN was founded to work with members and partners around the world. This network of marine specialists within IUCN is dedicated to help in creating an international framework for protection of marine biodiversity. The goal of the GMPP is to achieve significant improvement in the conservation of marine and polar biodiversity, and sustainable use of natural resources in marine, coastal and polar ecosystems, throughout the world. From local fishermen and decision-makers right up to the United Nations General Assembly, the GMPP works in partnership to deal with tourism, offshore renewable energy and to minimize the environmental impacts from coastal oil and gas operations. Additionally, the GMPP works with several commissions such as the World Commission on Protected Areas-Marine Division, the Species Survival Commissions-Specialists Groups, the Marine Conservation sub-committee, and the Commission on Environmental Law-Oceans Law.

Conservation of the Seacow *Dugong Dugon*

The sea cow or pig fish, *Dugong dugon* (Fig. 10.4) is the only species of the Dugongidae family in the Order Sirenia. It is the large mammal distantly related to elephants. It is an herbivore, feeding mainly on seagrasses during night hours. It inhabits coastal and near shore waters of 37 countries ranging from east Africa to Australia, through south and south-east Asia.

In India, information about the animal is restricted to sighting, stranding and mortality records. Dugongs are listed as vulnerable on the IUCN Red List. The animals have large size, slow movement and dependence on coastal habitats and hence they are vulnerable to human impact. Its oil is used for preventing the attachment of wood borers on wooden materials. Historically dugongs were abundant in the Andaman and Nicobar islands, but hunting and habitat loss has led to the decline in their numbers. Fishing nets are a major cause of population decline, since Dugongs are unable to hold their breath for about 12 minutes, and therefore easily drown once entangled.

Fig.10.4: *Dugong dugon*
(http://www.nevillecoleman.com.au/dive-travel-destinations--scuba-diving---snorkeling/vanuatu-diving.aspx)

Whale shark: A success story of conservation in India

The whale shark (*Rhincodon typus*) migrates from Australia to Saurashtra coastal waters in Gujarat, India. But, a large number of them were killed every year by fishermen. Therefore, Govt. of India put a ban on killing the whale shark in the year 2001. But it was in vain. The people continued to harvest the shark. At the time, a conservationist touched the sentiment of the people by saying that the whale sharks are like our daughters, coming from Australia to India for delivery at motherland. Moved by this story, the people stopped killing the sharks.

10.1.4.3. WTO (World Trade Organization)

The World Trade Organization (WTO) deals with the rules of trade between 153 nations at a global level. It is an organization for liberalizing trade, a forum for governments to negotiate trade agreements and a place to settle trade disputes. It operates a system of trade rules. The WTO began life on 1 January 1995, but its trading system is half a century older. Since 1948, the General Agreement on Tariffs and Trade (GATT) had provided the rules for the system, mainly dealt with trade in goods including marine-derived products, whereas the WTO covers trade in services and in traded inventions, creations and designs (intellectual property).

10.1.4.4. CITES (Convention on International Trade in Endangered Species of Wild Fauna and Flora)

CITES was drafted and initially signed in 1973 in Washington. It entered into force in 1975. The treaty now has 166 parties. Some marine species whose trade is regulated by CITES are beaked whales, marine turtles, coelacanths, dugongs, Irrawaddy river dolphin, surgeon fish, fur seal, marine dolphin, porpoise, black corals, giant clam, stony coral species, basking shark, Great white sharks, hump-head wrasse, queen conch, seahorses, and whale shark.

10.1.4.5. UNEP (United Nations Environment Programme)

UNEP (United Nations Environment Programme) promotes the use of science to apply ecosystem management to address the factors causing decline of ecosystem services. UNEP's Marine and Coastal Strategy focuses on priority issues for maintaining marine ecosystems and their services for human well-being. The vision of UNEP is "prosperous and healthy oceans and coasts where conservation, productivity and resource use are sustainable". This vision will be achieved through a long-term plan to improve marine and coastal environments and ultimately reduce human impact through four objectives:

- *Land-Ocean Connections:* to integrate the management of coastal watersheds, the coastal area (including cities) and the marine environment to optimize the ecosystem services and resilience of marine and coastal systems.
- *Ecosystems for Human Well-Being:* to identify, assess and value the status, key drivers of change and services of marine and coastal ecosystems and their link to human well-being.
- *Reconciling Use and Conservation:* Management tools, capacity and options are available for regions, countries, communities and the private sector to effectively engage in the sustainable management of marine and coastal ecosystems.
- *Vulnerable People and Places:* to strengthen the ecological, economic and social security of vulnerable communities and places including Small Island Developing States to adapt and respond to natural disasters and climate change, through enhancing the resilience of marine and coastal ecosystems and social capital and improved access and benefit sharing.

10.1.4.6. CMS (Convention on the Conservation of Migratory Species of Wild Animals)

CMS (also known as Bonn Convention) aims to conserve terrestrial, marine and avian migratory species throughout their range. It is an intergovernmental treaty, included under the aegis of the United Nations Environment Programme (UNEP), concerned with the conservation of wildlife and habitats on a global scale. Its membership includes 115 parties (as of 1 March 2011). CMS is the only global convention specialising in the conservation of migratory species, their habitats and migration routes. CMS co-operates with a number of other international organizations, NGOs and partners in the media as well as in the corporate sectors. CMS Parties strive towards strictly protecting these animals, conserving or restoring the places where they live, mitigating obstacles to migration and controlling other factors that might endanger them. CMS makes agreements with its partners. The agreements may range from legally binding treaties (called 'Agreements') to less formal instruments, such as Memoranda of Understanding. The agreements are adapted to the requirements of particular regions.

10.1.4.7. IOC (Intergovernmental Oceanographic Commission) of UNESCO

IOC is a competent organization of the United Nations in the field of marine sciences and services. Its purpose is to promote international cooperation and to coordinate programmes in research, services and capacity-building. Its purpose is also to learn more about the nature and resources of the ocean and coastal areas and to apply that knowledge for the improvement of management, sustainable development and protection of the marine environment as well as decision-making processes of its member state.

10.1.4.8. LOICZ (Land-Ocean Interactions in the Coastal Zone)

LOICZ is a core project of the International Geosphere-Biosphere Programme (IGBP) and the International Human Dimensions Programme on Global Environmental Change (IHDP). LOICZ is an international research project involving scientists from across the globe. This has been investigating changes in the biology, chemistry and physics of the coastal zone since 1993. Since 2003, LOICZ has expanded its areas of research to include social, political and economic sciences in order to address the human dimensions of the coastal zone. The research results are used to explore the role of humans to play in the coastal zone, their vulnerability to changing environments, and the options to protect coasts for future generations. The main goal of LOICZ is "to provide the knowledge, understanding and prediction needed to allow coastal communities to assess, anticipate and respond to the interaction of global change and local pressures which determine coastal change."

10.1.4.9. SCOR (Scientific Committee on Oceanic Research)

SCOR was the first interdisciplinary body formed by the International Council for Science (ICSU) in 1957. SCOR is the leading international non-governmental organization for the promotion and coordination of international oceanographic activities. SCOR does not have the resources to fund research directly. Therefore, SCOR science activities focus on promoting international cooperation in planning and conducting oceanographic research and solving methodological and conceptual problems. Scientists from 36 nations participate in SCOR working groups and scientific steering committees for the large-scale ocean research projects. SCOR promotes capacity building for marine scientists in developing countries and also provides travel grants to visiting scholars. As a non-governmental organization, individuals are representing the national SCOR committees, not national governments. Each national committee can be represented by three individual ocean scientists who serve as nominated members.

10.1.4.10. WWF (World Wild Fund)

WWF is the world's leading organization for conservation of nature. WWF works in 100 countries and is supported by 1.2 million members in the United States and close to 5 million globally. WWF works at every level from local to global, and ensures the delivery of innovative solutions that meet the needs of both people and nature. WWF's mission is the conservation of nature through the following ways.

- Protecting natural areas and wild populations of plants and animals, including endangered species;
- Promoting sustainable approaches to the use of renewable natural resources; and,
- Promoting more efficient use of resources and energy and the maximum reduction of pollution.

10.1.4.11. SCAR (Scientific Committee on Antarctic Research)

SCAR is an inter-disciplinary committee of the International Council for Science (ICSU). SCAR is charged with initiating, developing and coordinating high quality international scientific research in the Antarctic region. The scientific business of SCAR is conducted by its Standing Scientific Groups, active in Antarctic research and report to SCAR. In addition to carrying out its primary scientific role, SCAR also provides objective and independent scientific advice to the Antarctic Treaty Consultative Meetings and other organizations on issues of science and conservation that affect the management of Antarctica and the Southern Ocean. In that role, SCAR has made recommendations on a variety of matters, most of which have been incorporated into Antarctic Treaty. SCAR has its secretariat at the Scott Polar Research Institute in Cambridge, England.

SCAR meets every two years to conduct its administrative business. At these meetings the members of SCAR, through their appointed delegates, are responsible for approving SCAR finances and formulating SCAR policy and strategy. SCAR also holds, prior to the delegates meeting a major open science conference to draw attention to Antarctic issues.

10.1.4.12. IMO (International Maritime Organization)

IMO is the United Nations specialized agency with responsibility for the safety and security of shipping and the prevention of marine pollution by ships. Shipping is the most important world's industry, serving more than 90% of global trade by carrying huge quantities of cargo cost effectively, cleanly and safely. The best way of improving safety at sea is by developing international regulations that are followed by all shipping nations. From the mid-19th century onwards, a number of such treaties were adopted. In 1948 an international conference in Geneva adopted a convention for formally establishing IMO.

10.1.4.13. ICES (International Council for the Exploration of the Sea)

ICES coordinates and promotes marine research on oceanography, marine environment, marine ecosystem and on living marine resources. Members of the ICES community include all coastal states bordering the North Atlantic and the Baltic Sea, with members in the Mediterranean Sea and southern hemisphere. Its vision is to be an international scientific community, concerning marine ecosystems and their relation to humanity. Its mission is to advance the scientific capacity to give advice on human activities affecting and affected by marine ecosystems. ICES is a network of more than 1600 scientists from 200 institutes linked by an intergovernmental agreement (the ICES Convention) to add value to national research efforts.

Scientists working through ICES gather information about the marine ecosystems. This information is also developed into non-political advice. The 20 member countries that fund and support ICES use this advice to help them to manage the North Atlantic Ocean and adjacent seas. ICES plans and coordinates marine research through its national delegates and through a large numbers of expert groups, symposia and an Annual Science Conference. Most meetings take place at the ICES Headquarters in Copenhagen, Denmark or in the member countries. ICES maintains some of the world's largest databases on marine fisheries, oceanography and marine environment. ICES publishes its scientific information and advice in reports, publications (including the *ICES Journal of Marine Science*), electronic media and on the ICES website.

- Individual countries collaborate with international agencies in conservation and management of marine environment through legally binding treaties (called 'Agreements') to less formal instruments, such as Memoranda of Understanding.
- The international agencies involved in marine management are Convention on Biological Diversity, IUCN *(International Union For Conservation of Nature)*, WTO (World Trade Organization), CITES (Convention On International Trade In Endangered Species Of Wild Fauna And Flora), UNEP (United Nations Environment Programme), CMS (Convention On The Conservation Of Migratory Species of Wild Animals), Intergovernmental Oceanographic Commission (IOC of UNESCO), Land-Ocean Interactions In The

Coastal Zone (LOICZ), SCOR, World Wild Fund, SCAR, International Maritime Organization (IMO), ICES and others.

10.2. PARTICIPATORY APPROACH FOR MARINE MANAGEMENT

People's participation is extremely important in conservation and management of marine resources. It is necessary to allow people to make decisions as they are responsible for the present and future status of the marine resources. This is broadly termed 'Co-management' or "Participatory management". The participatory management involves partnership arrangements, in which government, community of local resource users, external agents (NGOs, academic and research institutions), and other stakeholders (fish traders, money lenders, tourism) share the responsibility and authority in decision-making on the management of natural resources.

10.2.1. Types of stakeholders

Stakeholders can be defined as those interested and dependent on the resources. They may vary in different places, and they largely include the following:

- *Local User Communities* – People of these communities live in and around marine habitats, directly using the resources such as fishes and other marine life.
- *Remote User Communities* – People of these communities hail from far away places and use resources.
- *Government Agencies* – They have the responsibility for managing and protecting marine resources *e.g.* fisheries, forest, tourism *etc.*
- *Supporters of Marine User Communities* – These belong to non-governmental organizations (NGOs) and voluntary agencies.
- *Research and Academic Institutions* – They monitor the status of health of the marine habitats and their resources on a continuous basis.

10.2.2. Types of participation

Participation of the stakeholders varies in intensity and approach. The degree of participation may be of three types: non participation, partial participation and genuine participation. In non-participation, the local people are ignored, and the decisions of conservation and management are taken by the officials and experts. In partial participation, the local people are not involved in decision-making, but are consulted or they may be allowed to participate for food, money and other incentives. In genuine participation, the stakeholders are involved in decision-making directly to participate in planning, implementation, and monitoring of resources. This is the most beneficial type of participation for successful management of natural resources.

10.2.3. Strategies for participation

No decision-making is complete without participation of local people whose livelihood depends upon resources. People have been using marine resources in a traditional manner since times immemorial. Both traditional and latest scientific technologies need to be blended to achieve long-term conservation goals. There are different steps of participation as briefed below, which may vary from place to place depending upon the type of resources:

- *Situation Analysis:* This is the first step that includes identification of stakeholders, resources available, resource utilization, degree of dependence by stakeholders, degradation pressures, causes of degradation pressures, government practices and socio-economic status of the local communities.
- *Partnership Development:* The stakeholders who are dependent on the natural resources are to be linked. They need to be motivated for co-operation and mutual benefits.
- *Community Issue Analysis:* For this, a common technique known as 'Participatory Rural Appraisal' (PRA) is followed. In this process, the stakeholders are consulted for the problems related to marine conservation and remedial measures. This helps in prioritization of actions.
- *Group Formation:* In order to share the responsibility of planning and implementation, different working groups are formed such as credit groups, women's groups and land-based alternative groups.
- *Micro-planning:* This is used for implementing a target-oriented action plan. It defines the problems, goals, targets to be achieved, implementation structure and time–bound activities.
- *Implementation and Monitoring:* This includes defining the implementation procedures for micro-plans, recording of the implementation activities and their outputs during monitoring activities.
- *Evaluation and Feedback:* It leads to the assessment and success of activities and identifying the drawbacks.
- *Refinement and Extension:* Based on the evaluation and feedback, the models can be refined and extended to other places.

The participatory approach is successful in the Sundarbans, India for improvement of relationship between people and the forest department, in saving the tigers, protecting the plantations and in voluntary patrolling to protect the coastal forest. Ecotourism generates employment for local people in the Sundarbans. Entry fees are collected from tourists entering the Sundarbans Tiger Reserve, with 25% of the revenue earned in the Reserve area is returned back to the eco-development committees. Using the revenue, the committees implement eco-development activities: (1) improvement of village roads, bridges and culverts, (2) construction of jetties, (3) supply of seedlings for forestry, fruit species for orchards, and fuel wood and fodder species, (4) provision of poultry and piggery units, (5) bee hives for apiculture, (6) solar street light and solar lanterns in remote villages, (7) artificial insemination and immunization of cattle, (8) conducting medical camps, (9) provision of smokeless cookers, nylon net and fencing nets and (10) offering training to people on crab fattening, sewing, apiculture, piggery, duckery, nursery techniques to improve their skills and to help them attain economic independence. The priority need of the people living there is fresh water, which is required for irrigation and drinking purpose. Under the co-management programme, the people are provided with deep tube wells for drinking water, irrigation tanks and canal for agriculture and ponds for fish culture.

10.2.4. Civil Society Awareness

Conservation and management of any natural resources will be successful on a long term basis, only when civil society awareness and people's participation in decision making are promoted. If these components are weak, the long term sustainability of natural resources is not possible. Eggs of the turtles are poached by humans, when the people do not aware of the value of the species in ecotourism. Therefore, environmental education and awareness is necessary.

Civil society awareness on the values of marine biodiversity can be created through many ways. Information materials have to be developed in local languages in an attractive manner. Individuals and community that contribute towards marine conservation should be recognized with awards and incentives. Public awareness programmes through mass media, websites and visuals have to be promoted. Awareness should be strengthened right from schooling through development of curricula on coastal and marine environment.

Marine environment in particular coastal areas are preferable sites for establishment of industrial developments, which cause concern of pollution. The industries are not aware of the values of the mangroves and seagrasses, which are efficient in carbon sequestration and in removal of heavy metal pollutants. The industries can be motivated to take up mangroves and seagrass plantations on a large scale. This will ultimately help in generating livelihood for the local people and income, besides coastal protection.

The land-based activities have a greater impact on marine ecology. However, the river based planners, land resource managers, and policy makers are largely not aware of their activities that affect the marine environment. For example, freshwater flow is required to feed the coastal water bodies, which are vital for the marine and fresh water fishes to complete part of their life cycles. The increased dam construction and water utility for irrigation and drinking purposes reduces the flow of water to the coastal water bodies. As a result the fishery resources and coastal habitats are shrinking, and this has resulted in loss of coastal livelihood.

10.2.5. Training and Capacity Building

Managers require much training on marine resource assessment, ecotourism, coastal rehabilitation, biodiversity conservation and coastal protection. There is also a need for training field officers on SCUBA diving, snorkeling, monitoring of corals, mangroves and seagrass ecosystems. The local communities are to be trained in skill development for alternate livelihood, mangrove restoration, coral reef rejuvenation and seagrass transplantation. The coastal managers should be educated with the latest developments around the world to tackle the management issues.

- The participatory management involves partnership arrangements, in which government, community of local resource users, external agents (NGOs, academic and research institutions), and other stakeholders (fish traders, money lenders, tourism) have to share the responsibility and authority in decision-making on the management of natural resources.
- To make the participatory management effective, civil society awareness and training and capacity building among the stakeholders are essential.

10.2.6. Coastal Rehabilitation

A huge human population living along the coast destroys the coastal ecosystems. Rehabilitation of the coastal ecosystems is vital for improving environmental conditions, and well-being of the human beings. This can be achieved better with people's participation. This involves two initial steps: (1) identification of potential areas for rehabilitation and natural regeneration of mangroves/seagrass/coral reefs; and, (2) assessment of the past and current restoration practices to find out the constraints and remedial measures for evolving better practices for coral rejuvenation, seagrass transplantation and mangrove restoration. People should be trained on the coastal rehabilitation practices, as well on impact assessment of the practices on ecological and fishery resources. In addition, people should be trained to breed the marine fishes of commercial and biodiversity values in captivity and to restock them in the natural environment for recovery of fish stock. It is also necessary to involve the people in establishment of 'gene banks' for genetic resources of corals, mangroves and seagrasses.

- Rehabilitation of the degraded coastal ecosystems can be achieved better with people's participation.
- This involves two initial steps: (1) identification of potential areas for rehabilitation and natural regeneration of mangroves/seagrass/coral reefs; and, (2) assessment of the past and current restoration practices to find out the constraints and remedial measures for evolving better practices for coastal restoration.

10.2.7. Coastal livelihood

Livelihoods are essential for income generation and sustenance of humans. The livelihood should be helpful for the benefit of coastal ecology and income generation of people. People should be empowered to take up different enterprises. In order to reduce pressure on marine ecosystems, it is necessary to develop alternate livelihood options for the people who are dependent on marine resources. In the absence of any alternative employment, the poor people depend largely on marine resources; they tend to resort to illegal practices such as over-fishing, poaching and felling. The people are collecting the tiger shrimp seeds in large numbers thereby causing damage to juvenile fish stocks. In the Sundarbans, each year about 500 honey collectors and woodcutters enter the forests in search of livelihood ignoring even the threat of attacks by tigers and crocodiles. Another example is camel herding, one of the activities practiced in the mangrove areas by a human community known as 'Maldharis' in Gujarat. The maldharis are in the habit of shifting along with their livestock to far away areas in search of fodder for their cattle. The degradation and non-access of mangroves has critically impacted their livelihoods. The local people should be trained and helped largely on alternate and supplement livelihoods to gain employment thereby reducing human pressure on mangrove resources.

Fishing is the major source of income for the coastal people. However, activities that integrate coastal rehabilitation with fishery development such as traditional canal fishing methods, crab fattening, culturing of oyster, mussel and clam may be promoted within the coastal water bodies. This will increase the income generation of the local community. Some examples of coastal livelihoods are: (i) mariculture practices in the open coastal water for farming of seaweeds, shell fish and finfish; (ii) integrated fish farming with cultivation of mangroves, other halophytes, fishes, prawns, mollusks, crabs; (iii) organic cultivation of vegetables along the non-saline or less saline landward areas of coast; (iv) honey production using apiculture practices, which will also help in better pollination opportunities for some mangrove species that are insect-pollinated; (v) cultivation of 'cash crops' such as cashew, palmyra, coconut etc., along the coastal sand dune; and (vi) ecotourism.

10.2.8. Income Generation and Financial Sustainability

This aspect is vital to ensure community participation in marine management activities. To cite an example from India, Gujarat Ecology Commission has been successful in restoration of mangroves in Gujarat with the help of India-Canada Environment Facility (ICEF) and State Government of Gujarat. With the involvement of local communities in regeneration of mangroves, about 4,000 hectare area has been brought under mangrove cover at ten different villages within a period of five years. Community participation has been institutionalized through formation of 10 Community Based Organization (CBO) at the village level. A corpus fund has been created in each village by depositing a part of daily wages in the bank account. Out of daily wage amount of ₹ 80, ₹ 56 is paid in cash, whereas ₹ 20 is deposited in the corpus fund. Remaining ₹ 4 is treated to be the contribution from the community towards the project. The 60% of fund along with the interest is reserved for maintenance of mangroves and other assets created under the project. Balance 40% amount is used as 'revolving fund' to give loans for income generation activities of CBO members.

- In order to reduce pressure on marine ecosystems, it is necessary to develop alternate livelihood options for the people who are dependent on marine resources. In the absence of any alternative employment, the poor people depend largely on marine resources; they tend to resort to illegal practices, such as over-fishing, poaching and felling.
- Fishing is the major source of income for the coastal people. However, activities that integrate coastal rehabilitation with fishery development such as traditional canals fishing methods, crab fattening, culturing of mussel, oyster and clam may be promoted within the coastal water bodies. This will increase the income generation of the local community.
- Income generation and financial sustainability are vital to ensure community participation in marine management activities.

10.2.9. Participatory Monitoring

The ecological and socio-economic assessment and monitoring are vital for understanding the changes in the habitats over time. Environmental evaluation is important to delineate the values of goods and services of marine ecosystems. It is of importance in evaluation of ocean management initiatives and laws/policies for their impacts on marine resources and management. The stakeholders should be trained in the monitoring processes on environmental evaluation. This should be a regular and continuous process. It is also necessary to involve local people and other stakeholders such as social scientists and ecologists in the exercises

(Fig. 10.5). In this process, local people have to be involved along with scientists especially social scientists. This participatory approach will greatly increase manpower to collect large amounts of data in a short time at a low cost. The effort will help in educating the local community and enhancing their feeling of ownership for the ecosystems. The team will be able to prepare the "biodiversity registers" for marine organisms at local levels linked to national and international data bank systems.

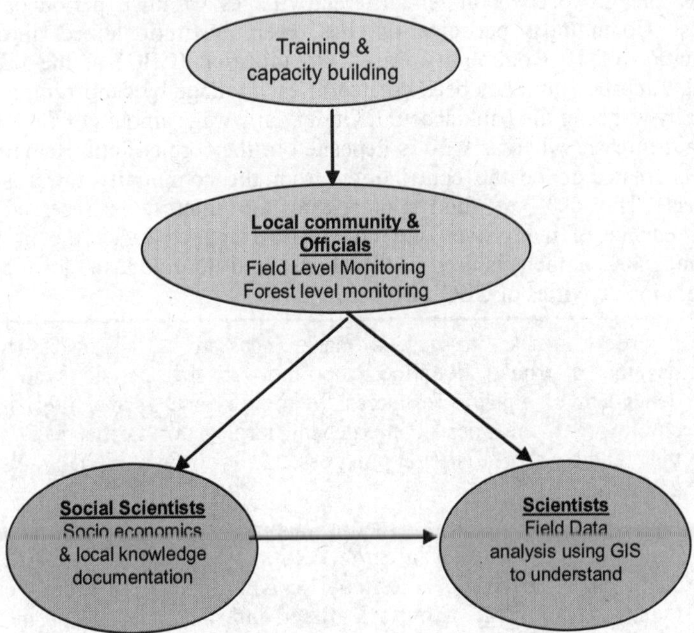

Fig. 10.5: Participatory monitoring of marine resources

- It is of importance in evaluation of ocean management initiatives and laws/policies for their impacts on marine resources and management.
- The stakeholders should be trained in the monitoring processes on environmental evaluation. This should be a regular and continuous process.
- It is also necessary to involve local people and other stakeholders such as social scientists and ecologists in the exercises.

10.2.10. Research and Development in Ocean Management

Research inputs are necessary for better preparation and implementation of management action plan. There are still many knowledge gap areas in the management of marine resources. Research activities are required to fill the knowledge gap areas, as briefed below for developing effective management action plan.

10.2.10.1. *Biodiversity Assessment*

Ocean is rich in biodiversity. World oceans host 31 of the 34 known animal phyla on the earth. New species are being discovered particularly in the deep sea. However, there was a decline of 14% of the marine species between 1970 and 2005. It is not clear about the actual number of species existing in the marine environment. Even the existing information is scattered and fragmentary; and, there is a need for comprehensive compilation of data on the coastal and marine biodiversity in site-specific manner.

10.2.10.2. *Marine Biological Research*

Biological studies especially for reproductive aspects are required on threatened species or commercial species in relation to environmental conditions. This will help in multiplication of the marine species. This will have bearing on conservation of rare or over-exploited commercial species through the 'sea ranching'/restocking practices and also on developing marketable items through aquaculture enterprises.

10.2.10.3. *Interrelationship between coastal ecosystems for augmentation of fish resources*

Coastal ecosystems are functionally linked. Each coastal ecosystem has peculiar species composition and interactions. Studies in this aspect will help in understanding the interdependence of species compositions to ensure better productivity. It has been proved experimentally in the Caribbean Sea that when coral habitat is connected with mangroves, the biomass of several commercially important fish species is more than doubled. Conversely, after removal of mangroves, the largest herbivorous fish in the Atlantic, *Scarus guacamaia* has suffered extinction. It is therefore necessary to study the connectivity among mangroves, coral reefs and seagrasses for augmentation of fish resources.

10.2.10.4. *Marine ecological studies*

Ecological variables including coastal processes (hydrodynamics, siltation, erosion, pollution) have significant impact on the marine living resources. The marine living resources, in turn, influence the ecological conditions of their environment. This two way interaction has to be understood for better management of the resources. The ecological variables are changing with climate change, and the biological species are also evolving to adapt to the changing conditions. This has a bearing on the long term management of the marine resources.

10.2.10.5. *Bioprospecting for high value products*

Bioprospecting is a search for high value products such as genes, medicines and chemicals. Marine organisms have been traditionally used for many purposes. However, there is not proper scientific validation for the traditional knowledge. Marine invertebrates and their associated microbes have been proved to be the potential source of life- saving drugs. An intensive search is necessary for extracting substances of pharmaceutical and industrial values from the sea. This effort may lead to development of patents, which can be a promising source of revenue and other opportunities.

- Research inputs are necessary for better preparation and implementation of management action plan.
- Research activities are required to fill the knowledge gap areas: biodiversity assessment, marine biological research, interrelationship between coastal ecosystems, marine ecological studies and bioprospecting for high value products.

Terms and Definitions

- Marine Protected Area (MPA) is any area of inter-tidal or sub-tidal terrain, together with its overlying water and associated flora, fauna, historical and cultural features, which has been reserved by law or other effective means to protect part or the entire enclosed environment.
- The coastal land up to 500m from the High Tide Line (HTL) and a stage of 100m along banks of creeks, estuaries, backwater and rivers subject to tidal fluctuations, is called the Coastal Regulation Zone (CRZ).
- The variability among living organisms from all sources including, terrestrial, marine and other aquatic ecosystems is known as Biological Diversity.
- Bioprospecting is finding useful biological substances, further developing them, and patenting them.

10.3. MANAGEMENT OF CORAL REEFS

Coral reefs are managed throughout the world. The management of coral reefs involves five important aspects: (1) protection, (2) monitoring, (3) research studies, (4) capacity building and awareness creation, and (5) coral restoration.

10.3.1. Protection of coral reefs

The coral reefs are legally protected under Marine Protected Areas. In India, four coral reef areas (Gulf of Kutch, Lakshadweep, Gulf of Mannar and Andaman) are protected and managed with 100% financial support by the Government of India. This effort has reduced man-made pressure on coral reef ecosystem and increased coral recruitment and live coral cover in many sites.

10.3.2. Monitoring of coral reefs

The coral reefs are monitored at regular intervals to detect changes on various aspects such as coral status (dead and live), coral recruitment, coral reef associated fishes, fisheries, destructive fishing practices, sea surface temperature, coral bleaching, sedimentation, incidence of diseases and water quality parameters including heavy metal concentrations. There is no simple methodology for routine reef monitoring. In this regard, the ratio of *Porites/Acropora* and growth patterns (vertical versus horizontal growth) is suggested for monitoring reef health and recovery.

10.3.3. Research studies in coral reef

Coral reef studies are mostly related to reproductive biology on gonad development, spawning processes, coral diseases and environmental monitoring. Also the coral reef surveys are being carried out in less explored and unexplored areas.

10.3.4. Capacity building and awareness creation

Fishing is a major livelihood for the people dependent on coral reef resources. The people use destructive fishing practices such as shore seine operation, push net operation, and trap fishing in reef areas. As the fish catches decline, fishermen resort to coral mining. The root cause of all these issues are lack of awareness among the people. Awareness creation is therefore, an essential component of management. Such efforts are promoted on the values of coral reefs and associated fishes, and the adverse impact of destructive fishing on decline of fish resources. In order to reduce the pressure on the coral reef resources, people are trained in alternate livelihood options such as crab and lobster fattening, development of value added products, cephalopod culture and vermi-composting. In India, SCUBA diving facilities are made available to scientists, students and field workers as well as public in all the major reef regions. Environmental education and awareness creation have been initiated as evident by involvement of local people in reef monitoring, and by increasing numbers of students who are interested in coral research.

Conservation of coral reef resources is not totally successful because of the large size of the areas and lack of facilities for enforcement. This problem is being resolved by people's participatory approach. In this regard, 'Gulf of Mannar Biosphere Trust' in India has established 252 grass root level organizations such as Village Marine Conservation and Eco-Development Committees within 10 kilometres landward zone. The committee members are taught about the coral reefs, coral conservation/restoration processes, sustainable utilization of coral fish resources, and building their capacity for alternative livelihoods, as well as capacity building on identification of endangered marine species under total judicial protection. The alternate livelihoods such as coral transplantation and captive breeding of ornamental fishes for restocking of coral reef areas are developed for conservation of coral reefs with people's participation.

10.3.5. Restoration and rehabilitation of coral reefs

Reef restoration is an act of bringing back the damaged reef back to its original conditions in structure and function. It involves manipulation of biotic or abiotic factors. In order to restore, it is necessary to reduce the abiotic causes of coral decline such as algal overgrowth, pollution, temperature and light extremes, damage by tourist, over-fishing, dynamite fishing and poison fishing. The damaged places must be repaired to provide stable sites for coral recolonisation. For biotic manipulation, different methods are used and they are artificial reef, coral transplantation, electrically stimulated coral growth, larval ranching, coral fragment reseeding, low-tech reef rehabilitation, and coral farming.

10.3.5.1. Artificial reefs

Any solid material placed in the sea becomes an 'artificial reef' (Fig. 10.6). It is the nature of marine species to settle on the substrates. In the method of artificial reef, substrates made of tires, bamboo, large concrete structures and mattresses are deployed on the seafloor to provide substrate for benthic organisms. The artificial structures have roles to play in restoration and conservation of marine biodiversity. They protect and shelter juvenile fish, add to habitat complexity and stability and provide new sites for larval settlement and recruitment. The value of the fish catches is significantly higher on artificial reefs than from adjacent non-reef areas. Breakwaters also provide excellent substrates for the build up of new communities. The walls of jetties act as breakwaters and provide sites for colonisation of small fish. Some of the largest fixed structures to be placed in our seas are pipes, jetties and oil or gas rigs and platforms. When located in mud or sand, they enhance complexity of marine habitats, and encourage the colonisation of both sessile and mobile species. The Caribbean *Panulirus argus* is a commercially important lobster species whose abundance can be significantly increased by the installation of artificial shelters, constructed out of PVC pipes and cement slabs. Using these shelters, many fold increase in juvenile density and biomass of the lobster can be achieved.

Fig. 10.6: An artificial reef
(http://pangaea-yep.com/blogs/entry/ARTIFICIAL-REEFS-Are-they-the-Answer-)

Fiber glass impregnated with sand and calcium carbonate to roughen the surface is also used as fiberglass artificial reef structures. These units are broad cone shape with 70 cm bottom diameter and 50 cm high with numerous regularly spaced holes for 3-5 diameters over the surface. These units are anchored down using metal stakes hammered into the seabed. The fiberglass has demerit of light in weight and hence concrete structures are used mostly as artificial reefs. A common form of reef structure is a cubical concrete structure of different dimensions, with holes of different sizes penetrating the block. Groups of concrete blocks are arranged into pyramids. Reef balls are the most widely used now-a-days as artificial reefs in the world. They are made in molds from microsilica concrete with rough surfaces and neutral pH. They are hollow with holes over the surface and are available in a variety of sizes up to 2 metres diameter weighing over 200 kilograms. They are used for mangrove planting, reef restoration and breakwater constructions. The reef balls containing concrete blocks show high fish abundance and species richness when compared to empty balls. Inner block material provides extra refuges and habitats for reef fish.

10.3.5.2. Transplantation of corals

In the method of coral transplantation, small fragments of corals are carefully removed as cuttings from healthy coral colonies, placed them in seawater, and allowed them to grow before introducing them to degraded reef sites. Two types of nursery structures are fabricated: one is suspended in seawater and another one is fixed on legs to the seabed. Coral fragments are pruned and super-glued to 2 centimetres plastic pins and attached to plastic frames suspended in mid-water at 6 metres depth off the coast.

Sometimes the corals are glued to ceramic tiles. Several factors may influence the success of the coral transplantation. Smaller fragments occupy less space and are easier to handle, but may be susceptible to damage, disease and death than large ones. Fragments of *Acropora* grow faster when collected below the topmost points of a donor coral reef branch. The corals of branching or foliose (leaf like) types have high growth in transplantation as compared to massive corals.

10.3.5.3. Other methods of coral restoration and rehabilitation

In the method of electrically stimulated coral growth, wire construction with a cathode and anode are connected to a power source like solar panels. In the method of larval ranching, the larvae are reared in aquaria tanks, after spawning processes, and then released in to the target site in net enclosures. In the method of coral fragment reseeding, the coral fragments are cut off from donor colonies, and deployed in the rehabilitation site with or without attached to substratum. In the method of low-tech reef rehabilitation, the coral fragments are tied together on a long polythene string and anchored at the transplantation site with stone at each end. In the method of coral farming, the fragments are cut off from the donor colonies and fixed on natural substrates and kept in farms until secondary basal discs are formed. Then the corals are deployed in the transplantation sites.

10.3.5.4. Stabilizing the substrate

One of the extreme forms of damage is dynamite blast fishing, especially in Southeast Asia. This blast fishing creates rubbles which are unstable substrate for the coral to settle on. Two methods are adopted here. One is to stabilize the substrate by laying out plastic nets with 2 centimetres mesh size on to the rubble and holding them in place with stakes hammered into the substrates. Another one is rock piles, in this one metre high pyramid shaped rock piles are made on land using reef rocks and placed on the net underwater. Coral recruitment is then allowed to occur naturally from adjacent healthy reefs. Within three years of rehabilitation the fish community can be restored.

10.3.5.5. Maintaining connectivity between coastal ecosystems

Coral reefs themselves can not exhibit closed community structure and assemblages without connectivity with other ecosystems such as seagrass and mangroves. Hence, restoration of these habitats is important to the reef. Reef restoration alone will be of no use, if juvenile reef fish have an obligate residence in seagrass or mangroves as young recruits or juveniles. Therefore, maintaining coastal connectivity between habitats of coastal wetlands such as mangroves, seagrasses, saltmarsh and mudflats, can enhance overall fish resources.

- The management of coral reefs involves five important aspects: (1) protection, (2) monitoring, (3) research studies, (4) capacity building and awareness creation, and (5) coral restoration.
- The coral reefs are monitored at regular intervals to detect changes on many aspects: coral status (dead and live), coral recruitment, coral diseases, coral reef associated fishes, fisheries, destructive fishing practices, sea surface temperature, coral bleaching, sedimentation, incidence of diseases, coastal water quality parameters including heavy metal concentrations.
- Coral reef studies are mostly related to reproductive biology on gonad development, environmental factors responsible for spawning, spawning cycle and to coral diseases at different depths of sea, percentages of infected colonies and pathogenic microorganisms. Also the coral surveys are carried out in less explored and unexplored areas.
- Fishing is a major livelihood for the people dependent on coral reef resources. The people use destructive fishing practices such as shore seine operation, push net operation and trap fishing in reef areas. As the fish catches decline, fishermen resort to coral mining. The root cause of all these issues are lack of awareness among the people.
- In order to restore, it is necessary to reduce the abiotic causes of coral decline such as algal overgrowth, pollution, temperature and light extremes, damage by tourist, over-fishing, dynamite fishing, and poison fishing. For biotic manipulation, different methods are used: artificial reef, coral transplantation, electrically stimulated coral growth, larval ranching, coral fragment reseeding, low-tech reef rehabilitation, stabilization of substrates, maintenance of connectivity among coastal ecosystems and coral farming.

10.4. CONSERVATION AND MANAGEMENT OF MANGROVE ECOSYSTEMS

Mangroves are spectacular ecosystems, offering many goods and services to humans, particularly through forest products, fisheries and coastal protection. Globally mangrove forests provide at least US $1.6 billion each year in ecosystem service and support coastal livelihoods. Realizing the importance of mangroves, the world nations have taken efforts on conservation and management of mangrove resources. However, mangroves are fast disappearing globally, but the the loss of mangroves varies between countries. This variation can be directly attributed to policy, legislation and management. For example, Indonesia has lost 31% of its mangroves, while Malaysia has lost 16% since 1980. Thus mangrove loss in Malaysia is 50% lower than that in Indonesia. This is due to the fact that mangrove forests in Malaysia are actively managed with good enforcement of legislation and most of the forests fall within a national forest reserve network. These efforts reduce the pressure on mangrove forests in Malaysia. In India, the mangrove forest cover has been well-maintained in the last two decades without any drastic change, as a result of effective conservation measures taken up by Government, in spite of growing threats by man and natural calamities such as cyclone, flood and tsunami. In fact, there was an increase of 58 km^2 in India between 2005 and 2007. The government of India has identified 38 mangrove areas in the country, protected and managed with 100% financial support, for the following activities related to mangrove ecosystems.

(i) Survey, assessment and demarcation;
(ii) Capacity building: staff training and skills;
(iii) Shelter belt development;
(iv) Protection and monitoring;
(v) Restoration and regeneration measures;
(vi) Alternate and supplementary livelihoods;
(vii) Community participation;
(viii) Mangrove afforestation/plantation on degraded areas and open mudflats;
(ix) Biodiversity conservation;
(x) Sustainable resource development;
(xi) Desilting;
(xii) Weed control;
(xiii) Pollution control;
(xiv) Environmental education and awareness;
(xv) Impact assessment; and,
(xvi) Concurrent and terminal evaluation.

10.4.1. Silviculture Management

Mangroves are managed actively in a number of large forests in Asia. Mangroves have been successfully managed for a very long time in the Sundarbans in Bangladesh and India since 1759; and in the Matang, Malaysia, since 1902. These forests are managed on a commercial basis with well-established plantations and mangrove-associated fisheries. Thinning of the trees for timber is done after several years, followed by a final clearing. Such cycles may be 30 years or more in duration.

10.4.2. Legal Protection

There is a strong legal frame works for the protection of mangroves in several countries: Brazil, Mexico, India, Cambodia, El Salvador and

Tanzania. In India, illegal destruction of mangroves is a violation of the Coastal Regulation Zone Notification-1991 under the Environment (Protection) Act 1986. Section 5 of the Act has provisions for closure of the unit, stoppage of electricity, or water to such units involved in illegal destruction of mangroves. Section 15 of the said Act provides imprisonment for a term of 5 years with a fine which may exceed to one lakh rupees, or both, and in case the failure continues, with additional fine which may extend to five thousand rupees for every day. If the failure continues beyond one year, the offender shall be punishable for a term which may extend to seven years. The above powers are delegated to the Coastal Zone Management Authorities of different states of India.

10.4.3. Marine Protected Areas

Marine Protected Areas (MPAs) are an important tool for conservation of marine ecosystems including mangroves. There are 1200 protected areas worldwide that include mangroves, covering about 25% of all mangrove areas and the remaining 75% of mangrove area is unprotected.

10.4.4. International Protection

There are three important agreements of global scope for mangrove management, and they are (1) the World Heritage Convention (Convention related to the protection of the world cultural and natural heritage, Paris, 1972), (2) the Ramsar Convention (Convention on Wetlands of International Importance especially as Water fowl habitat, Ramsar, Iran, 1971) and (3) the UNESCO's Man and the Biosphere (MAB) programme. The designation of mangrove sites under any of these international agreements offers prestige, support and collaboration.

10.4.4.1. The World heritage convention

This agreement aims to protect places of great value. At the end of the year 2009, there were 31 World Heritage sites in 18 countries that included mangroves. A few examples are the Sundarbans in India and Bangladesh, the Wet Tropics of Queensland in Australia, the Belize Barrier Reef system, and the Lorenz National Park in Indonesia.

10.4.4.2. The Ramsar convention

It is aimed to provide a framework for national action and international co-operation for the conservation and wise-use of wetlands and their resources. Of 1800 Ramsar sites, 215 sites are colonized with mangroves in 65 countries in the year 2009.

10.4.4.3. UNESCO's Man and the Biosphere

It is not a legally binding convention, but more of a co-operative venture. It empowers local people to manage their resources. The sites are zoned into core areas for protection, buffer areas and transition areas for human settlement and resource use. At the end of 2008, there were 501 sites in 105 countries, of which 34 in 21 countries are colonized with mangroves.

- Mangroves have been managed for a very long time in Matang, Malaysia and Sundarbans in India and Bangladesh.
- There are 1200 protected areas worldwide that include mangroves, covering about 25% of all mangrove areas and the remaining 75% of mangrove area is unprotected.
- There are three important international agreements for mangrove management, and they are (1) the world heritage convention, (2) the Ramsar Convention, and (3) the UNESCO's Man and the Biosphere (MAB) programme. The designation of sites under any of these international agreements offers prestige, support and collaboration.

10.4.5. Restoration and Afforestation

Restoration of an ecosystem can be defined as the act of bringing it back nearly to its original conditions. Several other terms used for this purpose are 'rehabilitation' or 'regeneration'. Afforestation refers to the plantation in new areas. In many places around the world, mangroves are planted. There are two types in restoration of degraded areas: (1) natural restoration, and (2) artificial restoration.

Natural restoration involves the natural process of establishment of 'seeds' of mangroves. This method is advantageous because it produces the forest pattern, similar to the original forest, and it is cheaper and less disturbing to the habitat. It may not be successful in places where the seeds are not readily available, and in places where there is excessive weed growth, debris, poor soil, and disturbed hydrodynamic conditions. In the natural restoration, hydrological manipulations are of great help. For example, in USA, mangrove restoration has been made in 500 hectares at a site in West Lake, Florida. This is successful even without any artificial planting, by simply using a combination of excavation of dredged material and hydrologic restoration. Another form of this hydrologic restoration is to reconnect dry mangrove areas to normal tidal influence.

Artificial restoration involves planting of seeds or seedlings in areas where there is inadequate availability of the planting materials. Hence, the seedlings are transplanted or seeds that are collected from other areas are planted to the new areas directly or after raising saplings in nurseries. The artificial restoration has its own merits: (i) species composition, distribution and density of seedlings can be controlled; (ii) genetically improved stocks can be developed; (iii) degraded areas can easily be restored; and, (iv) plantation can be made according to the objectives of restoration such as coastal protection, biodiversity conservation, and augmentation of forestry and fishery resources.

The restoration strategy is adopted based on the tidal amplitude. Consequently, the coastal area for restoration of mangroves is divided in to two categories *viz.,* high tidal amplitude area and low tidal amplitude area. In the high tidal amplitude areas, direct seed sowing and seedling planting in the mud flats can be practiced. In the low tidal amplitude areas, a planting technique called "Canal Bank Planting' with 'Fish bone' design model is successful for restoration of mangroves (Fig. 10.7). This technique involves digging of the feeder canal (with dimensions of 3 metres-top, 1 metre- bottom and 1 metre-deep), and the distribution canals (with dimensions of 2 x 0.75 x 0.75 metre) in the mud flats, and then planting of propagules directly in the inter-tidal zone of the canals.

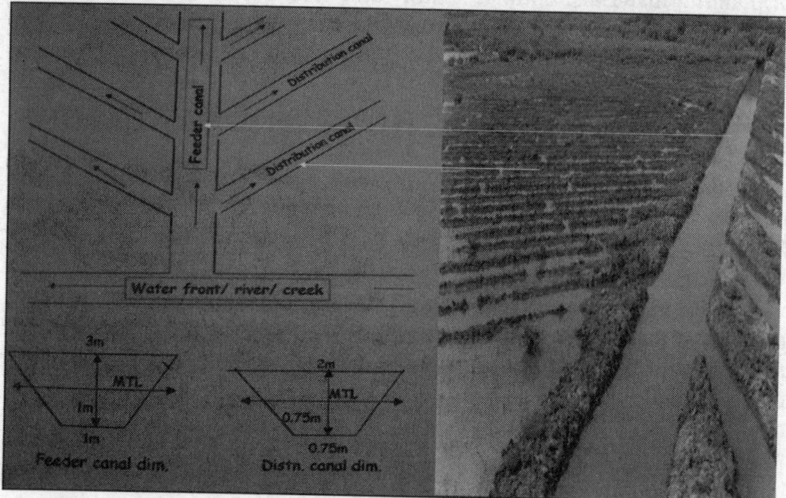

Fig.10.7.: "Canal bank planting" and fish bone design: a preferred restoration technique for low tidal amplitude areas

10.4.5.1. *Techniques of Mangrove Plantation*

Plantation techniques are of two types:

(1) Direct planting of seeds or propagules in the muddy areas; and,

(2) Planting of seedlings obtained from nurseries.

In the first type of direct planting type, the propagules are used without any nursery development. In the second type of planting the nursery-derived seedlings is used in the areas where seeds are available seasonally and in small quantities. In such case, the nursery is developed in polythene bags at the upper intertidal areas, under shady conditions for six months to one year and then transplanted to the field according to their zonation patterns.

Plantation of the lengthy propagules of Rhizophoraceae members (*Rhizophora, Kandelia, Ceriops, Bruguiera*) are planted directly, whereas relatively small seeds or propagules of *Avicennia, Sonneratia* and *Excoecaria* are raised in the nursery, and then these are transplanted to the field. The following aspects are to be considered in mangrove plantation programme.

Zonation patterns: It is necessary to do planting according to the natural zonation pattern of mangrove species. In general, the plants follow a pattern according to the length and size of the seeds. For example, lengthy propagules colonize along the water front, and smaller ones establish towards landward intertidal areas.

Seasons: Planting seasons differ with different species, the availability of planting materials, optimal salinity and other conditions. The extreme seasons of high salinity, high waves and strong winds are not suitable for planting mangroves. In general, the post-monsoon season with moderate salinity is the ideal period for plantation.

Working hours: Proper planning of time, duration, number of seedlings, man power, and transportation facilities is necessary for planting operations. The ideal time is at low tide. Seedlings have to reach the planting sites before the low tide so as to be ready for planting. For direct planting, it is possible to plant seeds as and when the tidal level permits workers to implant the seeds at arm's length.

Depth and spacing of planting: The propagules are to be planted in general, at their pointed tip in the muddy soil. The depth of planting can increase with the length of propagules, and the seeds of small size can be planted gently in the soil. The spacing for the planting normally is maintained at 1 to 2 metres. A long line rope can be used to make the plantation at appropriate spacing in between the seedlings.

Soil conditions: Generally, mangroves prefer soft, clay mud for their growth, but not the calcified or sandy areas. The frequency of high tidal flooding of the soil varies with species. Species with longer propagules require greater flooding frequencies and those with shorter propagules need lower flooding frequencies of the soil.

Gap filling: Mangrove seedlings require protection against grazing pressure and with the operation of fishing nets. Damaged seedling should be removed and the gaps thus created should be filled with fresh seedlings.

Storage: The propagules of *Rhizophora, Avicennia, Bruguiera, Ceriops,* but not *Kandelia* can be stored for a week in brackish water. However, fruits of *Sonneratia* and *Xylocarpus* can be stored for a long period by partially immersing their tip regions in low saline water.

Monitoring: Plantations are to be closely monitored so as to ensure proper regeneration of mangroves. It is necessary to identify the causes of the following risk factors and take appropriate remedial measures.

1. *Algal growth*: It is often a serious issue due to overgrowth of filamentous algae such as *Enteromorpha* and *Chaetomorpha*

that cover the leaves of seedlings and topple them into the water. This situation occurs during summer and post-monsoon seasons. This can be prevented by hand picking and erecting bamboo fencing with fishing net.

2. *Menace by water hyacinth*: This aquatic weed accumulates heavily during the monsoon season through freshwater inflow, and bends the seedlings down. This weed can be removed by hand picking.

3. *Infestation by crab and barnacles*: Several species of crabs are crasping plant shoots. The crab holes can be blocked with stones. A heavy load of barnacles gets attached at the low and middle tidal levels of the stem of seedlings. It is prevented by scraping with the help of knife carefully without damaging the seedlings. The areas where the infestations by barnacle and crabs are serious, it is ideal to use nursery-raised taller seedlings for plantation.

4. *Infestation by insects*: Moth worms and other insects create problems to the plantation sites. The insects can be prevented by removing and killing them manually or by spraying 5% parathion solution over them as a last option as it is hazardous to environment, or by simply washing the seedlings with seawater.

5. *Siltation*: It occurs during the monsoon period. The silt deposited on leaves and stems very often leads to death of plants. It can be prevented by washing the seedlings with seawater.

6. *Cattle grazing*: It is a very serious problem, besides grazing the cattle stampede the young seedlings. This can be prevented by erecting thick fence of iron wire or by planting biotoxic *Excoecaria* and spiny *Acanthus ilicifolius* for protection.

7. *Trash*: Solid waste materials that are dumped in the waters clog the mangrove habitats. This can be prevented by erecting bamboo fences and water gates to trap the trash at their entry points.

8. *Strong currents/high waves/strong winds*: These create dangerous situations. To overcome such problems, strong fences should be erected, or planting should be done deeper, and the seedlings can be supported with bamboo poles.

- There are two types of mangrove restoration in degraded areas: (1) natural restoration and (2) artificial restoration.

- Natural restoration involves the natural process of establishment of 'seeds' of mangroves, through hydrological manipulations by simply using a combination of excavation of dredged material and hydrologic restoration. Another form of this hydrologic restoration is to reconnect dry mangroves to normal tidal influence.
- Artificial restoration involves planting of two types: (1) Direct planting of seeds or propagules; and, (2) Planting of seedlings obtained from nurseries in the areas where seeds are available seasonally and in small quantities.
- Plantation of the lengthy propagules can be planted directly, whereas relatively small seeds or propagules can be raised in the nursery, and then these are transplanted to the field.
- Plantation should consider ideal natural zonation patterns, season, working hours, depth and spacing of planting, soil conditions, gap filling and storage of seeds.
- The risk factors for mangrove plantation programme are filamentous algal growth, menace by water hyacinth, infestation by crab and barnacles, infestation by insects, siltation, cattle grazing, trash and strong currents/high waves/strong winds.

10.5. MANAGEMENT OF CORAL REEF RESILIENCE AND RESISTANCE TO CLIMATE CHANGE (CORAL BLEACHING)

Climate change notably coral bleaching (Fig. 10.8) affects coral reefs. In order to overcome the disturbances, coral reefs exhibit resistance and resilience. Resistance is the ability of a coral reef to withstand the disturbances, while resilience is the ability of a coral reef to recover from the disturbances. The resistance factors are of two types: resistance and avoidance. Resistance factors are acclimatization, adaptive leaching hypothesis, thermal tolerance variation in zooxanthellae strains, whereas avoidance factors are upwelling, water currents, shading and screening. The resilience factors are (1) species and functional diversity, (2) reproduction and connectivity. Tools and strategies for enhancing coral reef resilience are (1) monitoring, (2) transplantation, (3) marine protected areas, (4) integrated coastal management, and (5) fisheries management. These factors are proposed by IUCN to promote resilience of coral reefs to climate change, as briefed below.

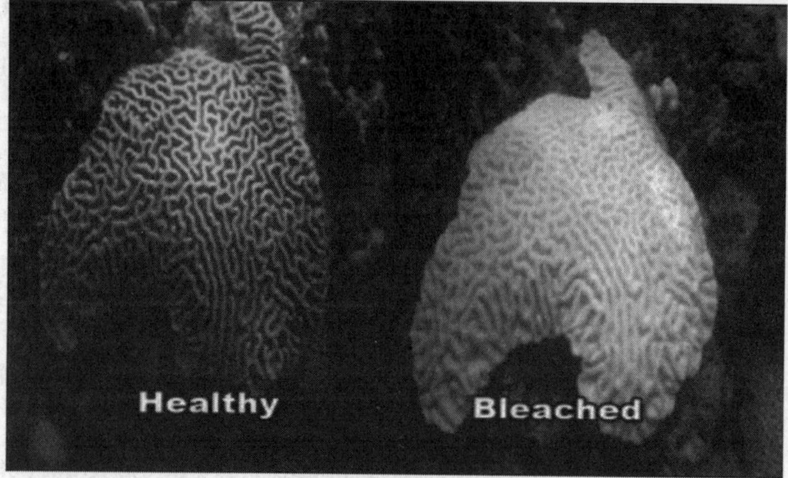

Fig.10.8.: Coral bleaching
(http://oceanicdefense.blogspot.in/2009/07/coral-bleaching-likely-in-caribbean.html)

10.5.1. Resistance Factors

10.5.1.1. Acclimatization

The coral reefs are often exposed to stressful environmental conditions and they get acclimatized. The coral reefs exhibit tolerance to high temperatures and solar UV-radiation. Small areas that are regularly heated during low tides often exhibit higher tolerance to coral bleaching. Corals on surface are more tolerant of high temperatures than corals at bottom. Corals expel their zooxanthellae during a bleaching event in order to replace them with more resistant strains.

10.5.1.2. Thermal tolerance variations in zooxanthellae

Zooxanthellae that live symbiotically in coral polyps play a role in bleaching tolerance. This tolerance varies with strains of the zooxanthellae. Coral morphology varies in temperature tolerance. In general, fast growing branching species of coral reefs (*e.g Acropora, Millepora* and *Pocillopora*) suffer higher bleaching mortality than slow growing massive species (*e.g. Favites, Favia, Turbinaria*). The thick tissue of slow growing corals could offer better protection for zooxanthellae against UV-radiation.

10.5.1.3. Local water movement

Coral reefs are protected against bleaching in upwelling areas. Upwelling is the process by which warm, less dense surface water is drawn away from a shore by offshore currents and replaced by cold, denser water brought up from the subsurface. During the process, the hot surface water is cooled with cooler deep waters because of tidal currents. In another process, tidal currents and fast water flow can protect coral reefs against bleaching by removing harmful oxygen radicals. More over the fast water flows inhibit the macroalgal settlement, and allowing the coral recruits to settle and grow.

10.5.1.4. Shading of corals

The shading of corals by clouds in the sky, rocks and high islands, or the screening of corals by suspended particulate matter, can protect them against solar UV-radiation, and allow them to prevent bleaching.

10.5.2. Resilience Factors

10.5.2.1. Species and functional diversity

The species and functional diversity increase the recovery of coral reefs from bleaching. It is necessary to have a balanced ecological community with sufficient species interactions for recovery of coral reefs from disturbances. The bleached coral reefs are deteriorated by over-growth of seaweeds. This will reduce the chance of coral recovery. In this context, the grazing animals such as herbivorous fish and sea urchins can enhance coral reef recovery by preventing fast growth of seaweeds.

10.5.2.2. Connectivity between and within reefs

This is one of the factors important for coral reef resilience. Healthy corals produce abundant, genetically diverse and healthy larvae which migrate to the degraded reefs, settle down and then grow. The larvae are poor swimmers, and they are to be transported quickly by water currents to settle on reefs within and between reefs of long distances. Furthermore, the dispersal of larvae depends on the mode of reproduction of the corals. Asexually reproducing corals disperse locally through coral fragments, while sexually-reproducing corals can disperse over larger distances through their larvae. The larval settlement depends on substrates. The ideal substrates are stable surrounded by calm water with salinity between 32 and 40 parts per thousand (ppt), sufficient light, less predation, limited sedimentation and less population of seaweeds. Dead corals or encrusting calcareous algae provide the best substrate for coral recruitment.

10.5.3. Tools for Enhancing Coral Reef Resilience

Man-made threat is crucial to resilience of coral reefs. Higher threat causes greater levels of stress to coral reefs, and lowers their resilience. In order to overcome the effect, IUCN has suggested two important tools such as coral reef monitoring and coral transplantation, and three important strategies such as marine protected areas, integrated coastal management, and fisheries management.

10.5.3.1. Monitoring

Ecological and socio-economic monitoring of coral reefs is an important management tool. Ecological monitoring is on the physical and biological parameters of coral reefs, while socio-economic monitoring is to understand how people use and interact with coral reefs. Good monitoring programmes can be used to improve coral resilience on different aspects such as sea surface tempe-rature, types of coral communities present, predators, herbivores, bleaching resistant corals, larval sources, connectivity patterns and threat factors.

10.5.3.2. Transplantation

In this process, juveniles from a healthy reef are introduced on to a degraded one. However, natural recruits survive better than transplants. This coral trans-plantation is better used tool to repair localized damage than to mitigating the impact of large scale events such as coral bleaching. However, transplantation is necessary for slow growing massive corals, better than fast growing branching corals for long-term success. Transplantation is highly important for the reason that the natural recruitment is often limited.

10.5.4. Strategies for Enhancing Coral Reef Resilience

10.5.4.1. Marine protected areas

Marine protected areas are the most accepted management strategy used to protect coral reefs. It is not protecting a coral reef from bleaching directly, but indirectly through improving resilience of coral reefs by reducing the man-made changes such as increased nutrient loads, pollution, diver and boat damage, sedimentation, and destructive over-fishing. This will make coral reefs healthy which will ensure rich larval output, essential for coral reef recovery.

10.5.4.2. Integrated coastal management

An important management strategy is integrated coastal management. This involves the management of coastal zone, as a single integrated system dealing with coral reefs and associated coastal ecosystems such as mangroves and seagrasses. Other issues are coastal development, fisheries, tourism, land-based resources of pollution and sedimentation, waste disposal, agriculture, forestry, mining, gas and oil industries, and shipping activities.

10.5.4.3. Fisheries management

The management of fisheries can play a crucial role in enhancing coral reef resilience. Coral mortality caused by bleaching can affect fish composition, fish catch and fishing activity. Over-fishing causes loss of biodiversity and reduces coral reef resilience. Destructive fishing practices such as blast or poison fishing can reduce coral reef resilience by decreasing coral cover by increasing the population of coral destroying crown-of-thorns starfish or depleting herbivores that graze down seaweeds.

Fisheries management involves creation of 'no-take' zones, the restriction of gear use, the imposition of fishing licenses, the implementation of legislation controlling destructive fishing techniques, the monitoring of catch compositions, the development of alternative livelihoods, and the regulation of the harvest of organisms for the aquarium trades. The Great Barrier Reef is managed for fisheries with 'no take' areas totaling 33%.

- Coral reefs exhibit resistance and resilience to combat disturbances in particular bleaching. Resistance is the ability of a coral reef to withstand the disturbances, and resilience is the ability of a coral reef to recover from the disturbances.
- Tools and strategies for enhancing coral reef resilience are (1) monitoring, (2) transplantation, (3) marine protected areas, (4) integrated coastal management, and (5) fisheries management.

10.6. MANAGEMENT OF MANGROVES FOR RESILIENCE TO CLIMATE CHANGE (SEA LEVEL RISE)

A growing threat to global mangrove ecosystem is the climate change, in particular changes in temperature, carbon dioxide, precipitation, hurricanes, storms and sea level. All these factors are synergistically acting upon the mangroves. To these changes, mangroves are likely to be one of the first ecosystems to be affected because of their location at the interface between land and sea. However, mangroves exhibit resistance and resilience to overcome the effects of climate change. Resistance of mangroves is the ability to withstand the disturbances, whereas resilience of mangroves is the ability to recover from the disturbances.

Once the mangroves are degraded, it may require long periods to recover from even minor disturbances. The habitat loss will result in loss of biodiversity of the mangrove ecosystems. However, there are a few genetically superior organisms in the mangrove ecosystems, which can overcome the climatic change. It is therefore, suggested as a long-term plan (i) to identify the mangrove geno-types and fauna which are tolerant to temperature and flooding, (ii) to propagate those genotypes, (iii) to create new hybrid species from those genotypes, for biodiversity enrichment and coastal protection against the climate change.

10.6.1. Impacts of Climate Change on Mangroves

10.6.1.1. *Temperature effects*

Mangroves may not be adversely affected by the projected increases in sea temperature of 2-6°C by 2100. Most mangroves produce maximal shoot density when mean air temperature rises to 25°C and the mangroves stop producing leaves when the temperature falls below 15°C. However, temperature greater than 35°C may alter root structure and seedling establishment. The photosynthesis may be affected at 40°C. Due to global warming, some species of mangroves will migrate to higher latitudes. This is evident that mangroves expand into saltmarshes in the US Gulf of Mexico, Australia and New Zealand. However, extreme cold conditions are likely to limit mangrove expansion into higher latitudes. At the same time, increased sediment temperature may increase growth rates of bacteria which are likely to increase recycling and regeneration of nutrients.

10.6.1.2. CO_2 effects

Most atmospheric carbon dioxide released from fossil fuels is absorbed into the ocean. The atmospheric carbon dioxide increased from 280 parts per million (ppm) in the year 1880 to 370 ppm in the year 2000. The increase in CO_2 may increase rate of net photosynthesis and growth rate of mangroves when the soil salinity is low. However, the photosynthesis and growth rate may be reduced, when the salinity increases in arid regions. One indirect effect of increase in temperature and CO_2 is the degradation of coral reefs due to mass bleaching and reduced growth. As a result, protection function of the coral reefs from wave action will be lost, thereby affecting the mangroves. The mangrove wetlands are efficient habitats for carbon burial, about 2.4 fold as high as salt marshes and 5.2 fold as high as seagrasses. The mangroves sequester as much as 4 times the amount of carbon in their sediment per hectare of tropical forest.

10.6.1.3. *Precipitation effect*

Precipitation may increase by 25% by 2050 due to global warming. However, both increases and decreases are predicted in different areas of the mangroves in the world. Changes in precipitation pattern may have significant effect on the growth, productivity and areal extent of mangroves. Decreased precipitation may change species composition, favouring more salt tolerant species with increase of hyper-saline flats.

10.6.1.4. *Hurricanes and storms effects*

There may be 5-10% increase in intensity of storms by the year 2050. The storm will increase in frequency and intensity due to climate change, posing threat to mangroves. This will affect mangrove health and species composition due to changes in salinity, recruitment, inundation, and changes in sediment budget. Storm surges can also flood the mangroves, and they may destroy mangroves in combination with sea-level rise. *Avicennia* species are more vulnerable than *Rhizophora* species. This is due to the difference in aerial root types: long stilt roots of *Rhizophora* stands above sea-level rise, while the short pneumatophores of *Avicennia* submerge under the sea-level rise.

10.6.1.5. Sea level rise effects

Sea level rise is the greatest challenge of mangroves to climate change. During the 21st century, mean sea-level predicts a range from 0.09 to 0.88 metre and this will accelerate over the coming decades. A recent study reveals that two islands namely Suparibhanga and Lohacharra in Indian part of Sundarbans disappeared that evacuated 10,000 people and 12 more islands in the southernmost part of the region are vulnerable for submergence in the seawater.

Mangroves colonise at the boundary between land and sea. Mangroves can adapt to sea level rise if it occurs slowly enough and if adequate expansion space exists. As the sea level rises, mangroves will tend to shift landward. Human encroachment at the landward boundary, however, makes this difficult. Consequently, the width of mangrove systems will decrease with the sea level rise. The ability of mangrove migration landward or seaward is also determined by local conditions such as infrastructure (roads, dikes, urbanization, and seawalls) and topography (steep slopes). Tidal range and sediment supply are two critical indicators of mangrove response to sea level rise. The mangroves with macro-tidal and sediment rich areas are able to survive sea level rise than those with micro-tidal and sediment starved areas.

The most vulnerable mangroves to sea-level change are located in areas with small islands, lack of rivers, carbonate settings, tectonic movements, ground-water extraction, underground mining, coastal development, steep topography micro-tidal and sediment-starved areas. The least vulnerable mangroves are situated in riverine areas, macro-tidal and sediment rich areas and dense mangrove forests.

10.6.2. Strategies to Mitigate Climate Change Effects

IUCN has proposed seven major strategies to promote resilience of mangroves to climate change, as briefed below.

10.6.2.1. Protection of species and habitats

Mangrove habitats can be identified based on local conditions such as the presence of sediment-rich, macrotidal environments and the availability of freshwater to reduce hyper-salinity. The mangrove habitats with abundant mature trees producing a healthy supply of seeds and propagules, along with dense epibiont communities such as oysters should be protected. This may serve as sources for colonizing new areas, and restoring the areas damaged by a disturbance. Such mangrove habitats should be protected under "Marine Protected Areas" or brought under integrated coastal management programmes. The mangrove species can be identified based on their potential of migrating to landward in response to sea level rise.

10.6.2.2. Management of man-made pressure

Man-made threats to mangroves are increasing with climate change. To promote resilience to global climate change, mangroves have to be protected from man-made pressures. In addition to this, land-based threats to mangroves have to be reduced through improving the land-use practices to decrease nutrient and sediment run-off, limit the unregulated felling, eliminate the use of chemical pollutants and to improve water quality. Community participatory programmes should be promoted by provision of alternative livelihood to the mangrove-dependent people. Alternative livelihoods such as charcoal production from coconut shells instead of from mangroves, bamboo extraction instead of mangrove wood, traditional honey making, and seaweed farming are income generating alternatives to mangrove destruction.

10.6.2.3. Establishment of green belts and buffer zones

Mangrove green belts can provide significant coastal protection from erosion. The green belts should be established along coastlines and river banks in the areas which experience damage from typhoons, tidal surges, cyclones and erosion. It is necessary to raise green belts from a minimum of 100 metres up to 1000 metres and preferably 500 metres at the open coast, 30-50 metres along river banks and lagoons and less than 10 metres on islands, creeks and channels. It is also necessary to maintain buffer zones bordering the seaward and landward margins of protected mangroves. This will help the mangroves for expansion on both landward and seaward areas.

Mangroves: A Fish Habitat

Mangroves contribute significantly to fish resources, on which a large human population is dependent for their livelihood. However, mangrove planting programmes concentrate only in raising the mangrove forest, ignoring its potential as fish habitat. To cite an example, in Andhra Pradesh, the forest area increased by 2.89% in two years from 1996 to 1998, with a reduction of water bodies by 0.83%. A scientific understanding of this important aspect is necessary.

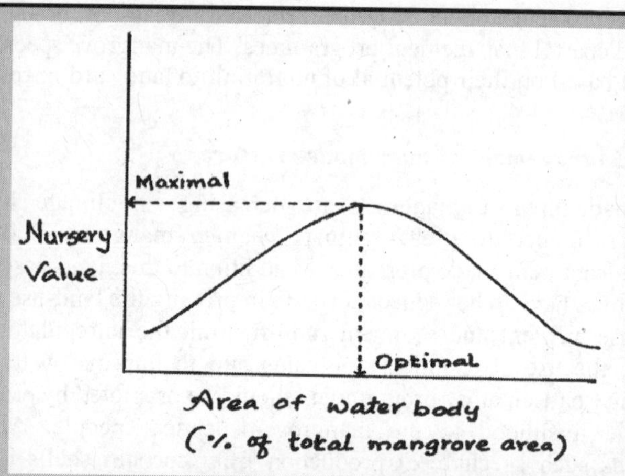

Fig. 10.9: A hypothetical model for the optimal area of water bodies in mangroves in relation to nursery value (Kathiresan, 2002)

Mangroves play an important role as 'nursery ground' for many fishes of commercial importance. In this regard, two variables and a balance between them are to be critically considered, and they are (1) the mangrove cover that supplies food for fish, and (2) the water bodies that provide shelter for fish. The nursery value of mangrove habitat increases with the water creeks. But the value decreases with decreasing mangrove forest area due to its lowering output of detritus food (Fig. 10.9). It is necessary to find an optimal extent of water creeks within a mangrove area for its maximum nursery potential. It is suggested here that the water creeks may be maintained at 40% of total mangrove cover in mangrove plantation programmes to achieve the potential 'nursery ground' for enhancing fish resources of the mangrove habitats and offshore areas (Fig.10.9). Much study and data are required especially on the engineering aspects of water creeks for their construction and maintenance for ultimately maximising the potential of mangroves as a fish habitat.

10.6.2.4. Restoration of degraded areas

One of the major causes of mangrove degradation in Southeast Asia and Latin America is conversion of large areas of mangroves into shrimp ponds. But the shrimp production has proven to be unsustainable in many regions, resulting in large areas of abandoned ponds. Restoration of degraded areas can be done by (i) hydrological manipulation through construction of creeks, thereby flushing the degraded areas with tidal waters, (ii) community participation, and (iii) integrated farming practices. This may help to create livelihoods for local communities and may also reduce the pressure on mangrove areas. The areas planted with multiple species of mangroves have greater ecological resilience.

10.6.2.5. Maintaining connectivity between mangroves and other coastal systems

This can be done by ensuring linkage between mangroves and their associated habitats such as coral reefs and seagrasses. Mangroves help the adjacent ecosystems by filtering sediments and pollutants from land run-off and improving the water quality for seagrasses, coral reefs and fish communities. Coral reefs reduce ocean currents and waves to create a sheltered environment for mangroves and seagrasses. Mangroves also enhance the biomass of coral reef fish species.

10.6.2.6. Baseline data development

It is required to establish baseline data on forestry structure, species richness, abundance and diversity of flora and fauna, primary production, nutrients and hydrological aspects for monitoring the response of mangroves to climate change.

10.6.2.7. Establishment of partnerships at local, regional and global scales

Building global, regional and local partnerships is important to overcome the financial constraints experienced in conservation for mitigating the effects of climate change.

- Resilience of mangroves will be threatened by climate change, in particular changes in temperature, carbon dioxide, precipitation, hurricanes, storms and sea level, in addition to man-made threats.
- Strategies to mitigate climate change effects are protection of species and habitats, management of man-made pressure, establishment of green belts and buffer zones, restoration of degraded areas, maintenance of connectivity between mangroves and other coastal ecosystems, base line data development and establishment of partnerships at local, regional and global scales.

Questions for Review/Exercise*

(1) Expand: WTO, CITES, IUCN, UNEP, CMS, ICES, IMO, SCAR, SCOR, LOICZ, CBD, IOC.

(2) What are the two principal approaches of ocean management? Explain

(3) 'Establishment of Marine Protected Areas is an important tool for conservation' - Discuss.

(4) Discuss about the marine fisheries Polices.
(5) Briefly describe the coastal management policies.
(6) List out the strategies for participatory approach of ocean management.
(7) Write a short note on research and development in ocean management.
(8) How are the coral reefs protected and conserved? Explain.
(9) Discuss the conservation and management of mangroves.
(10) What are the resistance, resilience factors and strategies for enhancing coral reef resilience to climate change? Explain.
(11) Explain briefly the potential impacts of climate change on mangrove ecosystems.
(12) What are the strategies for enhancing mangrove resilience to climate change? Explain
(13) Explain various techniques for mangrove restoration.
(14) Explain various techniques for coral rehabilitation.

* Collect current information through internet and newspapers about the marine environmental issues and the policies being adopted and modified to overcome the issues. Discuss about the management of marine resources with the persons connected with it.

Bibiliography

Abijit M and Kakoli B 2005 Living Resources of the sea: Focus Indian Sundarbans WWF, India, 120 pp

Abril G, Etcheber H, Delille B, Frankignoulle M and Borges AV 2003 Carbonate dissolution in the turbid and eutrophic Loire estuary *Marine Ecology Progress Series,* 259: 129-138

Aceves-Medina G, Saldierna-Martinez R, Hinojosa-Medina A, Jimenez-rosenberg SPA, Hernandez-Rivas, ME and Morales-Avila 2008 Vertical structure of larval fish assemblages during diel cycles in summer and winter in the southern part of Bahia de LaPaz, Mexico *Estuarine, Coastal Shelf Science*, 76(4) : 889-901

Ainsworth TD, Kvennefors EC, Blackall LL, Fine M and Hoegh-Guldberg O 2007 Disease and cell death in white syndrome of Acroporid corals on the Great Barrier Reef *Marine Biology*, 151(1): 19-29

Almany GR, Berumen ML, Thorrold SR, Planes S and Jones GP 2007 Local replenishment of coral reef fish populations in marine reserve *Science,* 316(5825): 742-744

Alongi DM 2002 Present state and future of the world's mangrove forests *Environmental Conservation*, 29(3): 331-349

Alvarez-Filip L, Dulvy NK, Gill JA, Cote IM and Watkinson AR, 2009 Flattening of Caribbean coral reefs: region-wide declines in architectural complexity *Proceedings of the Royal Society B-Biological Sciences,* 276(1669): 3019-3025

Andersen JH and Conley DH 2009 Eutrophication in costal ecosystems *Hydrobiologia,* 629: 269 pp

Arena PT, Jordan LKB and Spieler RE 2007 Fish assemblages on sunken vessels and natural reefs in southeast Florida, USA *Hydrobiologia,* 580: 157-171

Armstrong CW and van den Hove S 2008 The formation of policy for protection of cold-water coral off the coast of Norway *Marine Policy*, 32(1): 66-73

Azeiteiro UM, Jenkinson I and PereiraMJ 2008 Plankton studies *Hydrobiology,* 614: 404 pp

Azzellino A, Gaspari S, Airoldi S And Nani B 2008 Habitat use and preferences of cetaceans along the continental slope and the adjacent pelagic waters in the western Ligurian sea *Deep-Sea Research Part I – Oceanographic Research papers,* 55(3) : 296-323

Bailey KM, Abookire AA and Duffy-Anderson JT 2008 Ocean transport paths for the early life history stages of offshore-spawning flatfishes: a case study in the Gulf of Alaska *Fish and Fisheries*, 9(1): 44-66

Baker AC, Glynn PW and Riegl B 2008 climate change and coral reef bleaching: An ecological assessment of long-term impacts recovery trends and future outlook *Estuarine, Coastal and Shelf Science,* 80(4): 435-471

Balakrishnan Nair N and Thampy DM 1980 A Text Book of Marine Ecology The Macmillan Company of India Ltd 352 pp

Balasubramanian H and Foster RS 2007 Species and space: role of volume in organizing coral reef fish assemblages in SE Cuba *Marine Ecology Progress Series,* 345: 229-236

Ballantine WJ and Langlois TJ, 2008 Marine reserves: the need for systems *Hydrobiologia,* 606: 35-44

Barnes DKA and Griffiths HJ 2008 Biodiversity and biogeography of southern temperate and polar bryozoans *Global ecology and Biogeography*, 17(1): 84-99

Barrett NS, Edgar GJ, Buxton CD and Haddon M 2007 Changes in fish assemblages following 10 years of protection in Tasmanian marine protected areas *Journal of Experimental Marine Biology and ecology,* 345(2): 141-157

Bates NR and Peters AJ 2007 The contribution of atmospheric acid deposition to ocean acidification in the subtropical North Atlantic Ocean *Marine Chemistry,* 107(4): 547-558

Beron-Vera B, Crespo EA, Raga JA and Fernandez M 2007 Parasite communities of common dolphins (*Delphinus delphis*) from Patagonia: The relation with host distribution and diet and comparison with sympatric hosts *Journal of Parasitology*, 93: 1056-1060

Biswas KP 2008 Hydrothermal vents: Limits to life *Everyman's science,* 13(6):339-348

Blachowiak-Samolyk K, Kwasniewski S, Dmoch K, Hop H and Falk-Petersen S 2007 Trophic structure of zooplankton in the fram strait in spring and autumn 2003 *Deep-Sea Research Part-II-Topical Studies in Oceanography,* 54(23-26): 2716-2728

Bosire JO, Dahdouh-Guebas F, Walton M, Crona BI, Lewis RR, Field C, Kairo JG and Koedam N, 2008 Functionality of restored mangroves: A review *Aquatic Botany*, 89(2): 251-259

Bouillon S, Borges AV, Castaneda-Moya E, Diele K, Dittma T, Duke NC, Kristensen E, Lee SY, Marchand C, Middleburg JJ, Rivera-Monroy VH, Smith III, TJ and Twilley RR 2008 Mangrove production and carbon sinks: A revision of global budget *Global Biogeochemical Cycles*, 22: 1-12

Bradbury, LR, Laurel, B, Snelgrove, EVR, Bentzen, P and Campana, SE, 2008 Global patterns in marine dispersal estimate: the influence of geography, taxonomic category and life history *Proceedings of the Royal Society B – Biological Sciences,* 275(1644) : 1803-1809

Briones-Fourzan P, Lozano-Alvarez E, Egrete-Soto F and Barradas-Ortiz, C 2007 Enhancement of juvenile Caribbean spiny lobsters: an evaluation of changes in multiple response variables with the addition of large artificial shelters *Oecologica,* 151(3): 401-416

Bronwyn M Gillanders, Kenneth W Able, Jennifer A Brown, David B Eggleston and Peter F Sheridan 2003 Evidence of connectivity between juvenile and adult habitats for mobile marine fauna: an important component of nurseries *Marine Ecology Progress Series*, 247: 281-295

Bruschetti M, Luppi T, Fanjul E, Rosenthal A and Iribarne O 2008 Grazing effect of the invasive reef-forming polychaete *Ficopomatus enigmaticus* (Fauvel) on phytoplankton biomass in a SW Atlantic coastal lagoon *Journal of Experimental Marine Biology and Ecology,* 354(2): 212-219

Burkepile DE and Hay ME 2007 Predator release of the gastropod *Cyphoma gibbosum* increases predation on gorgonian corals *Oecologia,* 154: 167-173

Busby TO and Plante CJ 2007 Deposit feeding during tidal emersion by the suspension-feeding polychaete, *Mesoshaetopterus taylori Southeastern Naturalist*, 6(2): 351-358

Cabaco S, Machas R, Vieira, V and Santos R 2008 Impacts of urban wastewater discharge on seagrass meadows (*Zostera noltii*) *Estuarine Coastal and Shelf Science*, 78(1): 1-13

Cabaitan PC, Gomez ED and Alino PM 2008 Effects of coral transplantation and giant clam restocking on the structure of fish communities on degraded patch reefs *Journal of Experimental Marine Biology and ecology,* 357(1): 85-98

Caplan P 2009 http://mafia-island-tanzaniagoldacuk/

Carpenter RC and Williams SL 2007 Mass transfer limitation of photosynthesis of coral reef algal turfs *Marine Biology*, 151(2): 435-450

Castellanos-Galindo GA and Giraldo A 2008 Food resource use in a tropical eastern Pacific tide pool fish assemblage *Marine Biology*, 153(6): 1023-1035

Chen Zhong, Yang Hua-Ping and Huang Chi-yue 2007 Characteristics of cold seeps and structures of chemoautosynthesis-based communities in seep sediments *Journal of Tropical Oceanography*, 26(6): 73-82

Chmura GL, Anisfeld SC, Cahoon DR and Lynch JC 2003 Global carbon sequestration in tidal, saline wetland soils *Global Biogeochemical Cycles,* 17 (4): 1-12

Christie, P and White, AT, 2007 Best practices for improved governance of coral reef marine protected areas *Coral Reefs*, 26(4): 1047-1056

Claudet J, Osenberg CW, Benedetti-Cecchi L, Domenici P, Garcia-Charton JA, Perez-Ruzafa A, Badalamenti F, Bayle-Sempere J, Brito A, Bulleri F, Culioli JM, Dimech M, Falcom JM, guala I, Milazzo M, Sanchez-Meca J, Somerfield PJ, Stobart B, Vandeperre F, Valle C and Planes S 2008 marine reserves: size and age do matter *Ecology Letters*, 11(5): 481-489

Clynick BG, 2008 Characteristics of an urban fish assemblage: Distribution of fish associated with coastal marinas *Marine Environmental Research*, 65(1): 18-33

Coastal Aquaculture Authority (Govt of India), 2006 Compendium of Act, Rules, Guidelines and Notifications Chennai127pp

COMPENDIUM On Consultants for Preparation of Project Reports on Fishing Harbours and Fish Landing Centres, Government of India, Ministry of Agriculture, New Delhi, *MARCH 2005 10pp*

Constantino R, Caspar MB, Tata Regala J, Carvalho S, Curdia J, Drago T, Taborda R and Monteiro CC 2009 Clam dredging effects and subsequent recovery of benthic communities at different depth ranges *Marine Environmental Research*, 67(2): 89-99

Convention on Biological Diversity 2010 In depth review of the implementation of the programme of work on marine and coastal biological diversity A note by Executive Secretary UNEP/CBD/SBSTTA/14/4 28th Jan 2010 17pp

Cooper KM, Frojan C, Defew E, Curtis M, Fleddum A, Brooks L and Paterson DM 2008 Assessment of ecosystem function following marine aggregate dredging *Journal of Experimental Marine Biology and Ecology,* 366(1-2): 82-91

Copley JTP, Flint HC, Ferrero TJ and Van Dover CL 2007 Diversity of meiofauna and free-living nematodes in hydrothermal vent mussel beds on the northern and southern East Pacific Rise *Journal of the Marine Biological Association of the United Kingdom*, 87(5): 1141-1152

Courrat A, Lobry J, Nicolas D, Laffargue P, Amara R, Lepage M, Giradin M And Le Pape O 2009 Anthropogenic disturbance on nursery function of estuarine areas for marine species *Estuarine Coastal and Shelf Science*, 81(2): 179-190

Cowen RK and Sponaugle S 2009 Larval dispersal and marine population connectivity *Annual Review of Marine Science*, 1: 443-466

Crabbe MJC, Martinez E, Garcia C, Chub J, Castro L and Guy J 2008 Growth modeling indicates hurricanes and severe storms are linked to low coral recruitment in the Caribbean *Marine Environmental Research*, 65(4): 364-368

Crain CM, Kroeker K and Halpern 2008 Interactive and cumulative effects of multiple human stressors in marine systems *Ecology Letters*, 11(12): 1304-1315

Croquer A and Weil E 2009 Spatial variability in distribution and prevalence of Caribbean scleractinian coral and octocoral diseases II Genera-level analysis *Diseases of Aquatic Organisms*, 83(3): 209-222

Daly B and Konar B 2008 Effects of macroalgal structural complexity on nearshore larval and post-larval crab composition *Marine Biology,* 153(6): 1055-1064

Daniel C Donato, Boone Kauffman J, Daniel Murdiyarso, Sofyan Kurnianto, Melanie Stidham and Markku Kanninen 2011 Mangroves among the most carbon-rich forests in the tropics *Nature Geoscience advance online publication* doi:101038/ngeo1123 pp 1-5.

Danielsen F, Sorensen MK, Olwing, MF, Selvam V, Parish F, Burgess ND, Hiraishi T, Karunagaran VM, Rasmussen MS, Hansen LB, Quadro A and Suryadiputra N 2005 The Asian Tsunami: A protective role for coastal vegetation *Science,* 310: 643

Danovaro R, Gambi C, Dell'Anno A, Corinaidesi C, Fraschetti S, Vanreusel S, Vincx M and Gooday AJ 2008 Exponential decline of deep-sea ecosystem functioning linked to benthic biodiversity loss *Current Biology*, 18(1): 1-8

Dalpadado P, Ellertsen B and Johannessen S 2008 Inter-specific variations in distribution, abundance and reproduction strategies of krill and amphipods in the Marginal Ice Zone of the Barents Sea *Deep-Sea Research Part II- Topical Studies in Oceanography*, 55(20-21): 2257-2265

Davenport J, BurnellG, Cross T, Emmerson M, Mcallen R, Ramasay R and Rogan E 2008 Challenges to marine ecosystems Springer, 211pp

David Obura and Gabriel Grimsditch 2009 Coral reefs, climate change and resilience an agenda for action from the IUCN World Conservation Congress in Barcelona, Spain *IUCN resilience Science Group Working Paper Series*: No 6, 38 pp de Castro IB, de Meirelles CAO, Matthews-Cascon H, Rocha-Barreira CA, Penchaszadeh P and Bigatti G, 2008 Imposex in endemic volutid from Northeast Brazil (Mollusca: Gastropoda) *Brazalian Archives of Biology and Technology*, 51(5): 1065-1069

De longh HH, Kiswara W, Kustiawan W and Loth PE 2007 A review of research on the interactions between dugongs (*Dugong dugon* Muller 1776) and intertidal seagrass beds in Indonesia *Hydrobiologia,* 591: 73-83

De Santo EM and Jones PJS 2007b Offshore marine conservation policies in the North East Atlantic: Emerging tensions and opportunities *Marine Policy*, 31(3): 336-347

De'ath G, Lough JM and Fabricius KE 2009 Declining coral calcification on the Great Barrier Reef *Science*, 323(5910): 116-119

Denton GRW, Morrison RJ, Bearden BG, Houk P, Starmer JA and Wood HR 2009 Impact of a coastal dump in tropical lagoon on trace metal concentrations in surrounding marine biota: A case study from Saipan, Commenwealth of the Northern Mariana Islands (CNMI) *Marine Pollution Bulletin*, 58(3): 424-431

Diaz RJ and Rosenberg R, 2008 Spreading dead zones and consequences for marine ecosystems *Science,* 321(5891): 926-929

Diez S and Bayona JM 2009 Butyltin occurrence and risk assessment in the sediments of the Iberian Peninsula *Journal of Environmental Management*, 90: 525-530

Diwan AP and Arora DK 1995 Marine Ecology : Recent advances in environmental ecology series – 4 Anmol Publications Pvt Ltd, New Delhi, 322 pp

Dorenbosch M, Verberk W, Nagelkerken I and van der Veldel G 2007 Influence of habitat configuration on connectivity between fish assemblages of Caribbean seagrass beds, mangroves and coral reefs *Marine Ecology Progress Series*, 334: 103-116

Dreanno C, Kirby RR and Clare AS 2007 Involvement of the barnacle settlement inducing protein complex(SIPC) in species recognition at settlement *Journal of experimental Marine Biology and Ecology,* 351(1-2): 276-282

Dryden J, Grech A, Moloney J and Hamann M 2008 Rezoning of the Great Barrier Reef World Heritage Area: does it afford greater protection for marine turtles? *Wildlife Research,* 35(5): 477-485

Duarte CM, Middleburg JJ and Caraco N 2005 Major role of marine vegetation on the oceanic carbon cycle *Biogeosciences*, 2 : 1-8

Dubilier N, Bergin C and Lott C 2008 Symbiotic diversity in marine animals: the art of harnessing chemosynthesis *Nature Reviews Microbiology*, 6(10): 725-740

Elliott M, Whitfield AK, Potter LC, Blaber SJM, Cyrus DP, Nordlie FG and Harrison TD 2007 The guild approach to categorizing estuarine fish assemblages: a global review *Fish and Fisheries*, 8(3): 241-268

Eline van Haastrecht and Marjanka Schaap 2003 A critical look at fisheries management practices: the 45-day ban in Tuticorin District, Tamil Nadu, India University of Amsterdam, 160 pp

Elizabeth McLeod and Rodney V Salm 2006 Managing mangroves for resilience to climate change, IUCN, Gland Switzerland, 64 pp

Elser JJ, Bracken MES, Cleland EE, Gruner DS, Harpole WS, Hillebrand H, Ngai JT, Seabloom EW, Shurin JB and Smith JE 2007 Global analysis of nitrogen and phosphorus limitation of primary producers in freshwater, marine and terrestrial ecosystems *Ecology Letters,* 10: 1135-1142

Fabinyi M, 2008 Diver tourism, fishing and marine protected areas in the Calamianes Islands, Philippines *Marine Policy*, 32(6): 898-904

FAO 2007 The role of coastal forests in the mitigation of Tsunami impacts RAP Publication, 30 pp

Fiksen O, Jorgensen C, Kristiansen T, Vikebo F and Huse G 2007 Linking behavioural ecology and oceanography: larval behaviour determines growth, mortality and dispersal *Marine Ecology Progress Series*, 347: 195-205

Freidenberg TL, Menge BA, Halpin PM, Webster M and Sutton-Grier A 2007 Cross-scale variation in top-down and bottom-up control of algal abundance *Journal of experimental Marine Biology and ecology*, 347(1-2): 8-29

Frid CLJ, Paramor OAL, Brockington S and Bremner J 2008 Incorporating ecological functions into the designation and management of marine protected areas *Hydrobiogia,* 606: 69-79

FSI (Forest Survey of India) 2009 State of Forest Report Forest Survey of India, Dehradun, Ministry of Environment & Forest, Government of India pp 27-32

Gabriel D Grimsditch and Rodney V Salm 2006 Coral Reef Resilience and Resistance to Bleaching IUCN, Gland, Switzerland, 52pp

Ganesh Kumar B, Datta KK, Joshi PK, Katiha PK, Suresh R, Ravisankar T, Ravindranath K and Menon M 2008 Domestic Fish Marketing in India – Changing Structure, Conduct, Performance and Policies *Agricultural Economics Research Review,* 21: 345-354

Garpe KC and Ohman MC 2007 Non-random habitat use by coral reef fish recruits in Mafia Island Marine Park, Tanzania *African Journal of Marine Science*, 29(2): 187-199

Gillanders BM, Able KW, Brown JA, Eggleston DB and Sheridan PF 2003 Evidence of connectivity between juvenile and adult habitats for mobile marine fauna: an important component of nurseries *Marine Ecology Progress Series*, 247: 281-295

Gilman EL, Ellison J, Duke NC and Field C 2008 Threats to mangroves from climate change and adaptation options: a review *Aquatic Botany*, 89: 237-250

Giomi Folco, Raicevich Sasa, Giovanardi Otello, Pranovi Fabio, Muro Paolo Di and Beltramini Mariano 2008 Catch me in winter! Seasonal variation in air temperature severely enhances physiological stress and mortality of species subjected to sorting operations and discarded during annual fishing activities *Hydrobiologia*, 606(1):195-202

Glardon CG, Walters LJ, Quintana-Ascencio PF, McCauley LA, Stam WT and Olsen JL 2008 Predicting risks of invasion of macroalgae in the genus Caulerpa in Florida *Biological Invasions*, 10(7): 1147-1157

Gravestock P, Roberts CM and Bailey A 2008 The income requirements of marine protected areas *Ocean & Coastal Management*, 51(3): 272-283

Griffin JN, De la Haye KL, Hawkins SJ, Thompson RC and Jenkins SR 2008 Predator diversity and ecosystem functioning : Density modifies the effect of resource partitioning *Ecology,* 89(2): 298-305

Gutowska MA and Melzner F 2009 Abiotic conditions in cephalopod (*Sepia officinalis*) eggs: embryonic development at low pH and high pCO_2 *Marine Biology,* 156 : 515-519

Halpern BS, Walbridge S, Selkoe KA, Kappel CV, Micheli F, D'Agrosa C, Bruno JF, Casey KS, Ebert C, Fox HE, Fujita R, Heinemann D, Lenihan HS, Madin EMP, Perry MT, Selig ER, Spalding M, Steneck R and Watson R 2008 A global map of human impact on marine ecosystems *Science,* 319(5865): 948-952

Hare CE, DiTullio Gr, Riseman SF, Crossley AC, Popels LC, Sedwick PN and Hutchins DA 2007 Effects of changing continuous iron input rates on a southern Ocean algal assemblage *Deep-Sea Research Part I – Oceanographic Research Papers*, 54(5): 732-746

Hilary B Moore 1958 Marine Ecology John Wiley & Sons, Inc 493 pp

Hiraishi T and Harada K 2003 Greenbelt Tsunami Prevention in South-Pacific Region, available at http://eqtapedmbosaigojp/useful_outputs/report/hiraishi/data/papers/greenbeltpdf

Huijbers CM, Mollee EM and Nagelkerken I 2008 Post-larval French grunts (*Haemulon flavolineatum*) distinguish between seagrass, mangrove and coral reef water: Implications for recognition of potential nursery habitats. *Journal of Experimental Marine Biology and Ecology,* 367(2): 134-139

Iglesias-Rodriguez MD, Halloran PR, Rickaby REM, Hall IR, Colmenero-Hidalgo E, Gittins JR, Green DRH, Tyrrell T, Gibbs SJ, von Dassow P, Rehm E, Armbrust EV and Boessenkool KP, 2008 Phytoplankton calcification in a high-CO_2 world *Science,* 320(5874): 336-340

Jeffrey S Levinton 1982 Marine Ecology Prentice – Hall Inc, Englewood Cliffs, New Jersey, 526 pp

Jessopp MJ and McAllen R 2008 Go with the flow: Tidal import and export of larvae from semi-enclosed bays *Hydrobiologia,* 606: 81-92

Jin X, Gruber N, Frenzel H, Doney SC and McWilliams JC 2008 The impact on atmospheric CO_2 of iron fertilization induced changes in the ocean's biological pump *Biogeosciences*, 5(2): 385-406

Joanne Wilson 2008 Preface *Hydrobiologia*, 608(1): 1-2

Jokiel PL, Rodgers KS, Kuffner IB, Anderson AJ, Cox EF and Mackenzie FT 2008 Ocean acidification and calcifying reef organisms: a mesocosm investigation *Coral Reefs,* 27(3): 473-483

Kathiresan K 2009 Coastal and marine biodiversity of India Final report, 201 pp

Kathiresan K 2000 A review of studies on Pichavaram mangroves, southeast coast of India, *Hydrobiologia,* 430 (1): 185-205

Kathiresan K 2002 Why are mangroves degrading? *Current Science,* 83: 1246-1249

Kathiresan K 2010 Globally threatened mangrove species in India *Current Science,* 98 (12):1551

Kathiresan K 2010 Conservation and Management Strategies: Restoration Technologies *In* Kathiresan, K and S Ajmal Khan (eds) *International Training Course on 'Coastal Biodiversity in Mangrove Ecosystems'– Manual,* Annamalai University, Parangi-pettai, India pp 636-660

Kathiresan K and Bingham BL 2001 Biology of mangroves and mangrove ecosystems *Advances in Marine Biology*, 40: 81-251

Kathiresan K and Rajendran N 2005 Coastal mangrove forests mitigated Tsunami *Estuarine Coastal and Shelf Science,* 65: 601-606

Kathiresan K and Qasim SZ 2005 Biodiversity of mangrove ecosystems Hindustan Publishing Corporation, New Delhi 251 pp

Kathiresan K and Faisal A M 2006 Managing Sundarbans for uncertainty and sustainability: in *International Conference and Exhibition on Mangroves of Indian and Western Pacific Oceans*, ICEMAN, Kuala Lumpur, Malaysia, pp1-31

Komiyama A, Ong JE and Poungparn S 2008 Allometry, biomass and productivity of mangrove forests: A review *Aquatic Botany*, 89: 128-137

Krauss Ken W, Lovelock Catherine E, McKee Karen L, Laura Lopez-Hoffman, Sharon ML Ewe and Wayne P Sousa 2008 Environmental drivers in mangrove establishment and early development: a review *Aquatic Botany,* 89:105-127

Kristensen E, Bouillon S, Dittmar T and Marchand C 2008 Organic carbon dynamics in mangrove ecosystems: A review *Aquatic Botany*, 89: 201-219

Lee PA, Saunders PA, de Mora SJ, Deibel D and Levasseur M 2003 Influence of copepod grazing on concentrations of dissolved dimethylsulfoxide and related sulfur compounds in the North Water, northern Baffin Bay *Marine Ecology Progress Series*, 255: 235 – 248

Li S W, Chan Benny KK and Tam Nora FY 2009 Barnacle fouling impedes the gaseous exchange and food production of the mangroves *Kandelia obovata*, a dominant mangrove species in Hong Kong and Taiwan *Hydrobiologia,* 618(1):199-203

Lirman D, Deangelo G, Serafy J, Hazra A, Smith Hazra D, Herlan J, Luo J, Bellmund S, Wang J and Clausing R 2008 Seasonal changes in the abundance and distribution of submerged aquatic vegetation in a highly managed coastal lagoon *Hydrobiologia* 596:105–120

Malakoff D 2003 Marine research – Scientists counting on census to reveal marine biodiversity *Science,* 302(5646): 773

Marañón E, Behrenfeld MJ, González N, Mouriño B and Zubkov MV 2003 High variability of primary production in oligotrophic waters of the Atlantic Ocean: uncoupling from phytoplankton biomass and size structure, *Marine Ecology Progress Series,* 257 : 1–11

Marianne Holmer and Nuria Marba 2010 Dynamics and functions of seagrass ecosystems *Hydrobiologia*, 649: 383pp

Mattice JS and Zittel HE 1976 Site-specific evaluation of power-plant chlorination *Journal Water Pollution Control Federation,* 48(10): 2284-2308

McCook LJ, Jpmpa J and Diaz-Pulido G 2001 Competition between corals and algae on coral reefs: a review of evidence and mechanisms *Coral Reefs,* 19(4): 400-417

McWillams JP, Cote IM, Gill JA, Siutherland WJ and Watkinson AR 2005 Accelerating impacts of temperature induced coral bleaching in the Caribeean *Ecology*, 86(8): 2055-2060

Mullineaux LS, Peterson CH, Micheli F and Mills SW 2003 Successional mechanism varies along a gradient in hydrothermal fluid flux at deep-sea vents *Ecological Monographs,* 73(4): 523-542

Mumby PJ, 2006 Connectivity of reef fish between mangroves and coral reefs: Algorithms for the design of marine reserves at seascape scales *Biological Conservation*, 128(2): 215-222

Munday PL, Jones GP, Pratchett MS and Williams AJ 2008 Climate change and the future for coral reef fishes *Fish and Fisheries*, 9(3): 261-285

Myers RA and Worm B 2003 Rapid worldwide depletion of predatory fish communities *Nature,* 423(6937): 280-283

Myhre S and Acevedo-Gutierrez A 2007 Recovery of sea urchin Diadema antillarum populations is correlated to increased coral and reduced macroalgal cover *Marine Ecology Progress Series*, 329: 205-210

Nakagawa S and Takai K 2008 Deep-sea chemoautotrophs: diversity, biochemistry and ecological significance *Fems Microbiology Ecology*, 65(1): 1-14

Nicoletti L, Marzialetti S, Paganelli D and Ardizzone GD 2007 Long-term changes in a benthic assemblage associated with artificial reefs *Hydrobiologia,* 580: 233-240

Nybakken James W 2001 Marine Biology and ecological approach Benjamin Cummings, an imprint of Addison Wesley Longman Inc, New York 516pp

Omori K, Doura, A Fukumoto T, Matsumoto J, Ohnishi H and Kathiresan K 2006 Sustainable reservation of mangrove forests on the basis of their ecological and socio-economical function against the global warming problem International Workshop on Prediction of greenhouse gas effect on global environment: Asian coastal ecosystem case study, North Vietnam, 1 p

Ong JE 1993 Mangroves – A carbon source and sink *Chemosphere*, 27(6): 1097-1107

Papush L and Danielsson A 2006 Silicon in the marine environment: disdolved silica trends in the Baltic Sea Estuarine, *Coastal and Shelf Science,* 67: 53-66

Pasquard S, Lobry J and Elie P, 2007 Facing the necessity of describing estuarine ecosystems: a review of food web ecology study techniques *Hydrobiologia*, 588: 159-172

Patnaik S 2008 Seafood exporters cry foul, The Hindu, November 14, 2008

Patterson JK, Jamila P and Wilhelmsson D 2008 Conservation and management of coral reefs and seagrasses of the Gulf of Mannar and Palk Bay, Southeastern India *SDMRI Special Research Publication* No13 35pp

Pedersen MF and Hansen PJ 2003 Effects of high pH on the growth and survival of six marine heterotrophic protists *Marine Ecology Progress Series,* 260: 33-41

Perillo GME 2009 Biological-physical interactions in estuaries, *Estuarine, Coastal and Shelf Science*, 85: v-vi

Polidoro BA, Carpenter KE, Collins L, Duke NC, Ellison AM, Ellison JC, Farnsworth EJ, Fernando ES, Kathiresan K, Koedam NE, Livingstone SR, Miyagi T, Moore GE, Vien Ngoc Nam, Ong J E, Primavera JH, Salmo III SG, Sanciangco JC, Sukardjo S, Wang Y and Yong JWH 2010 The loss of species: Mangrove extinction risk and geographic areas of global concern *PLoS ONE*, 5 (4): 1-10

Portman ME 2007 Zoning design for Ross-border marine protected areas: The Red Sea Marine Peace Park Case Study *Ocean and Coastal Management*, 50: 499-522

Potter IC, Chuwen BM, Hoeksema SD and Elliott M 2010 The concept of an estuary: A definition that incorporates systems which can become closed to the ocean and hypersaline *Estuarine, Coastal and Shelf Science,* 87: 497-500

Pramod G 2010 Illegal, Unreported and Unregulated Marine Fish Catches in the Indian Exclusive Economic Zone FIELD REPORT *Policy and Ecosystem Restoration in Fisheries Fisheries Centre, University of British Columbia, BC, Vancouver, Canada,* April 2010

Pratchett MS 2005 Dynamics of an outbreak population of Acanthaster planci at Lizard Island, northern Great Barrier Reef (1995-1999) *Coral Reefs,* 24(3): 453-462

Priede IG, Froese R, Bailey DM, Bergstad OA, Collins MA, Dyb JE, Henriques C, Jones EG and King N 2006 The absence of sharks from abyssal regions of the world's oceans *Proceedings of the Royal Society B –Biological Sciences,* 273(1592): 1435-1441

Qasim SZ 2004 Handbook of Tropical estuarine biology Narendra Publishing House, Delhi131pp

Raghubanshi AS, Rai LC, Gaur JP and Singh JS 2005 Invasive alien species and biodiversity in India *Current Science*, 88 (4): 540

Rober Llafite, Josette Garnier and Victor N de Jonge 2007 Consequences of estuarine management on hydrodynamics and ecological functioning *Hydrobiologia,* 588:302pp

Roberts CM, Andelman S, Branch G, Bustamante RH, Castilla JC, Dugan J, Halpern BS, Lafferty KD, Leslie H, Lubchenco J, McArdle D, Possingham HP, Ruckelshaus M and Warner RR 2003 Ecological criteria for evaluating candidate sites for marine reserves *Ecological Applications,* 13(1): S199-214

Ronald Semyalo, Juliet K Nattabi and Petter Larsson 2009 Diel Vertical Migration of zooplankton in a eutrophic bay of Lake Victoria *Hydrobiologia*, 635(1): 383-394

Salas-Leiva, Dayana Elizabeth, Mayor-Duran, Victor Manuel and Toro-Perea Nelson 2009 Genetic diversity of black mangrove (*Avicennia germinans*) in natural and reforested areas of Salamanca Island Parkway, Colombian Caribbean *Hydrobiologia,* 620(1): 17-24

Saudamini Das 2004 Mangroves – A natural defense against cyclones: An investigation from Orissa, India *Sandee Policy Brief*, No 24-07

Schallenberg M, Hall CJ and Burns CW 2003 Consequences of climate induced salinity increases on zoo- plankton abundance and diversity in coastal lakes *Marine Ecology Progress Series,* 251: 181-189

Seaman W 2007 Artificial habitats and the restoration of degraded marine ecosystems and fisheries *Hydrobiologia,* 580: 143-155

Soong K and Chen TA 2003 Coral transplantation: Regeneration and growth of Acropora fragments in a nursery Restoration Ecology, 11(1): 62-71

Spalding, M, Kainuma, M and Collins, L, 2010 World Atlas of Mangroves Earthscan, London, Washington, DC 319pp

Speight, M and Henderson, P, 2010 Marine Ecology concepts and applications Wiley – Blackwell, 276 pp.

Stæhr, PA, Waite, AM and Markager, S, 2008 'Effects of sewage water on bio-optical properties and primary production of coastal systems in West Australia' *Hydro-biologia*, 620(1): 191-205.

Stoner, AW, 2003 What constitutes essential nursery habitat for a marine species? A case study of habitat form and function for queen conch *Marine Ecology Progress Series,* 257: 275-289.

Subramanian B R 2011 Long-term data on coastal pollution and dissemination to academic and research communities *Current Science,* 100: 49-51.

Svavarsson J 2000 Imposex in the dogwhelk (*Nucella lapillus*) due to TBT contamination: Improvement at high latitudes *Marine Pollution Bulletin*, 40(11): 893-897.

Swaminathan MS 2004 Committee Report on "Revised Coastal Regulation Zone Report", submitted on July 2004 to Ministry of Environment & Forests, Government of India, New Delhi Pp3 (http://envfor.nic.in/).

Tan Kim Hooi, Ong JE and Primavera JH 2006 Proceedings of the International Conference and exhibition on mangroves of Indian and western Pacific oceans Maritime Institute of Malaysia, Kuala Lumpur, 21-24 August 2006, 320pp.

Tateda Y (Ed) 2007 Greenhouse and carbon balances in Mangrove Coastal Ecosystems Central Research Institute of Electric Power Industry, Japan 257pp.

Thompson JR and Flower RJ 2009 Environmental science and management of coastal lagoons in the Southern Mediterranean

Region: key issues revealed by the MELMARINA Project *Hydrobiologia,* 622: 221-232.

Tomlinson PB 1986 "The Botany of Mangroves" Cambridge University Press Cambridge, UK 413 pp.

Van Alstyne KL and Houser LT 2003 Dimethyl sulfide release during macroinvertebrate grazing and its role as an activated chemical defense *Marine Ecology Progress Series,* 250: 175-181.

Van Langevelde F and Prins H 2007 Resilience and restoration of soft-bottom nearshore ecosystems *Hydrobiologia*, 591: 1-4.

Venkataraman K 2006 Coral reefs in India National Biodiversity Authority, Chennai, India, 70 pp.

Venkataraman K 2007 Conservation and management of coastal and marine ecosystems in India National Biodiversity Authority, Chennai, 91pp.

Vivekanandan E 2011 Climate change and Indian Marine Fisheries CMFRI Special Publication No 105 Central Marine Fisheries Research Institute, Kochi 90 pp.

Volkenborn N, Hedtkamp SIC, van Beusekom JEE and Reise K 2007 Effects of bioturbation and bioirrigation by lugworms (*Arenicola marina*) on physical and chemical sediment properties and implications for intertidal habitat succession *Estuarine Coastal and Shelf science*, 74(1-2): 331-343.

Wafar M and Wafar S 2001 101 Questions on corals towards awareness National Institute of Oceanography, Goa and Administration of the Union Territory of Lakshadweep 79pp.

Wai TC and Willaims GA 22005 Therelative importance of herbivore-induced effects on productivity of crustose coralline algae: Sea urchin grazing and nitrogen excretion *Journal of Experimental Marine Biology and Ecology,* 324(2): 141-156.

White JW 2007 Spatially correlated recruitment of a marine predator and its prey shapes the large-scale pattern of density-dependent prey mortality *Ecology Letters,* 10: 1054-1065.

Williams PJL 1998 The balance of plankton respiration and photosynthesis in the open oceans *Nature,* 394(6688): 55-57.

Young CR, Fujio S and Vrijenhoek RC 2008 Directional dispersal between mid-ocean ridges: deep-ocean circulation and gene flow in *Ridgeia piscesae Molecular Ecology,* 17(7): 1718-1731.

Yunkai Li, Yong Chen, Derek Olson, Na Yu and Liqiao Chen 2009 Evaluating ecosystem structure and functioning of the East China Sea Shelf ecosystem, China *Hydro-biologia*, 636(1): 331-351.

Subject Index

Ocean and Coastal Ecology

Series : 21st Century Biology and Agriculture

K. Kathiresan

Life on the Earth without ocean is impossible. The oceans are of great ecological significance to have direct effect on global environment and human life; and of significant economic importance as the biggest storehouse of mineral, food and energy resources. This book entitled "Ocean and Coastal Ecology" under the series of 21st Century Agriculture and Biology has 10 chapters aimed to provide a clear understanding on scientific principles of marine ecology in order to manage the marine environment and its bio-resources in view of deterioration of marine ecosystems, natural calamities and climate change. This book would be of immense help to students, scholars and all others interested in marine ecology.

Contents

978-81-7233-891-6 | 2014 | PB |